Interpretation and Implementation of the Evaluation Requirements for Classified Protection of Cybersecurity

Editor in Chief: Guo Qiquan(郭启全)
Associate Editor in Chief: Liu Jianwei(刘建伟)
Wang Xinjie(王新杰)

北京航空航天大学出版社

图书在版编目(CIP)数据

网络安全等级保护测评要求应用指南 = Interpretation and Implementation of the Evaluation Requirements for Classified Protection of Cybersecurity：英文 / 郭启全主编. -- 北京：北京航空航天大学出版社，2022.5
ISBN 978-7-5124-3774-6

Ⅰ.①网… Ⅱ.①郭… Ⅲ.①计算机网络—网络安全—评价—指南—英文 Ⅳ.①TP393.08-62

中国版本图书馆 CIP 数据核字(2022)第 066348 号

版权所有，侵权必究。

Interpretation and Implementation of the Evaluation Requirements for Classified Protection of Cybersecurity
Editor in Chief：Guo Qiquan(郭启全)
Associate Editor in Chief：Liu Jianwei(刘建伟)
Wang Xinjie(王新杰)

策划编辑　董宜斌　　责任编辑　张　凌

*

北京航空航天大学出版社出版发行

北京市海淀区学院路 37 号(邮编 100191)　http://www.buaapress.com.cn
发行部电话：(010)82317024　传真：(010)82328026
读者信箱：copyrights@buaacm.com.cn　邮购电话：(010)82316936
三河市华骏印务包装有限公司印装　各地书店经销

*

开本：787×1 092　1/16　印张：26　字数：666 千字
2022 年 5 月第 1 版　2022 年 5 月第 1 次印刷
ISBN 978-7-5124-3774-6　定价：159.00 元

若本书有倒页、脱页、缺页等印装质量问题，请与本社发行部联系调换。联系电话：(010)82317024

本书编委会

主　编：Guo Qiquan（郭启全）
副主编：Liu Jianwei（刘建伟）　Wang Xinjie（王新杰）
编　者：Zhu Guobang（祝国邦）
　　　　Fan Chunling（范春玲）
　　　　Pan Wenbo（潘文博）
　　　　Wang Lianqiang（王连强）
　　　　Yang Yuzhong（杨玉忠）

Foreword

The Cybersecurity Law of the People's Republic of China was officially implemented on June 1, 2017. In this fundamental law in the field of cybersecurity, it is clearly stipulated that China implements the classified system of classified protection of cybersecurity. On December 1, 2019, *Information Security Technology Network Security—Evaluation Requirements for Classified Protection of Cybersecurity* GB/T 28448—2019 (hereinafter referred to as "Evaluation Requirements"), the National Standard of the People's Republic of China, was implemented.

The Evaluation Requirements is the core standard that guides the test and evaluation agencies to carry out the evaluation for the classified protection of cybersecurity. The correct understanding and use of this standard is the prerequisite for the smooth implementation for the classified protection of cybersecurity.

In order to better understand and comprehend the "Evaluation Requirements" and further improve the evaluation capabilities of test and evaluation agencies, the Cybersecurity Bureau under the Ministry of Public Security, the Zhongguancun Information Security Evaluation Alliance, and the Information Security Rating Center of the Ministry of Public Security jointly organized and compiled the "Guidelines for the Application of Evaluation Requirements for Classified Protection of Cybersecurity".

For each evaluation unit in the Evaluation Requirements, this book focuses on the determination of evaluation targets, the key points and methods of evaluation implementation, so as to better guide the classified test and evaluation agencies, the operation and using organizations of classified protection objects and the competent authorities to carry out the evaluation work for classified cybersecurity protection.

This book is divided into 8 chapters. Chapter 1 is the basic concept, which explains the terms or concepts related to the evaluation of classified cybersecurity protection, mainly including classified test and evaluation, evaluation targets and selection, evaluation index and selection, the mapping relationship between evaluation targets and evaluation indicators, and non-applicable evaluation index, evaluation intensity, evaluation method, singular evaluation, overall evaluation and evaluation conclusion, etc. Chapter 2 is the general introduction of the Evaluation Requirements, elaborating on the meaning of general requirements for security evaluation and extended requirements for security evaluation. Chapter 3 is the application interpretation of the general evaluation requirements at Level Ⅲ and Level Ⅳ. Chapter 4 is the application and interpretation of the extended requirements of cloud computing security evaluation. Chapter 5 is the application and interpretation of the extended security evaluation requirements of mobile Internet. Chapter 6 is the application and interpretation of the extended security evaluation requirements of Internet of Things. Chapter 7 is the application and interpretation of the extended security evaluation requirements of industrial control systems, and Chapter 8 is the application and interpretation of the extended security evaluation requirements of big data. The content of interpretation includes the evaluation targets, the main points and methods of the evaluation implementation, etc. , and the security protection level of the evaluation metric is identified by the evaluation unit number.

The editor in chief of this book is Guo Qiquan, the associate editor in chief are Liu Jianwei and Wang Xinjie, and other main contributors are Zhu Guobang, Fan Chunling, Pan Wenbo, Wang Lianqiang, Yang Yuzhong.

Due to the limited knowledge of the authors, there are inevitably some inadequacies in this book. Please feel free to kindly provide your feedback and correction.

the Author
March, 2022

Contents

Chapter 1 Basic Concepts ··· 1

    1.1 Classified Evaluation ··· 1

    1.2 Evaluation Targets Selection ·· 1

    1.3 Evaluation Index Selection ··· 2

    1.4 The Mapping Relationship Between Evaluation Targets and Evaluation Index ·· 3

    1.5 Non-Applicable Evaluation Index ··· 4

    1.6 Evaluation Strength ··· 5

    1.7 Evaluation Method ··· 6

    1.8 Singular Evaluation ··· 6

    1.9 Overall Evaluation ··· 7

    1.10 Evaluation Conclusion ·· 7

Chapter 2 General Introduction of the Evaluation Requirements ············ 9

    2.1 Relevant Description of the Evaluation Requirements ················ 9

    2.2 Text Structure of the Evaluation Requirements ························ 11

    2.3 General Requirements and Extended Requirements of Security Evaluation ··· 12

Chapter 3 Application and Interpretation of the General Security Evaluation Requirements at Level III and Level IV ······························· 13

    3.1 Security Physical Environment ·· 13

        3.1.1 Selection of Physical Location ···································· 13

        3.1.2 Physical Access Control ·· 14

        3.1.3 Anti-Theft and Anti-Vandal ·· 15

        3.1.4 Lightning Prevention ··· 16

        3.1.5 Fire Protection ··· 17

        3.1.6 Water Resistance and Moisture Resistance ················· 19

        3.1.7 Static Electricity Prevention ······································ 20

        3.1.8 Temperature and Humidity Control ····························· 20

        3.1.9 Power Supply ·· 21

        3.1.10 Electromagnetic Protection ······································ 23

- 3.2 Security Communication Network ……………………………………………………… 24
  - 3.2.1 Network Architecture ……………………………………………………… 24
  - 3.2.2 Communication Transmission ……………………………………………… 27
  - 3.2.3 Trusted Verification ………………………………………………………… 29
- 3.3 Security Area Boundary ………………………………………………………………… 31
  - 3.3.1 Boundary Protection ………………………………………………………… 31
  - 3.3.2 Access Control ………………………………………………………………… 34
  - 3.3.3 Intrusion Prevention ………………………………………………………… 37
  - 3.3.4 Malicious Code and Spam Prevention …………………………………… 39
  - 3.3.5 Security Audit ………………………………………………………………… 40
  - 3.3.6 Trusted Verification ………………………………………………………… 43
- 3.4 Security Computing Environment …………………………………………………… 44
  - 3.4.1 Network Device and Security Device …………………………………… 44
  - 3.4.2 Server and Terminal ………………………………………………………… 55
  - 3.4.3 Application System …………………………………………………………… 67
  - 3.4.4 Data Security ………………………………………………………………… 79
- 3.5 Security Management Center ………………………………………………………… 87
  - 3.5.1 System Management ………………………………………………………… 87
  - 3.5.2 Audit Management …………………………………………………………… 88
  - 3.5.3 Security Management ………………………………………………………… 89
  - 3.5.4 Centralized Management and Control …………………………………… 90
- 3.6 Security Management Systems ……………………………………………………… 94
  - 3.6.1 Security Strategy …………………………………………………………… 94
  - 3.6.2 Management Systems ………………………………………………………… 95
  - 3.6.3 Development and Release …………………………………………………… 96
  - 3.6.4 Review and Revision ………………………………………………………… 97
- 3.7 Security Management Organization ………………………………………………… 98
  - 3.7.1 Post Setting …………………………………………………………………… 98
  - 3.7.2 Staffing ………………………………………………………………………… 99
  - 3.7.3 Authorization and Approval ……………………………………………… 101
  - 3.7.4 Communication and Cooperation ………………………………………… 102
  - 3.7.5 Review and Inspection ……………………………………………………… 104
- 3.8 Security Management Personnel ……………………………………………………… 106
  - 3.8.1 Staff Recruitment …………………………………………………………… 106
  - 3.8.2 Staff Dismissal ……………………………………………………………… 108
  - 3.8.3 Security Awareness Education and Training …………………………… 109
  - 3.8.4 External Visitor Access Management …………………………………… 110

3.9　Security Development Management ··· 113
　3.9.1　Grading and Filing ··· 113
　3.9.2　Security Scheme Design ··· 114
　3.9.3　Product Procurement and Usage ··· 116
　3.9.4　Software Self-Development ··· 118
　3.9.5　Outsourcing Software Development ··· 122
　3.9.6　Security Engineering Implementation ··· 123
　3.9.7　Test and Acceptance ··· 125
　3.9.8　System Delivery ··· 126
　3.9.9　Classified Security Evaluation ··· 127
　3.9.10　Service Provider Management ··· 128
3.10　Security Operation and Maintenance Management ··· 130
　3.10.1　Environment Management ··· 130
　3.10.2　Asset Management ··· 132
　3.10.3　Media Management ··· 133
　3.10.4　Device Maintenance Management ··· 134
　3.10.5　Vulnerability and Risk Management ··· 136
　3.10.6　Network and System Security Management ··· 137
　3.10.7　Malicious Code Prevention Management ··· 142
　3.10.8　Configuration Management ··· 143
　3.10.9　Password Management ··· 144
　3.10.10　Change Management ··· 145
　3.10.11　Backup and Recovery ··· 146
　3.10.12　Security Incident Handling ··· 148
　3.10.13　Contingency Plan Management ··· 150
　3.10.14　Outsourcing Operation and Maintenance Management ··· 152

Chapter 4　Application and Interpretation of the Extended Security Evaluation Requirements of Cloud Computing ··· 155

4.1　Overview of Cloud Computing ··· 155
　4.1.1　Basic Concepts ··· 155
　4.1.2　Characteristics of Cloud Computing System ··· 156
　4.1.3　Deployment Model of Cloud Computing ··· 157
　4.1.4　Service Model of Cloud Computing ··· 157
　4.1.5　Cloud Computing Evaluation ··· 158

4.2 Application and Interpretation of the Extended Security Evaluation Requirements of Cloud Computing at Level Ⅲ and Level Ⅳ ……………… 162
    4.2.1 Security Physical Environment ……………………………………… 162
    4.2.2 Security Communication Network …………………………………… 163
    4.2.3 Security Area Boundary ……………………………………………… 168
    4.2.4 Security Computing Environment …………………………………… 175
    4.2.5 Security Management Center ………………………………………… 186
    4.2.6 Security Management System ………………………………………… 190
    4.2.7 Security Management Organization ………………………………… 190
    4.2.8 Security Management Personnel …………………………………… 190
    4.2.9 Security Development Management ………………………………… 190
    4.2.10 Security Operation and Maintenance Management ……………… 194

Chapter 5 Application and Interpretation of the Extended Security Evaluation Requirements of Mobile Internet ……………………………… 196

5.1 Basic Concepts ……………………………………………………………… 196
    5.1.1 Mobile Interconnection ………………………………………………… 196
    5.1.2 Mobile Terminals ……………………………………………………… 196
    5.1.3 Wireless Access Gateway …………………………………………… 196
    5.1.4 Mobile Application Software ………………………………………… 196
    5.1.5 Mobile Terminal Management System ……………………………… 197
5.2 Application Interpretation of Extended Security Evaluation Requirements of Mobile Internet at Level Ⅲ and Level Ⅳ ……………… 197
    5.2.1 Security Physical Environment ……………………………………… 197
    5.2.2 Security Communication Network …………………………………… 198
    5.2.3 Security Area Boundary ……………………………………………… 199
    5.2.4 Security Computing Environment …………………………………… 206
    5.2.5 Security Management Center ………………………………………… 230
    5.2.6 Security Management System ………………………………………… 230
    5.2.7 Security Management Organization ………………………………… 231
    5.2.8 Security Management Personnel …………………………………… 231
    5.2.9 Security Development Management ………………………………… 231
    5.2.10 Security Operation and Maintenance Management ……………… 233

Chapter 6 Application and Interpretation of the Extended Security Evaluation Requirements of IoT ……………………………………………… 235

6.1 Overview of IoT System …………………………………………………… 235

  6.1.1 Characteristics of IoT System 235
  6.1.2 Composition of IoT System 235
  6.1.3 Overview of Extended Requirements for IoT Security 237
  6.1.4 Basic Concepts 238
 6.2 Application and Interpretation of the Extended Security Evaluation Requirements of IoT at Level Ⅲ and Level Ⅳ 239
  6.2.1 Security Physical Environment 239
  6.2.2 Security Communication Network 241
  6.2.3 Security Area Boundary 245
  6.2.4 Security Computing Environment 253
  6.2.5 Security Management Center 268
  6.2.6 Security Management System 277
  6.2.7 Security Management Organization 277
  6.2.8 Security Management Personnel 277
  6.2.9 Security Development Management 277
  6.2.10 Security Operation and Maintenance Management 277

Chapter 7 Application and Interpretation of the Extended Security Evaluation Requirements of Industrial Control System 280

 7.1 Overview of Industrial Control System 280
  7.1.1 Characteristics of Industrial Control System 280
  7.1.2 Functional Hierarchy Model of Industrial Control System 282
  7.1.3 Evaluation Target and Index of Industrial Control System 284
  7.1.4 Typical Industrial Control System 287
 7.2 Application and Interpretation of the Extended Security Evaluation Requirements of Industrial Control System at Level Ⅲ and Level Ⅳ 290
  7.2.1 Security Physical Environment 290
  7.2.2 Security Communication Network 291
  7.2.3 Security Area Boundary 300
  7.2.4 Security Computing Environment 317
  7.2.5 Security Management Center 343
  7.2.6 Security Management System 343
  7.2.7 Security Management Organization 343
  7.2.8 Security Management Personnel 344
  7.2.9 Security Development Management 344
  7.2.10 Security Operation and Maintenance Management 345

Chapter 8　Application and Interpretation of the Extended Security Evaluation Requirement of Big Data ⋯⋯ 346

　8.1　Basic Concepts ⋯⋯ 346

　　8.1.1　Big Data ⋯⋯ 346

　　8.1.2　Targets of Big Data Classification Protection ⋯⋯ 347

　8.2　Extended Security Requirements and Best Practices ⋯⋯ 348

　8.3　Application and Interpretation of the Extended Security Evaluation Requirements of Big Data at Level Ⅲ and Level Ⅳ ⋯⋯ 358

　　8.3.1　Security Physical Environment ⋯⋯ 358

　　8.3.2　Security Communication Network ⋯⋯ 358

　　8.3.3　Security Area Boundary ⋯⋯ 363

　　8.3.4　Security Computing Environment ⋯⋯ 364

　　8.3.5　Security Management Center ⋯⋯ 382

　　8.3.6　Security Management System ⋯⋯ 385

　　8.3.7　Security Management Organization ⋯⋯ 387

　　8.3.8　Security Management Personnel ⋯⋯ 390

　　8.3.9　Security Development Management ⋯⋯ 392

　　8.3.10　Security Operation and Maintenance Management ⋯⋯ 396

Chapter 1 Basic Concepts

1.1 Classified Evaluation

The classified evaluation refers to the activities of evaluation agencies to test and evaluate the status of cybersecurity classified protection that does not involve state secrets in accordance with the provisions of the national cybersecurity classified protection regime and according to relevant administrative norms and technical standards. The classified evaluation includes standard conformity evaluation activities and risk assessment activities, that is, a scientific, fair and comprehensive evaluation process for the cybersecurity protection capabilities of the network by following a specific methodology in accordance with national standards or industry standards for cybersecurity classified protection.

The evaluation agency conducts the classified security evaluation in accordance with relevant national standards such as GB/T 28448—2019 *Information Security Technology — Evaluation Requirement for Classified Protection of Cybersecurity* (hereinafter referred to as "Evaluation Requirements") and GB/T 28449—2018 *Information Security Technology— Evaluation Process Guide for Classified Protection of Cybersecurity*, then issues the evaluation report according to the "*Report Templates for Classified Protection of Cybersecurity* (2019 Edition)" uniformly formulated by the Ministry of Public Security.

Through the classified evaluation, the security issues of the tested system (the classified protection object) can be discovered, the security status of the tested system can be understood, the potential dangers and weaknesses of the system can be sorted out, and the rectification needs of the system security development can be determined. Furthermore, it can be measured whether the system's security protection management measures and technical measures meet the baseline requirements of classified protection, and whether it has equipped with the corresponding security protection capabilities.

1.2 Evaluation Targets Selection

The evaluation target is the direct work object of the classified evaluation and is the specific system component of the security function corresponding to the specific evaluation indicator. Therefore, selecting and determining the evaluation target is a necessary step for the development of a evaluation plan and an critical link in the entire evaluation work. Appropriate selection of the type and quantity of evaluation targets is an important guarantee for the entire classified evaluation work to obtain sufficient evidence and understand the true security protection status of the object under test.

The determination of the evaluation targets generally adopts the method of spot check, that is, the representative components are randomly selected as evaluation targets.

When determining the evaluation targets, the following principles shall be followed:

(1) Importance. The servers, databases, and network devices that are important to the tested object shall be spot-checked.

(2) Security. The network boundaries exposed to the outside shall be checked randomly.

(3) Shareability. The spot-checking shall cover shared devices and data exchange platform/devices.

(4) Comprehensiveness. The spot-checking shall cover the various device types, operating system types, database system types and application system types in the system.

(5) Conformity. The selected device, software system, etc. shall be able to meet the evaluation intensity requirements of the corresponding level.

The following steps can be referred to when determining the evaluation targets:

(1) Classify the system components based on the granularity, such as the Client (mainly considering operating system), server (including operating system, database management system, application platform and business application software system), network interconnection device, security device, security-related personnel and security management documents, and further refinement can be made on the basis of the above classification.

(2) For each type of system components, the importance analysis shall be carried out based on the survey results, and the server operating system, database system, network interconnection device, security device, security related personnel, and security management documents with high importance shall be selected.

(3) For the selection results obtained in step (2), the security, sharability and comprehensiveness analysis shall be conducted to further improve the set of evaluation targets.

1.3 Evaluation Index Selection

The evaluation index are derived from the security requirements in GB/T 22239—2019 *Information Security Technology—Baseline for Classified Protection of Cybersecurity* (hereinafter referred to as "*Baseline*"). Security requirements can be further broken down into 3 types according to the "*Baseline*": the information security requirements (abbreviated as S) to protect data from being leaked, destroyed, and from unauthorized modification during storage, transmission, and processing, the assurance requirements (abbreviated as A) to protect the continuous and normal operation of the system from unauthorized modification and destruction that may cause the unavailability of the system, and other security protection requirements (abbreviated as G), all security management requirements and extended security requirements are marked as G.

For classified protection objects with the determined level, the selection method for evaluation index selection is as follows: according to the grading result of the evaluated system, including the business information security protection level and the system service security protection level, the system service assurance (Type A) security requirements, business information security (Type S) security requirements, general security requirements (Type G) and extended security requirements (Type G) of the tested system are obtained. Based on the combination of Type A, Type S, and Type G security requirements of the evaluated system, select the corresponding level of security requirements from the "*Baseline*" as the evaluation index. According to the characteristics of different industries or different objects, analyze the possible special security protection capabilities in some aspects, and select higher-level security requirements or supplementary security requirements of other standards.

1.4 The Mapping Relationship Between Evaluation Targets and Evaluation Index

On the basis of determining the evaluation targets and evaluation index of the evaluated system, the evaluation index and evaluation targets are combined, that is, mapping the evaluation index to each evaluation target, then combining with the characteristics of the evaluation target to illustrate the the evaluation method adopted by each evaluation target, thereby constituting a singular evaluation content that can be specifically implemented. The content of the evaluation is the basis for the evaluation personnel to develop the evaluation guide.

In view of the complexity and particularity of the evaluated system, certain evaluation index may not be suitable for all evaluation targets, and such evaluation index belong to non-applicable evaluation index. Except for non-applicable evaluation index, there is a one-to-one or one-to-many relationship between other evaluation index and evaluation targets, that is, one evaluation metric can be mapped to one or more evaluation targets. Taking the control point of identity authentication in the security computing environment as an example, the mapping relationship between evaluation index and evaluation targets is shown in Table 1.1.

Table 1.1 The Mapping Relationship Between Evaluation Index and Evaluation Targets (Example)

Security Item	Control Points	Evaluation Index	Evaluation Targets
Security Cloud Computing	Identify Authentication	a) Any users with login attempts shall be identified and authenticated. The identification has the character of uniqueness, while the identity authentication information has complexity requirements and is regularly replaced	operating systems in devices such as terminals and servers (including host and virtual equipment operating systems), network devices (including virtual network devices), security devices (including virtual security devices), business application systems, database management systems, etc.
		b) It shall have the function of handling login failures, configure and enable related measures such as ending the session, limiting the number of illegal logins, and automatically logging out when the login connection times out	operating systems in devices such as terminals and servers (including host and virtual equipment operating systems), network devices (including virtual network devices), security devices (including virtual security devices), business application systems and database management systems, etc.
		c) When performing remote management, necessary measures shall be taken to prevent the authentication information from being intercepted during network transmission	operating systems in devices such as terminals and servers (including host and virtual equipment operating systems), network devices (including virtual network devices), security devices (including virtual security devices), business application systems and database management systems, etc.
		d) The combination of two or more authentication technologies such as passwords, cryptography techniques, and biometric techniques shall be used for authenticating users' identity, and at least one of the authentication techniques shall be implemented by using cryptography	operating systems in devices such as terminals and servers (including host and virtual equipment operating systems), network devices (including virtual network devices), security devices (including virtual security devices), business application systems and database management systems, etc.

1.5 Non-Applicable Evaluation Index

In view of the complexity and particularity of the evaluated system, some requirements in the "*Baseline*" may not be applicable to all evaluation targets. For these non-applicable items, the reasons of non-applicability shall be explained in the evaluation report.

1.6 Evaluation Strength

Evaluation strength is the intensity of the evaluation work implemented in the classified evaluation process, which is reflected in the actual degree of commitment in the evaluation work, which is specifically reflected by the breadth and depth of the evaluation. The greater the breadth of the evaluation, the greater the scope of the evaluation implementation, and the more evaluation targets included in the evaluation implementation. The deeper the evaluation, the more detailed it is necessary to carry out the evaluation, and the more rigorous the evaluation, therefore, the greater the need for more work involvement. The more commitment, the stronger the evaluation intensity, thus the more guaranteed the evaluation effect.

In order to verify whether the classified protection objects with different levels have the corresponding level of security protection capabilities and whether they meet the corresponding level of protection requirements, it is necessary to implement the evaluation that is compatible with the level of security protection, and make the corresponding work commitment to achieve the required evaluation strength. The breadth and depth of the evaluation are implemented in three different evaluation methods: interview, verification and test, which can reflect the different levels of work involvement in interview, verification and test during the implementation of the evaluation, as shown in Table 1.2.

Table 1.2 The Evaluation Strength Requirements for Classified Protection Objects with Different Levels

Evaluation Strength	Evaluation Method	Level I	Level II	Level III	Level IV
Breadth	Interview	evaluation targets are sampled in type and quantity, and the type and quantity are relatively small	evaluation targets are sampled in type and quantity, and there are many types and quantities	evaluation targets are sampled in quantity and basically covered in type	evaluation targets are sampled in quantity and covered in all types
	Verification				
	Test				
Depth	Interview	Brief	Adequate	Relatively Comprehensive	Comprehensive
	Verification				
	Test	Functional Test	Functional Test	Functional Test and Test Verification	Functional Test and Test Verification

From Table 1.2, it can be seen that when evaluating classified protection objects of different levels, the types and numbers of selected evaluation targets are different. As the security protection level of the classified protection object becomes higher, the type and number of evaluation targets for random inspection also increases.

1.7 Evaluation Method

The classified evaluation method generally includes three types: interview, verification and test.

Interview refers to the process by which the evaluation personnel guides the relevant personnel of the classified protection object to have a purposeful (objected) communication to help the assessor understand, clarify or obtain evidence. In terms of the scope of interviews, different levels of grading objects have different requirements during evaluation. Generally, all types of safety-related personnel shall be basically covered, and samples shall be taken in number.

Verification refers to the process by which testers observe, inspect and analyze the test objects (such as system documents, various equipment and related safety configurations, etc.) to help testers understand, clarify, or obtain evidence. Verification can be subdivided into several specific methods such as document review, field inspection, and configuration verification.

Testing refers to the process in which testers use predetermined methods/tools to make test objects (various types of equipment or safety configurations) produce specific results, and compare the running results with the expected results. Testing generally uses technical tools to test the system, including vulnerability scanning based on network detection and host auditing, penetration testing, functional testing, performance testing, intrusion detection and protocol analysis.

1.8 Singular Evaluation

Singular evaluation is an evaluation for each security requirement item, supporting the repeatability and reproducibility of evaluation results. The singular evaluation in the "Evaluation Requirements" standard consists of evaluation index, evaluation targets, evaluation implementation and unit judgment results. Each specific evaluation implementation requirement item in the singular evaluation corresponds to the requirement item (evaluation index) included under the security control point. When evaluating each requirement item, three evaluation methods, interview, verification, and test may be used, or one or two of them may be used.

The content of the evaluation implementation completely covers the evaluation requirements of all the requirement items in the "*Baseline*". When using it, the evaluation requirements for each requirement item in the "*Baseline*" shall be extracted from the evaluation implementation of the singular evaluations, and the evaluation guidebook shall be developed in accordance with these evaluation requirements to standardize and guide the classified evaluation activities.

1.9　Overall Evaluation

The overall evaluation is based on the singular evaluation to judge the overall security protection capabilities of the classified protection objects. For the non-conformity items and partial-conformity items of the singular evaluation results, the method of one-by-one judgment shall be adopted, and the specific result of the overall evaluation are provided from the perspective of inter-points and inter-levels.

The inter-points evaluation refers to the singular evaluation item for "partial-conformity" and "non-conformity" requirement of the evaluation targets, analyzing whether other evaluation items related to the evaluation item in the same area and the same category can be associated with it, and what kind of association relationships occurs, and whether the effects of these association relationships can "make up for" the shortcomings of the evaluation item or "weaken" the protection capability achieved by the evaluation item, and whether the evaluation result of this evaluation item will affect the evaluation results of other evaluation items that are related to it.

Inter-levels evaluation refers to the singular evaluation item for "partial-conformity" and "non-conformity" requirement of the evaluation target, analyzing whether other evaluation targets related to the evaluation item can be associated with it, and what kind of association relationship occurs, and whether the effects of these relationships can "make up for" the shortcomings of the evaluation item or "weaken" the protection capability achieved by the evaluation item, and whether the evaluation result of the evaluation item will affect the evaluation results of other evaluation items that are related to it.

1.10　Evaluation Conclusion

The classified evaluation conclusion reflects the degree to which the classified protection object meets the corresponding level of protection requirements, which is determined by the risk analysis result of security issues and the comprehensive score.

After overall evaluation, the risk analysis and evaluation shall be carried out for the non-conforming items or partially conforming items in the singular evaluation results. Analyze the possibility of security issues (non-conforming items or partially conforming items) being exploited, determine the degree of impact on business information security and system service security after being exploited, and comprehensively evaluate the security risks caused by these non-conforming items or partially conforming items to the classified objects, and the result of the security problem risk analysis is obtained.

The comprehensive scoring algorithm is as follows:

$$100\left[1-\frac{1}{q}\sum_{j=1}^{q}\frac{\sum_{k=1}^{p(j)}\left(\sum_{i=1}^{m(k)}\text{weight of non-conforming}+\frac{1}{2}\sum_{i=1}^{m(k)}\text{weight of partially conforming}\right)}{\sum_{k=1}^{p(j)}\sum_{i=1}^{m(k)}\text{weight of evaluation items}}\right]$$

Among them, q is the security type involved in the tested object, $p(j)$ is the total number of evaluation items corresponding to a certain security type, excluding non-applicable evaluation items, and $m(k)$ is the number of evaluation targets corresponding to the evaluation item k.

The classified evaluation conclusions are divided into four categories: "excellent", "good", "medium" and "poor". The judgment basis of the classified evaluation conclusion is as follows:

Excellent: There exists security problems in the tested object, but it will not cause the tested object to face medium and high-level security risks, and the system gets a comprehensive score of 90 points or more (including 90 points).

Good: There exists security problems in the tested object, but it will not cause the tested object to face high-level security risks, and the system gets a comprehensive score of 80 points or more (including 80 points).

Medium: There exists security problems in the tested object, but it will not cause the tested object to face high-level security risks, and the system gets a comprehensive score of 70 points or more (including 70 points).

Poor: There exists security problems in the tested object, and it will cause the tested object to face high-level security risks, or the comprehensive score of the tested object is lower than 70 points.

Chapter 2　General Introduction of the Evaluation Requirements

2.1　Relevant Description of the Evaluation Requirements

1. Technical Framework of Classified Evaluation

The technical framework of classified evaluation includes singular evaluation and overall evaluation.

The singular evaluation is the evaluation aimed at each security requirement item, supporting the repeatability and reproducibility of evaluation results. The individual evaluation in this standard consists of evaluation index, evaluation targets, evaluation implementation and unit judgment.

The overall evaluation is based on the singular evaluation to judge the overall security protection capabilities of the classified protection objects. The content of overall evaluation includes security control points evaluation, inter-points evaluation and inter-levels evaluation.

In addition, in order to better enable the evaluation personnel of the test and evaluation agencies to determine the objects of the evaluation work, the evaluation targets are added to the evaluation unit. The evaluation target refers to the object upon which different evaluation methods are implemented in the classified evaluation process, mainly involving with supporting system documents, equipment and facilities, and personnel.

2. Numbering of Evaluation Unit

In order to facilitate the understanding and use of the "Evaluation Requirements", Appendix C provides the evaluation unit coding rules and special abbreviations. The evaluation unit is numbered as three sets of data in the format XX-XXXX-XX. The meanings and coding rules of each group are as follows:

The 1st Group consists of 2 characters, the first digit is the letter L, and the second digit is the number, where the number 1 stands for Level Ⅰ, 2 for Level Ⅱ, 3 for Level Ⅲ, 4 for Level Ⅳ, and 5 for Level Ⅴ.

The 2nd Group consists of 4 characters, the first 3 digits are letters and the 4th digit is the number. The letter represents different types: PES stands for Physical Environment

Security, CNS is Communication Network Security, ABS is Area Boundary Security, CES is Computing Environment Security, SMC is Security Management Center, PSS is Policy and System Security, ORS is Organization and Resource Security, and HRS is Human Resources Security, CMS is Construction Management Security, and MMS is Maintenance Management Security. The number represents different application scenarios: 1 is the general requirements for security evaluation, 2 is the extended requirements for cloud computing security evaluation, 3 is the extended requirements for mobile internet security evaluation, 4 is the extended requirements for Internet of Things(IoT) security evaluation, and 5 is the extended requirements for industrial control system security evaluation.

The 3rd Group consists of two numbers, which are numbered sequentially according to the required items in the general requirements.

Example: The evaluation unit number is L1-PES1-01, which represents the first metric that of the physical environment security of the general requirements security assessment for Level Ⅰ.

The evaluation unit numbering for referable big data security evaluation method consists of three sets of data in the format of XXX-XX-XXX, and the meaning and coding rules of each group are as follows:

The 1st Group consists of 3 characters. The BDS represents referable big data security evaluation method.

The 2nd Group consists of 2 characters, the first digit is the letter L, and the second digit is the number, where the number 1 stands for Level Ⅰ, 2 for Level Ⅱ, 3 for Level Ⅲ, 4 for Level Ⅳ, and 5 is the 5th Level.

The 3rd Group consists of two numbers, which is numbered sequentially in accordance with the security controls in the basic requirements.

Example: The evaluation unit number is BDS-L1-01, which represents the first metric of the referable big data security evaluation method for Level Ⅰ.

3. Level Differences in Evaluation Requirements

The evaluation work of different levels mainly reflects the level differences of the evaluation requirements from the following 3 aspects:

Different evaluation methods are used at different levels: In the actual on-site evaluation implementation process, the evaluation methods at the technical level are mainly involved with configuration check and test verification, with almost no interviews. At the management level, interviews can be used for evaluation. The use of different evaluation methods at different levels can reflect the different evaluation strengths of interviews, verification and tests during the implementation of the evaluation.

The scope of evaluation targets at different levels is different: the scope of the evaluation targets of Level Ⅰ and Level Ⅱ focus on critical devices, Level Ⅲ involves with main devices, and Level Ⅳ includes devices. The scope of evaluation targets with different levels is different, which can reflect the different evaluation breadths of interviews, verification and tests during the implementation of the evaluation.

The on-site evaluation and implementation work for different levels is different: Level Ⅰ and Level Ⅱ mainly check the security mechanism, and Level Ⅲ and Level Ⅳ firstly check the security mechanism, then check the effectiveness of the security strategy.

2.2　Text Structure of the Evaluation Requirements

There are 12 Chapters and 3 Appendices in the "Evaluation Requirements", and the content of each chapter is described in below:

Chapters 1, 2 and 3 are general descriptions of the Standard, including scope, normative references, terms and definitions.

Chapter 4 is abbreviations.

Chapter 5 outlines the evaluation methods of classified security protection, the composition of individual evaluations and overall evaluations.

Chapters 6,7,8 and 9 describes the evaluation requirements at Level Ⅰ, Level Ⅱ, Level Ⅲ and Level Ⅳ respectively, and each level follows the framework of the "*Baseline*" describing how to implement the evaluation work. Each level includes five parts, including general requirements of security evaluation, extended requirements for cloud computing security evaluation, extended requirements for mobile Internet security evaluation, extended requirements for IoT security evaluation and extended requirements for industrial control systems security evaluation. Among them, the technical level is rolled out from five aspects: security physical environment, security communication network, security area boundary, security computing environment and security management center; while the management level is rolled out from security management system, security management organization, security management personnel, security development management and security operations and maintenance, forming a mutually contrasted, harmonious and unified standard text structure with the "*Baseline*".

Chapter 10 omits the evaluation requirements at Level Ⅴ.

Chapter 11 describes the overall evaluation method of the system, considering how to conduct a systematic evaluation from the perspective of the system as a whole on the basis of the singular evaluation. It is described from three aspects: security control points, inter-points and inter-area evaluation, analyzing the content that needs to be considered in the overall evaluation of the system.

Chapter 12 summarizes the method of drawing the conclusion of the evaluation and the main content of the evaluation conclusion.

Appendix A describes the evaluation intensity of various evaluation methods, and specifically describes the evaluation intensity for classified protection objects with different levels.

Appendix B describes the referenced security evaluation methods for big data.

Appendix C describes the coding rules of evaluation index and special abbreviations.

2.3 General Requirements and Extended Requirements of Security Evaluation

Due to factors such as different business objectives, different technologies used, and different application scenarios, classified protection objects with different levels appear in different patterns. The manifestation may be basic information networks, information systems (including systems using mobile Internet and other technologies), cloud computing platform/system, big data platform/system, IoT, industrial control system, etc. Different types of protection objects face different threats, and their security protection requirements are also different. In order to facilitate the realization of universal and personalized protection of classified protection objects with different levels and different forms, the classified protection requirements are divided into general security requirements and extended security requirements. Correspondingly, the requirements for classified security protection evaluation are also divided into general requirements for security evaluation and extended requirements for security evaluation.

Regardless of the form and pattern of the classified protection objects, the the corresponding level of general security requirements must be achieved according to their security protection levels, and the security test and evaluation of these universal protection measures shall be carried in accordance with the general requirements of security evaluation. For cloud computing, mobile Internet, IoT and industrial control systems, it is necessary to selectively implement extended security requirements based on security protection levels and specific technologies used or specific application scenarios, and perform security test and evaluation on these personalized protection measures in accordance with extended requirements of security evaluation. The general requirements for security evaluation and the extended requirements for security evaluation together constitute the security evaluation requirements for the classified protection objects.

Chapter 3　Application and Interpretation of the General Security Evaluation Requirements at Level Ⅲ and Level Ⅳ

3.1　Security Physical Environment

3.1.1　Selection of Physical Location

[Standard Requirements]

This control point for Level Ⅲ includes the evaluation units of L3-PES1-01, L3-PES1-02, and Level Ⅳ includes that of L4-PES1-01, L4-PES1-02.

[L3-PES1-01/L4-PES1-01 Interpretation and Description]

The main evaluation targets of the evaluation metric that "The location of the equipment room shall be chosen in a building with capabilities of earthquake prevention, wind proof and rain resistance" are the equipment room, the design/ acceptance document of the equipment room, including the description of the equipment room's performance of earthquake prevention, wind proof and rain resistance.

The key points of the evaluation implementation include: checking whether the building where the equipment room is located has the building seismic fortification approval document, and whether the design/acceptance document of the equipment room contains design requirements or the acceptance conclusion for the equipment room's quakeproof, windproof and rainproof capabilities; checking whether there is rainwater leakage in areas such as the roof, walls or windows of the equipment room; checking whether the doors and windows in the equipment room are full of dust caused by wind; checking whether the roof, walls, doors, windows and ground of the equipment room are damaged or cracked.

If the test results show that the building where the equipment room is located has a building seismic fortification approval document, and the design/acceptance document of the equipment room contains design requirements or the acceptance conclusion for the equipment room's capabilities in earthquake resistance, wind resistance and rain resistance, and there is no dust covered all over the doors and windows of the equipment room, and the roof, walls, doors, windows and ground of the equipment room are not damaged or cracked, the result is conforming; otherwise it is non-conforming or partially conforming.

[L3-PES1-02/L4-PES1-02 Interpretation and Description]

The main evaluation targets of the evaluation metric that "The equipment room shall avoid being located at the top floor or basement of the building. Otherwise, water-proof and moisture-proof measures shall be strengthened" are the equipment room, the design/acceptance document of the equipment room, including the description of strengthening earthquake-resistant, water-proof and moisture-proof measures in the equipment room.

The key points of the evaluation implementation include: checking whether the building where the equipment room is located has a building seismic fortification approval document, and whether the design/acceptance document of the computer room contains design requirements or the acceptance conclusion for the earthquake-resistant, wind-proof and rain-proof capabilities of the equipment room; checking whether there exists rainwater leakage on the roof, walls or windows of the equipment room.

If the test results show that the building where the equipment room under test is located has a building seismic fortification approval document, and the equipment room's design/acceptance document contains design requirements or the acceptance conclusion for earthquake-resistant, wind-proof and rain-proof capabilities, and there is no rainwater leakage on the roof, walls or windows of the equipment room, the doors and windows of the equipment room are not severely dusty due to wind, and the roof, walls, doors and windows and the ground of the equipment room are not damaged or cracked, the result is conforming; otherwise it is non-conforming or partially conforming.

3.1.2 Physical Access Control

[Standard Requirements]

This control point for Level III includes the evaluation unit of L3-PES1-03, and Level IV includes that of L4-PES1-03, L4-PES1-04.

[L3-PES1-03/L4-PES1-03 Interpretation and Description]

The main evaluation targets of the evaluation metric that "The entrance and exit of the equipment room shall be equipped with the electronic access control system to control, identify and record the personnel entering" are the electronic access control system of the equipment room, the control and authentication records of the electronic access control system.

The key points of the evaluation implementation include: checking whether the entrance and exit of the equipment room are equipped with the electronic access control system; checking the opening and closing records of the electronic access control system to confirm that the electronic access control system is operating normally and has been in the enabled status; checking whether the electronic access control system can identify and record the information of personnel entering, and checking whether the historical records are complete.

If the test results show that the entrance and exit of the tested equipment room are equipped with the electronic access control system, and the electronic access control system can identify and record the information of entering personnel and keep historical records, the result is conforming; otherwise it is non-conforming or partially conforming.

[L4-PES1-04 Interpretation and Description]

The main evaluation targets of the evaluation metric that "Important areas shall be equipped with a second electronic access control system to control, identify and record the entering personnel" include the second electronic access control system in the equipment room, the control and identification records of the second electronic access control system.

The key points of the evaluation implementation include: check whether the second electronic access control system is equipped and activated at the entrances and exits of important areas; checking whether the second electronic access control system can identify and record the information of incoming personnel, and checking related records; checking whether the first electronic access control system and the second electronic access control system are separately authorized.

If the test results show that the second electronic access control system is equipped and activated at the entrances and exits of important areas in the tested computer room, and the second electronic access control system can identify and record the information of the entering personnel, and keep the historical records, and the first electronic access control system and the second electronic access control system are separately authorized, the result is conforming; otherwise it is non-conforming or partially conforming.

3.1.3 Anti-Theft and Anti-Vandal

[Standard Requirements]

This control point for Level III includes the evaluation units of L3-PES1-04, L3-PES1-05, L3-PES1-06, and Level IV includes that of L4-PES1-05, L4-PES1-06, L4-PES1-07.

[L3-PES1-04/L4-PES1-05 Interpretation and Description]

The main evaluation targets of the evaluation metric that "The equipment or main components shall be fixed, and obvious signs that are not easy to remove shall be set up" are the equipment room or main components, racks, etc.

The key points of the evaluation implementation include: checking whether the equipment or main components in the equipment room are fixed on the rack; checking whether the equipment or main components in the equipment room have obvious signs that are difficult to remove, including information of asset name, asset number, asset model, responsible personnel, responsible department, asset location, etc.

If the test results show that the equipment or main components in the equipment room are fixed on the rack, and the equipment or main components in the equipment room are affixed with obvious signs that are difficult to remove, the result is conforming; otherwise it is non-conforming or partially conforming.

[L3-PES1-05/L4-PES1-06 Interpretation and Description]

The main evaluation targets of the evaluation metric that "Communication cables shall be laid in a concealed and safe place" is the communication cables in the equipment room.

The key points of the evaluation implementation include: checking whether the communication cables in the equipment room are laid in a concealed and safe place, such as in the bridge above the cabinet or the electrostatic floor under the cabinet, which is not exposed to the ground or outside the cabinet.

If the test results show that the communication cable in the equipment room is laid in a concealed and safe place and is not exposed to the ground or outside the cabinet, the result is conforming; otherwise it is non-conforming or partially conforming.

[L3-PES1-06/L4-PES1-07 Interpretation and Description]

The main evaluation targets of the evaluation metric that "The equipment room anti-theft alarm system shall be set up or the video surveillance system with the dedicated personnel shall be equipped" are anti-theft alarm system and video surveillance system in the equipment room.

The key points of the evaluation implementation include: checking whether the equipment room is equipped with the anti-theft alarm system or the dedicated video surveillance system; checking whether the anti-theft alarm system or video surveillance system is enabled, and whether there is a dedicated person on duty to keep track of the video surveillance system's records.

If the test results show that the tested computer room is equipped with the anti-theft alarm system or the video surveillance system with a dedicated person on duty, the systems are activated, and there is the record of the dedicated person taking care of the video monitoring system, the result is conforming; otherwise it is non-conforming or partially conforming.

3.1.4 Lightning Prevention

[Standard Requirements]

This control point for Level I includes the evaluation units of L3-PES1-07, L3-PES1-08, and Level II includes that of L4-PES1-08, L4-PES1-09.

[L3-PES1-07/L4-PES1-08 Interpretation and Description]

The main evaluation targets of the evaluation metric that "All kinds of cabinets, facilities and equipment shall be securityly grounded through the grounding system" are cabinets, facilities and equipment in the equipment room, the design/acceptance documents of the equipment room, and instructions for safe grounding of various cabinets, facilities and equipment through the grounding system.

The key points of the evaluation implementation include: checking whether the design/ acceptance document of the equipment room contains the safe grounding design or acceptance conclusion of the cabinet, facilities and equipment; checking whether the equipment in the cabinet and the cabinet are connected to the earthing busbar with the grounding wire, and whether the earthing busbar is connected to the grounding body in the earth through the grounding wire; checking whether the deployment of grounding system is consistent with the equipment room design/acceptance documents.

If the test results show that the design/acceptance document of the equipment room contains the design requirements or acceptance conclusions for safe grounding of cabinets, facilities and equipment, and the actual situation is consistent with the design/acceptance document, the result is conforming; otherwise it is non-conforming or partially conforming.

[L3-PES1-08/L4-PES1-09 Interpretation and Description]

The main evaluation targets of the evaluation metric that "Measures shall be taken to prevent induced lightning strike, such as installing lightning protection devices or overvoltage protection devices" include the lightning protection facilities in the equipment room and the technical inspection report of the lightning protection device.

The key points of the evaluation implementation include: checking whether the equipment room is equipped with induced lightning protection measures, such as whether the power supply surge protector is installed in the power distribution cabinet of the equipment room, etc. ; checking the technical test report of the lightning protection device to check whether the lightning protection device has passed the acceptance or the technical inspection of the relevant state authorities; checking relevant records and documents to confirm whether surge protectors are regularly inspected by professional maintenance companies (inspection records shall be available).

If the test results show that the tested equipment room is equipped with induced lightning protection measures, has a technical inspection report of the lightning protection device, the lightning protection device has passed the acceptance or the technical inspection by the relevant state authorities, the surge protector is regularly inspected by the professional maintenance company and the inspection record is available, the result is conforming; otherwise it is non-conforming or partially conforming.

3.1.5　Fire Protection

[Standard Requirements]

This control point for Level Ⅲ includes the evaluation units of L3-PES1-09, L3-PES1-10, L3-PES1-11, and Level Ⅳ includes that of L4-PES1-10, L4-PES1-11, L4-PES1-12.

[L3-PES1-09/L4-PES1-10 Interpretation and Description]

The main evaluation targets of the evaluation metric that "The equipment room shall be equipped with an automatic fire-fighting system, which can automatically detect, alarm and extinguish the fire" are fire-fighting facilities in the equipment room and the technical inspection report of the automatic fire-fighting system.

The key points of the evaluation implementation include: checking whether the automatic fire protection system is installed in the equipment room, checking the temperature, smoke and automatic gas fire-fighting system on site; checking the functions and performance indicators of the automatic fire-fighting system in the equipment room's design/acceptance documents, and checking whether the automatic fire-fighting system is activated; checking the design/acceptance documents of the equipment room to confirm whether the automatic fire-fighting system has passed the acceptance or the technical inspection by the relevant state authorities.

If the test results show that the automatic fire-fighting system has passed the acceptance or technical inspection of relevant state authorities and can automatically detect, alarm and extinguish the fire, the result is conforming; otherwise it is non-conforming or partially conforming.

[L3-PES1-10/L4-PES1-11 Interpretation and Description]

The main evaluation target of the evaluation metric that "The equipment room, relevant work rooms and auxiliary rooms shall use building materials with fire-resistance rating" is the design/acceptance document of the equipment room.

The key points of the evaluation implementation include: checking the equipment room acceptance document to conform whether fire-resistant building materials are adopted for the equipment room and relevant auxiliary rooms, and clarify the fire-resistant rating of relevant building materials; checking the equipment room and related work rooms and auxiliary rooms on site to conform whether the building materials used meet the required requirements.

If the test results show that the tested equipment room and related work rooms and auxiliary rooms use fire-resistant materials, and the acceptance document includes the design requirements and acceptance conclusions of the building materials, the result is conforming; otherwise it is non-conforming or partially conforming.

[L3-PES1-11/L4-PES1-12 Interpretation and Description]

The main evaluation targets of the evaluation metric that "The equipment room shall be divided into areas for management, and isolation and fire prevention measures shall be set up between areas" are the equipment room administrator and the equipment room.

The key points of the evaluation implementation include: interviewing with the equipment room administrator to find out whether the equipment room has been divided into areas and inspecting it on the spot; checking whether fire prevention measures are taken to isolate each area, such as fire-proof rolling shutter doors.

If the test results show that the equipment room is divided into areas and fire prevention measures are taken to isolate different areas, the result is conforming; otherwise it is non-conforming or partially conforming.

3.1.6 Water Resistance and Moisture Resistance

[Standard Requirements]

This control point for Level Ⅲ includes the evaluation units of L3-PES1-12, L3-PES1-13, L3-PES1-14, and Level Ⅳ includes that of L4-PES1-13, L4-PES1-14, L4-PES1-15.

[L3-PES1-12/L4-PES1-13 Interpretation and Description]

The main evaluation targets of the evaluation metric that "Measures shall be taken to prevent rainwater from penetrating through windows, roofs and walls of the equipment room" are the design/acceptance documents of the equipment room, and the instructions on preventing rainwater from penetration.

The key points of the evaluation implementation include: reviewing the equipment room design/acceptance documents to check the acceptance conclusion of rainwater prevention facilities; and checking the windows, roof and walls of the computer room on the site for signs of rainwater infiltration.

If the test results show that the design/acceptance document of the tested equipment room contains the rainwater infiltration prevention design or acceptance conclusion, and there is no trace of rainwater infiltration by inspecting on the spot, the result is conforming; otherwise it is non-conforming or partially conforming.

[L3-PES1-13/L4-PES1-14 Interpretation and Description]

The main evaluation target of the evaluation metric that "Measures shall be taken to prevent water vapor condensation in the equipment room and prevent the transmission and penetration of underground water" is the equipment room.

The key points of the evaluation implementation include: checking whether measures are taken to prevent water vapor condensation in the equipment room, such as equipping with the special precision air conditioners and setting up the floor drain around the ground in the hidden water leakage area in the equipment room; checking whether measures have been taken to drain underground water in the equipment room and prevent underground water from infiltrating.

If the test results show that measures are taken to prevent water vapor condensation in the equipment room, and measures are taken to drain underground water and prevent underground water from infiltrating in the equipment room, the result is conforming; otherwise it is non-conforming or partially conforming.

[L3-PES1-14/L4-PES1-15 Interpretation and Description]

The main evaluation target of the evaluation metric that "Water-sensitive testing instruments or components shall be installed to perform waterproof detection and generate alarms in the equipment room" is the water leakage detection facility in the equipment room.

The key points of the evaluation implementation include: checking whether a water-sensitive detection device is installed in the equipment room; checking the strategies and records of the waterproof detection and alarm devices; checking whether the waterproof detection and alarm devices are enabled and working properly.

If the test results show that the tested equipment room is equipped with water-sensitive detection devices, the waterproof detection and alarm devices are activated and there are related disposal records, etc., the result is conforming; otherwise it is non-conforming or partially conforming.

3.1.7 Static Electricity Prevention

[Standard Requirements]

This control point for Level III includes the evaluation unit of L3-PES1-15, L3-PES1-16, and Level IV includes that of L4-PES1-16, L4-PES1-17.

[L3-CNS1-15/L4-PES1-16 Interpretation and Description]

The main evaluation target of the evaluation metric that "Anti-static floor or ground and necessary grounding anti-static measures shall be adopted" is the equipment room.

The key points of the evaluation implementation include: checking whether anti-static floor or ground is installed in the equipment room; checking whether anti-static measures such as grounding, overlap and shielding are adopted in the equipment room.

If the test results show that the anti-static floor or ground is installed in the equipment room, and anti-static measures such as grounding, overlap and shielding are adopted in the equipment room, the result is conforming; otherwise it is non-conforming or partially conforming.

[L3-PES1-16/L4-PES1-17 Interpretation and Description]

The main evaluation target of the evaluation metric that "Measures shall be taken to prevent the generation of static electricity, such as the use of static eliminators, wearing anti-static bracelets" is the equipment room.

The key points of the evaluation implementation include: checking whether the equipment room is equipped with the anti-static equipment, such as using static eliminators, wearing anti-static bracelets, etc.

If the test results show that the tested equipment room is equipped with the anti-static equipment, including using the static eliminator, wearing the anti-static bracelet, etc., the result is conforming; otherwise it is non-conforming or partially conforming.

3.1.8 Temperature and Humidity Control

[Standard Requirements]

This control point for Level III includes the evaluation unit of L3-PES1-17, and Level IV includes that of L4-PES1-18.

[L3-PES1-17/L4-PES1-18 Interpretation and Description]

The main evaluation targets of the evaluation metric that "Automatic temperature and humidity adjustment facilities shall be set up so that the temperature and humidity changes in the equipment room are within the allowable range for equipment operation" are the temperature and humidity adjustment facilities in the equipment room.

The key points of the evaluation implementation include: checking whether the equipment room is equipped with precision air conditioners; checking whether the current temperature and humidity in the equipment room are within the allowable range for the equipment operation, usually the suitable temperature range in the equipment room is 20—24℃, and the air humidity range is 45%—65%; checking whether the temperature and humidity history records of the equipment room are within the allowable range for equipment operation; checking whether the temperature and humidity of each main area are within the allowable range for the equipment operation for large-area or area-managed equipment rooms.

If the test results show that the tested computer room is equipped with precision air conditioners, and the current temperature and humidity of the computer room are within the allowable range for equipment operation, the result is conforming; otherwise it is non-conforming or partially conforming.

3.1.9 Power Supply

[Standard Requirements]

This control point for Level III includes the evaluation units of L3-PES1-18, L3-PES1-19, L3-PES1-20, and Level IV includes that of L4-PES1-19, L4-PES1-20, L4-PES1-21, L4-PES1-22.

[L3-PES1-18/L4-PES1-19 Interpretation and Description]

The main evaluation targets of the evaluation metric that "Voltage stabilizer and over-voltage protection device shall be installed on the power supply wires of the equipment room" are the power supply facilities in the equipment room and the technical inspection report of power supply facilities in the equipment room.

The key points of the evaluation implementation include: checking whether voltage stabilizer and over-voltage protection devices are installed on the power supply wires of the equipment room, including surge protectors, self-compound over-voltage and under-voltage protectors, vacuum circuit breakers, over-voltage absorbers, etc.; checking whether the voltage stabilizer and over-voltage protection device have passed the acceptance or the technical inspection of the relevant state authorities or checking whether the certificate of conformity is available.

If the test results show that the power supply wires of the tested equipment room is equipped with the voltage stabilizer and over-voltage protection equipment, which have passed the acceptance or the technical inspection of the relevant state authorities, the result is conforming; otherwise it is non-conforming or partially conforming.

[L3-PES1-19/L4-PES1-20 Interpretation and Description]

The main evaluation targets of the evaluation metric that "Short-term standby power supply shall be provided to at least meet the normal operation requirements of equipment in the case of power outage" are power supply facilities of the equipment room, the design/acceptance documents of the equipment room, including the description of capabilities of the power supply facilities in the equipment room.

The key points of the evaluation implementation include: checking whether the equipment room is equipped with the uninterruptible power supply (UPS) or the standby power supply system. According to the design requirements or acceptance conclusions of the power supply facilities of the equipment room in the design/acceptance documents, checking whether the backup power system such as UPS meets the normal operation requirements of the equipment in the case of power failure.

If the test results show that the tested equipment room is equipped with the backup power system such as UPS, and the backup power system such as UPS meets the normal operation requirements of the equipment in the case of power failure, the result is conforming; otherwise it is non-conforming or partially conforming.

[L3-PES1-20/L4-PES1-21 Interpretation and Description]

The main evaluation targets of the evaluation metric that "The redundant or parallel power cable lines shall be set to provide power to the computer system" include the equipment room administrator, the equipment room, the equipment room design/acceptance documents, including instructions on setting up redundant or parallel power cable lines in the equipment room to provide power to the computer system, etc.

The key points of the evaluation implementation include: interviewing with the equipment room administrator to understand whether the power supply of the equipment room comes from two different substations; checking whether the design document, the construction plan and other related records of the equipment room's power supply wiring include the relevant requirements and content of the redundant or parallel power supply of the equipment room; checking the power distribution facilities and power supply wiring of the equipment computer room on site to confirm whether the equipment room supplies power to the computer system through redundant or parallel power cable lines.

If the test results show that the power supply of the evaluated equipment room comes from two different substations, and the related facilities of dual power supply are deployed in the power distribution cabinet of the equipment room, the result is conforming; otherwise it is non-conforming or partially conforming.

[L4-PES1-22 Interpretation and Description]

The main evaluation targets of the evaluation metric that "The emergency power supply facilities shall be provided" include the emergency power supply facilities in the equipment room, the design/acceptance documents of the equipment room, including instructions on

Chapter 3 Application and Interpretation of the General Security Evaluation Requirements at Level III and Level IV

the equipment room to be equipped with emergency power supply facilities.

The key points of the evaluation implementation include: checking whether the emergency power supply facilities or backup power supply systems (such as diesel generators, or third-party backup power supply services) are equipped; checking whether emergency power supply facilities are available; reviewing the equipment room design/acceptance documents to check whether the rated capacity of emergency power supply facilities meet the continuous power supply needs of the loading equipment.

If the test results show that the evaluated equipment room is equipped with emergency power supply facilities or backup power supply system, which can meet the continuous power supply needs of the loading equipment and can be operated normally, the result is conforming; otherwise it is non-conforming or partially conforming.

3.1.10 Electromagnetic Protection

[Standard Requirements]

This control point for Level III includes the evaluation units of L3-PES1-21, L3-PES1-22, and Level IV includes that of L4-PES1-23, L4-PES1-24.

[L3-PES1-21/L4-PES1-23 Interpretation and Description]

The main evaluation target of the evaluation metric that "Power cables and communication cables shall be laid in isolation to avoid mutual interference" is the cable in the equipment room.

The key points of the evaluation implementation include: checking whether power cables and communication cables in the equipment room are laid separately. For example, power cables and communication cables in the equipment room are laid in different bridge frames and pipes.

If the test results show that power cables and communication cables in the equipment room are laid in isolation, the result is conforming; otherwise it is non-conforming or partially conforming.

[L3-PES1-22 Interpretation and Description]

The main evaluation targets of the evaluation metric that "The electromagnetic shielding shall be implemented for critical devices" are critical devices in the equipment room.

The key points of the evaluation implementation include: checking whether electromagnetic shielding is implemented for critical devices in the equipment room, such as electromagnetic shielding instruments, electromagnetic shielding nets, etc.; checking whether the electromagnetic shielding device has passed the acceptance or the technical inspection by the relevant state authorities or whether it has a certificate of conformity.

If the test results show that critical devices in the evaluated equipment room are equipped with qualified electromagnetic shielding apparatus, the result is conforming; otherwise it is non-conforming or partially conforming.

【L4-PES1-24 Interpretation and Description】

The main evaluation targets of the evaluation metric that "The electromagnetic shielding shall be implemented for critical devices or critical areas" are critical devices or critical areas in the equipment room.

The key points of the evaluation implementation include: checking whether electromagnetic shielding is implemented for critical devices or critical areas in the equipment room, such as electromagnetic shielding instruments, electromagnetic shielding nets, etc.; checking whether the electromagnetic shielding device has passed the acceptance or the technical inspection by the relevant state authorities or whether it has a certificate of conformity.

If the test results show that critical areas or critical devices in the evaluated equipment room are equipped with qualified electromagnetic shielding apparatus, the result is conforming; otherwise it is non-conforming or partially conforming.

3.2 Security Communication Network

3.2.1 Network Architecture

【Standard Requirements】

This control point for Level Ⅲ includes the evaluation units of L3-CNS1-01, L3-CNS1-02, L3-CNS1-03, L3-CNS1-04, L3-CNS1-05, and Level Ⅳ includes that of L4-CNS1-01, L4-CNS1-02, L4-CNS1-03, L4-CNS1-04, L4-CNS1-05, L4-CNS1-06.

【L3-CNS1-01/L4-CNS1-01 Interpretation and Description】

The main evaluation targets of the evaluation metric that "It shall be ensured that the business processing capacity of the network devices meet the needs during the peak period of the business" are routers, switches, wireless access devices, firewalls and other devices or related components that provide network communication functions.

The key points of the evaluation implementation include: interviewing with the network administrators/business administrator, checking recent statistics on the network management platform, determining the business peak hour of the evaluated system, and checking whether the business processing capabilities of the main network devices meet the needs; if it cannot accurately know the service peak and there is no equipment breakdown ever, if conditions pernut, such methods as pressure testing can be used to analyze whether the service processing capability of the network equipment is redundant. The evaluation process shall focus on checking whether key network devices (such as core switches, firewalls, etc.) have experienced downtime and restart problems caused by insufficient device performance. If there exists such a problem, special attention shall be paid during risk analysis.

If the test results show that the peak usage of the CPU and memory of the network device is not greater than 70%, and there is no device downtime or abnormal restart alert log in the network management platform, the result is conforming; otherwise it is non-conforming or partially conforming.

[L3-CNS1-02/L4-CNS1-02 Interpretation and Description]

The main evaluation target of the evaluation metric that "It shall be ensured that the bandwidth of each part of the network meets the needs of during the peak period of the business" is the integrated network management system.

The key points of the evaluation implementation include: interviewing with the network administrator to understand the peak period of business, and checking the current and historical resource utilization of important network devices (backplane bandwidth, concurrency, throughput, etc.) during the business peak, and whether there is a long-term utilization rate above 80%; checking the historical resource utilization of devices from the integrated network management system during the peak period of business if the integrated network management system is deployed in the evaluated system and evaluation; checking the switch interface bandwidth usage during peak business hours if the integrated network management system is not deployed; checking the log of the integrated network management system or the running time of devices to confirm whether there have been insufficient bandwidth or device downtime due to insufficient device processing capacity.

If the test results show that the bandwidth of each part of the evaluated network meets the needs of the peak period of business, the result is conforming; otherwise it is non-conforming or partially conforming.

[L3-CNS1-03/L4-CNS1-03 Interpretation and Description]

The main evaluation targets of the evaluation metric that "Different network areas shall be divided, and addresses shall be assigned to each network area in accordance with the principle of convenient management and control" include routers, switches, wireless access devices and firewalls that provide network communication functions or related components.

The key points of the evaluation implementation include: interviewing with the network administrator to understand the principle of network area division, including physical network division and logical division (for example, division according to the principles of different security levels, different functional areas, etc.); checking the relevant configuration to confirm whether it is consistent with the network area (or subnet) planning.

If the test results show that the evaluated network is divided into different network areas based on factors such as importance and department, and the relevant network device configuration files are checked to verify that the divided network areas are consistent with the division principles the result is conforming; otherwise it is non-conforming or partially conforming.

【L3-CNS1-04/L4-CNS1-04 Interpretation and Description】

The main evaluation target of the evaluation metric that "important network areas shall avoid being deployed at the boundary, and reliable technical isolation measures shall be adopted between important network areas and other network areas" is the network topology.

The key points of the evaluation implementation include: interviewing with the network administrator and reviewing the network topology, checking whether the actual network area division is consistent with the network topology diagram, and checking whether important network areas are deployed at the boundary. For example, the core exchange area is directly exposed at the network boundary, checking whether ACLs are installed or firewalls, gatekeepers, one-way isolation devices are deployed between important network areas and other network areas, and verifying the effectiveness of technical isolation measures.

If the test results show that the important network area is not deployed at the network boundary, the reliable technical isolation method is adopted between important network areas and other network areas, the result is conforming; otherwise it is non-conforming or partially conforming.

【L3-CNS1-05/L4-CNS1-05 Interpretation and Description】

The main evaluation targets of the evaluation metric that "Hardware redundancy for communication lines, critical network devices and critical computing devices shall be provided to ensure the availability of the system" include the network topology, communication lines, critical network devices and critical computing devices, etc.

The key points of the evaluation implementation include: checking whether the network topology of the evaluated system and evaluation adopts the redundant design; checking whether critical network devices and critical computing devices adopt high availability measures such as hardware stacking, virtualization, active-standby or active-active; checking whether the communication lines are provided by different network operators with at least dual links.

If the test results show that the communication lines, critical network devices and key computing devices of the evaluated system and evaluation have hardware redundancy, the result is conforming; otherwise it is non-conforming or partially conforming.

【L4-CNS1-06 Interpretation and Description】

The main evaluation targets of the evaluation metric that "Bandwidth shall be allocated according to the importance of business services, and priority shall be given to guaranteeing important services" include routers, switches, and flow control devices that provide bandwidth control functions or related components.

The key points of the evaluation implementation include: checking whether the bandwidth control devices are configured according to the importance of business services and the bandwidth strategies are enabled.

Chapter 3 Application and Interpretation of the General Security Evaluation Requirements at Level III and Level IV

If the test results show that network devices have bandwidth control strategy configuration commands like PQ, CQ, WFQ, CBWFQ, LLQ, WRED, CAR, NBAR, GTS, etc., the result is conforming; otherwise it is non-conforming or partially conforming.

3.2.2 Communication Transmission

【Standard Requirements】

This control point for Level III includes the evaluation units of L3-CNS1-06 and L3-CNS1-07, and Level IV includes that of L4-CNS1-07, L4-CNS1-08, L4-CNS1-09, L4-CNS1-10.

【L3-CNS1-06 Interpretation and Description】

The main evaluation targets of the evaluation metric that "Verification techniques or cryptography techniques shall be adopted to ensure the integrity of the data during the communication process" are devices or components that provide verification technology or cryptographic technology functions, such as cipher machine, SSL/IPsec VPN, and so on.

The key points of the evaluation implementation include: interviewing with the security administrator and checking the design/acceptance documents to understand the measures to be taken to protect the data integrity in the communication process; testing to verify the effectiveness of the integrity protection measures: by tampering with the data message of the communicating party, checking whether the other party can identify it.

If the test results show that verification technologies or cryptographic technologies are used in the data transmission process to ensure its integrity, the result is conforming; otherwise it is non-conforming or partially conforming.

【L4-CNS1-07 Interpretation and Description】

The main evaluation target of the evaluation metric that "Cryptographic technology shall be adopted to ensure the integrity of data in the communication process" is the overall system network under text and evaluation. In the evaluation implementation, devices or components that provide cryptographic technology functions shall be selected according to the actual situation of the information system for verification. Typical evaluation targets include cipher machines, vertical encryption devices, and SSL/IPsec VPN, etc.

The key points of the evaluation implementation include: checking whether cryptographic technology is adopted to protect data in the communication process, and checking the specific ways in which protection measures are implemented. Generally speaking, it can be implemented either at the network level (such as deploying the cipher machine) or at the application level (such as using the HTTPS protocol), realizing either one of the two levels shall suffice; the effectiveness of integrity protection measures shall be tested and verified. If conditions permit, the data message of one party of the communication can be tampered with to check whether the other party of the communication can identify it.

If the test results show that the evaluated system uses cryptographic technology to ensure its integrity during the data transmission process, the result is conforming; otherwise it is non-conforming or partially conforming.

[L3-CNS1-07/L4-CNS1-08 Interpretation and Description]

The main evaluation targets of the evaluation metric that "cryptography techniques shall be used to ensure the confidentiality of data during the communication process" are cipher machines, SSL/IPsec VPN and other devices or components that provide cryptographic technology functions.

The key points of the evaluation implementation include: understanding the types of data communicated and transmitted in the evaluated system, such as authentication data, important business data, personal information data, etc., and analyze their confidentiality and security requirements; checking whether cryptographic technology is used to Security the data processing during the communication process, and check the specific ways in which the protection measures are implemented. Generally speaking, it can be implemented either at the network level (such as deploying cipher machines, SSL/IPsec VPN, etc.), or through the application layer (such as using HTTPS, etc.), realizing either one of the two levels shall suffice; testing shall be performed to verify confidentiality protection The effectiveness of the protection measures shall be analyzed and verified by capturing communication messages to analyze whether sensitive information such as user passwords and session keys are encrypted.

If the test results show that the evaluated system and evaluation uses cryptographic technology to ensure the confidentiality of the data in the communication process, the result is conforming; otherwise it is non-conforming or partially conforming.

[L4-CNS1-09 Interpretation and Description]

The main evaluation targets of the evaluation metric that "Both parties of the communication shall be verified or authenticated based on cryptography techniques prior to communication" are cipher machines, SSL/IPsec VPN and other devices or components that provide cryptographic technology functions.

The key points of the evaluation implementation include: interviewing with the network administrator and checking design/acceptance documents, checking whether the two parties to the communication are verified or authenticated based on cryptographic technology prior to communication, checking whether the cryptographic technology used is correct and effective, analyzing and verifying it by capturing communication data packets.

If the test results show that the two parties in the communication are verified or authenticated based on cryptographic technology before the communication, the result is conforming; otherwise it is non-conforming or partially conforming.

[L4-CNS1-10 Interpretation and Description]

The main evaluation targets of the evaluation metric that "The cryptographic computation and key management shall be performed for important communication processes based on hardware cryptographic modules" are devices or components that provide cryptographic technology functions, such as cipher machines, vertical encryption devices, SSL/IPsec VPNs, etc.

The key points of the evaluation implementation include: understanding the types of data communicated and transmitted in the evaluated system, such as authentication data, important business data, personal information data, etc., and analyzing their confidentiality requirements; checking whether the key is generated and cryptographic computing is performed based on the hardware cryptographic modules. Typical products generally include cipher machines, vertical encryption devices, SSL/IPsec VPN, etc.; checking whether the relevant product has obtained the valid inspection report or the cryptographic product model certificate specified by the state cryptography administration.

If the test results show that the cryptographic computation and key management for important communication processes are based on hardware cryptographic modules, the result is conforming; otherwise it is non-conforming or partially conforming.

3.2.3 Trusted Verification

[Standard Requirements]

This control point for Level III includes the evaluation unit of L3-CNS1-08, and Level IV includes the evaluation unit of L4-CNS1-11.

[L3-CNS1-08 Interpretation and Description]

The main evaluation targets of the evaluation metric that "The trusted verification of the system boot loader, the system program, important configuration parameters and communication application of the communication device can be performed on the basis of trusted root, and dynamic trusted verification shall be performed in the key execution link of the application. After detecting that the credibility is damaged, an alarm shall be issued and the verification result shall be formed into an audit record and sent to the security management center" are devices or components that provide trusted verification, and systems that provide centralized audit functions.

The key points of the evaluation implementation include: interviewing with the network administrator to understand what kind of trusted verification architecture and trusted root are used in the system, and understanding whether the communication device contains chip with trusted root and reviewing the technical white papers of communication devices; checking whether the system boot loader, the system program, important configuration parameters and communication application program can be measured and verified step by step through the algorithm and keys in the chip with trusted root and the integrated special micro-controller during the boot process of related communication device; checking whether the communication devices use trusted verification technologies such as the application behavior whitelist to dynamically obtain their operating characteristics and perform dynamic measurement and verification in the key execution links of the application, and judge whether the key execution links are operating normally according to rules or model analysis; checking whether the communication device can detect the credibility of the system boot loader, system program, important configuration parameters and communication application program is damaged and then generate an alarm; checking whether the trusted verification result is formed into an audit record and sent to the security management center.

If the test results show that all the requirements of the evaluation metric are met after the verification of the above steps, the result is conforming; otherwise it is non-conforming or partially conforming.

〖L4-CNS1-11 Interpretation and Description〗

The main evaluation targets of the evaluation metric that "The trusted verification of the system boot loader, the system program, important configuration parameters and communication application of the communication device can be performed on the basis of trusted root, and dynamic trusted verification shall be performed in all execution links of the application. After detecting that the credibility is damaged, an alarm shall be generated and the verification result shall be formed into an audit record and sent to the security management center for dynamic correlation and situation awareness" are devices or components that provide trusted verification, and systems that provide centralized audit functions.

The key points of the evaluation implementation include: interviewing with the network administrator to understand what kind of trusted verification architecture and trusted root are used in the system, and understanding whether the communication device contains chip with trusted root and reviewing the technical white papers of communication devices; checking whether the system boot loader, the system program, important configuration parameters and communication application program can be measured and verified step by step through the algorithm and keys in the chip with trusted root and the integrated special micro-controller during the boot process of related communication device; checking whether the communication devices use trusted verification technologies such as the application behavior whitelist to dynamically obtain their operating characteristics and perform dynamic measurement and verification in the key execution links of the application, and judge whether the key execution links are operating normally according to rules or model analysis; checking whether the communication device can detect the credibility of the system boot loader, system program, important configuration parameters and communication application program is damaged and then generate an alarm; checking whether the trusted verification result is formed into an audit record and sent to the security management center for dynamic correlation and situation awareness.

If the test results show that the above checking results meet the requirements of the evaluation metric, the result is conforming; otherwise it is non-conforming or partially conforming.

3.3 Security Area Boundary

3.3.1 Boundary Protection

[Standard Requirements]

This control point for Level Ⅲ includes the evaluation units of L3-ABS1-01, L3-ABS1-02, L3-ABS1-03, L3-ABS1-04, and Level Ⅳ includes that of L4-ABS1-01, L4-ABS1-02, L4-ABS1-03, L4-ABS1-04, L4-ABS1-05, L4-ABS1-06.

[L3-ABS1-01/ L4-ABS1-01 Interpretation and Description]

The main evaluation targets of the evaluation metric that "It shall be ensured that any cross-boundary access and data flow is communicated through a controlled interface provided by the boundary device" are devices or related components that provide access control functions such as gatekeepers, firewalls, routers, switches and wireless access gateway devices.

The key points of the evaluation implementation include: interviewing with the network administrator and obtaining the network topology diagram consistent with the actual operation condition, and identifying the external boundaries of the system and the boundaries between network areas within the system in the network topology diagram; combined with the network devices list and the boundary conditions identified in the network topology, checking whether boundary access control devices are deployed at the network boundary; checking the device connected to the physical interface of the boundary access control device, VLAN division, router configuration; checking whether the data flow across the boundary is through the designated port for network communication, and whether the designated port is configured and the security strategy is enabled; check whether there is any violation of the establishment of wireless Wi-Fi hot spots.

If the test results show that boundary access control devices are deployed on the external network boundary and internal network area boundary of the evaluated system and evaluation, the access and data flow across the boundaries are all through the designated port of the boundary device for network communication, and the designated port is configured and the security strategy is enabled, there is no violation of the establishment of Wi-Fi hot spots in the network, the result is conforming; otherwise it is non-conforming or partially conforming.

[L3-ABS1-02/ L4-ABS1-02 Interpretation and Description]

The main evaluation targets of the evaluation metric that "It shall be able to check or restrict unauthorized devices from privately connecting to the internal network" are endpoint management systems, routers, switches, security devices and other related equipment.

The key points of the evaluation implementation include: interviewing with the network administrator to understand the inspection or restriction measures to prevent unauthorized devices from privately connecting to the internal network; checking the configuration strategies of related systems or devices to see if the devices connected to the network are authorized; with the consent and cooperation of the network administrator, trying to use unauthorized devices to connect to the internal network to verify the effectiveness of related technical control measures; checking whether all idle ports of routers, switches and other related devices are closed, and checking whether the IP/MAC address binding strategy is configured for the connected device in the network.

If the test results show that the system under evaluation has taken technical measures to control the access of unauthorized devices to the internal network, and all routers, switches and other related devices have closed idle ports, and the connected devices have been configured with IP/MAC address binding strategy, the result is conforming; otherwise it is non-conforming or partially conforming.

[L3-ABS1-03/ L4-ABS1-03 Interpretation and Description]

The main evaluation targets of the evaluation metric that "It shall be able to detect or limit the behavior of internal users' unauthorized access to the external network" are endpoint management system, routers, switches, security devices, business endpoints and management endpoints related devices.

The key points of the evaluation implementation include: interviewing with the system administrator and network administrator to understand network areas for business endpoints and system management endpoints, network resources accessed by various endpoints, checking the routing configuration information of network areas where endpoints are located, and judging whether the routing configuration is correct or not, and whether there is redundant routing configuration; interviewing with the network administrator to find out whether the terminal illegal external connection monitoring device is deployed, and whether the device can check or restrict the external connection behavior of endpoints. With the cooperation of the network administrator, test and verify whether the function of the illegal external connection monitoring device is effective.

If the test results show that the evaluated system and evaluation can check or restrict the unauthorized connection of internal users to the external network, the result is conforming; otherwise it is non-conforming or partially conforming.

[L3-ABS1-04/ L4-ABS1-04 Interpretation and Description]

The main evaluation targets of the evaluation metric that "The use of the wireless network shall be restricted to ensure that the wireless network accesses the internal network through controlled boundary devices" are network topology diagram and wireless network devices. If the evaluated system does not involve a wireless network, the evaluation metric is not applicable.

The key points of the evaluation implementation include: interviewing with the network administrator and reviewing the network topology diagram to confirm the boundary between the wireless network and the internal network; checking whether access control devices are deployed at the boundaries; checking the physical interface of the boundary protection device, VLAN division, routing strategy configuration and other information, and checking whether the wireless network can only access the internal wired network through the border protection device.

If the test results show that the wireless network of the evaluated system and evaluation is connected to the internal wired network after the separate networking, and border protection devices are deployed between the internal wired network, and the wireless network can only be connected to the internal wired network through the border protection device, the result is conforming; otherwise it is non-conforming or partially conforming.

[L4-ABS1-05 Interpretation and Description]

The main evaluation targets of the evaluation metric that "It shall be able to effectively block unauthorized access to the internal network by unauthorized devices or unauthorized access to the external network by internal users" are endpoint management systems, routers, switches, security devices, business endpoints, management endpoints and other related devices.

The key points of the evaluation implementation include: interviewing with the network administrator to understand the deployment of network access control device, and with the cooperation of the network administrator, testing whether the device can block unauthorized device from accessing; interviewing with the network administrator to understand the deployment of illegal external connection monitoring devices, with the cooperation of the network administrator, selecting the endpoint with illegal external connection monitoring, inserting the USB wireless network card to check whether the operation is blocked, or connecting its network interface to the external network to verify whether its access to the external is blocked.

If the test results show that the network access control device and the illegal external connection monitoring device are deployed in the tested network, which can block unauthorized devices from connecting to the internal network privately or internal users from connecting to the external network without authorization, the result is conforming; otherwise it is non-conforming or partially conforming.

[L4-ABS1-06 Interpretation and Description]

The main evaluation targets of the evaluation metric that "Trusted verification mechanism shall be adopted to perform trusted authentication on devices connected to the network to ensure that devices accessing the network are authentic and reliable" are endpoints, network devices and security devices.

The key points of the evaluation implementation include: checking whether the trusted verification mechanism is used in the network to verify the credibility of the devices connected to the network; carrying out the field test to verify the credibility of the devices connected to the internal network.

If the test results show that the evaluated system uses the trusted verification mechanism to perform trusted verification on the devices connected to the network, the result is conforming; otherwise it is non-conforming or partially conforming.

3.3.2 Access Control

[Standard Requirements]

This control point for Level III includes the evaluation units of L3-ABS1-05, L3-ABS1-06, L3-ABS1-07, L3-ABS1-08, L3-ABS1-09, and Level IV includes that of L4-ABS1-07, L4-ABS1-08, L4-ABS1-09, L4-ABS1-10, L4-ABS1-11.

[L3-ABS1-05/ L4-ABS1-07 Interpretation and Description]

The main evaluation targets of the evaluation metric that "The access control rules shall be set between network boundaries or areas according to the access control strategy. By default, the controlled interface shall reject all communication except the allowed communications" include the network topology and devices or related components that provide access control functions such as gatekeepers, firewalls, routers, switches, and wireless access gateways.

The key points of the evaluation implementation include: interviewing with the network administrators, obtaining the latest network topology diagram to understand network boundaries and regional boundaries, determining the devices that provide boundary access control measures and understanding the communication status of various types of data between different network boundaries; checking whether the boundary access control device is configured and enabled with the access control strategy, and the access control strategy specifies the source address, destination address, source port, destination port and protocol, etc.; focusing on checking whether the last access control strategy of the boundary access control device prohibits all network communications. If this access control strategy is not explicitly configured, it is necessary to analyze whether the device rejects all communications by default, or it can be verified by constructing the test data packet.

If the test results show that the boundary access control device is configured and the access control strategy is enabled, and the last item of the access control strategy is that the controlled interface denies all communications except for allowed communications, the result is conforming; otherwise it is non-conforming or partially conforming.

[L3-ABS1-06/ L4-ABS1-08 Interpretation and Description]

The main evaluation targets of the evaluation metric that "The redundant or invalid access control rules shall be deleted, the access control list shall be optimized, and it shall be ensured that the number of access control rules are minimized" include the network topology and devices or related components that provide access control functions such as gatekeepers, firewalls, routers, switches, and wireless access gateways.

The key points of the evaluation implementation include: checking the access control rules of gatekeepers, firewalls, routers, switches, etc., and checking whether there are redundant or invalid access control rules (such as temporary rules for testing, debugging, emergency, etc., rules that are obviously different from the actual network topology, rules that cannot explain their roles, etc.). Analyzing the access control strategy as a whole to check whether the logical relationship between different access control rules and the sequence of rules are reasonable.

If the test results show that the logic of the access control rules for the relevant device is clear and specific, there are no redundant or invalid rules, and the administrator regularly optimizes the access control rules, the result is conforming; otherwise it is non-conforming or partially conforming.

[L3-ABS1-07/ L4-ABS1-09 Interpretation and Description]

The main evaluation targets of the evaluation metric that "The source address, destination address, source port, destination port and protocol shall be checked to allow/deny packets in and out" include the network topology and devices or related components that provide access control functions such as gatekeepers, firewalls, routers, switches, and wireless access gateways.

The key points of the evaluation implementation include: checking whether the access control strategy of the access control device specifies the source address, destination address, source port, destination port and protocol. Generally speaking, in addition to the deny access control strategy, the source address and destination address of other access control strategies are forbidden to be arbitrary at the same time. If one of them is arbitrary, its rationality shall be judged based on the business scenario; logging into the relevant device to view the access control rules, and checking whether the range restriction of the source address, destination address, source port and destination port in the access control rule is reasonable. Generally speaking, the range restriction shall not be too excessive, otherwise there will be greater security risks.

If the test results show that the access control rules specify the source address, destination address, source port, destination port and protocol, etc., and there is no excessive restriction, the result is conforming; otherwise it is non-conforming or partially conforming.

[L3-ABS1-08/ L4-ABS1-10 Interpretation and Description]

The main evaluation targets of the evaluation metric that "It shall be able to provide explicit permission/denial of access for incoming and outgoing data streams based on session status information" include the network topology and devices or related components that provide access control functions such as gatekeepers, firewalls, routers, switches, and wireless access gateways.

The key points of the evaluation implementation include: interviewing with the network administrator, determine the devices that use session authentication and other mechanisms to provide clear permission/deny of access for incoming and outgoing data streams, and conducting on-site checking. For example, the status detection firewall can track the connection session status through the status detection checklist, and making a comprehensive judgment based on the relationship between the data packets before and after, and then decide whether to allow the data packet to pass through; it shall be tested to verify whether the ability to explicitly allow/deny access for incoming and outgoing data stream is provided.

If the test results show that the relevant device has the status detection function and the status detection function is enabled, the result is conforming; otherwise it is non-conforming or partially conforming.

[L3-ABS1-09 Interpretation and Description]

The main evaluation targets of the evaluation metric that "The access control based on application protocol and application content shall be realized for the data streams in and out of the network" include the network topology, second-generation firewalls, application firewalls, and other devices or related components that provide application layer access control functions.

The key points of the evaluation implementation include: checking whether devices or related components with application layer access control functions are deployed at key network nodes, and typical devices include second-generation firewalls, application firewalls, etc. ; logging into the devices such as application firewalls to check which servers in the area are under the protection of the application firewall, then further checking which kinds of protections are activated, whether they include SQL injection, XSS attack, cross-site request forgery, system command injection and other types of protections, checking the actions of applying protection rules. Generally speaking, behaviors with a risk level of medium or higher shall be intercepted; it shall be tested to verify whether the relevant device access control strategy can achieve the access control based on the application protocol and application content on the data flow entering and leaving the network. Penetration testing can be used to verify whether the device can intercept related statements and characters that may cause SQL injection and cross-site scripting.

If the test results show that the relevant device protection rules are configured properly, and the penetration test results show that the application firewall can block attacks such as SQL injection and cross-site scripting, the result is conforming; otherwise it is non-conforming or partially conforming.

[L4-ABS1-11 Interpretation and Description]

The main evaluation targets of the evaluation metric that "the data exchange shall be carried out at the network boundary by means of communication protocol conversion or communication protocol isolation" include the network topology and devices that provide communication protocol conversion or communication protocol isolation functions such as gatekeepers and FGAP.

The key points of the evaluation implementation include: interviewing with the network administrator to understand the method of data exchange at the network boundary, and checking whether gatekeepers, FGAP, and other devices are deployed at the network boundary to perform communication protocol conversion or communication protocol isolation for data entering and leaving the network boundary; logging into the gatekeeper, FGAP and other devices with the cooperation of the administrator to check whether the device is operating normally and whether relevant strategies are properly configured, confirming that the device has stripped and rebuilt relevant general protocols; the gatekeeper shall be tested and verified through test methods such as sending data with general protocols, testing and verifying whether GAP and FGAP devices can effectively block them.

If the test results show that the gatekeeper device is deployed at the network boundary to perform communication protocol conversion or communication protocol isolation, and the gatekeeper device is configured correctly and operates normally, the result is conforming; otherwise it is non-conforming or partially conforming.

3.3.3 Intrusion Prevention

[Standard Requirements]

This control point for Level III includes the evaluation units of L3-ABS1-10, L3-ABS1-11, L3-ABS1-12, L3-ABS1-13, and Level IV includes that of L4-ABS1-12, L4-ABS1-13, L4-ABS1-14, L4-ABS1-15.

[L3-ABS1-10/ L4-ABS1-12 Interpretation and Description]

The main evaluation targets of the evaluation metric that "Network attack behaviors initiated from the external threat actors shall be detected, prevented or restricted at critical network nodes" include anti-APT attack system, network backtracking system, threat intelligence detection system, anti-DDoS attack system, intrusion protection system or related components.

The key points of the evaluation implementation include: interviewing with the network administrator and checking the network topology to determine whether devices or related components that provide intrusion detection functions are deployed at key network nodes of the network (such as the Internet boundary, core data exchange nodes, etc.), and checking whether the scope of detection covers key network nodes, and focusing on checking whether it covers the boundaries connected to external networks; logging into the relevant devices or components to check the update time of the rule base version and the threat intelligence library, and checking whether its configuration information or security strategy is reasonable; adopting penetration testing and other methods to simulate external network attack behaviors to verify whether the configuration information or security strategy of related devices or components are effective.

If the test results show that the evaluated system has deployed intrusion detection devices or related components at key nodes, and the security protection strategy is properly configured and the strategy database is updated in a timely manner, which can detect, prevent or limit external network attacks, the result is conforming; otherwise it is non-conforming or partially conforming.

【L3-ABS1-11/ L4-ABS1-13 Interpretation and Description】

The main evaluation targets of the evaluation metric that "Network attack behaviors initiated from the internal threat actors shall be detected, prevented or restricted at critical network nodes" include anti-APT attack system, network backtracking system, threat intelligence detection system, anti-DDoS attack system, intrusion protection system or related components.

The key points of the evaluation implementation include: interviewing with the network administrator and checking the network topology to determine whether devices or related components that provide intrusion detection functions are deployed at key network nodes of the network (such as the Internet boundary, core data exchange nodes, etc.), and checking whether the scope of detection covers key network nodes; logging into the relevant devices or components to check the update time of the rule base version and the threat intelligence library, and checking whether its configuration information or security strategy is reasonable; adopting penetration testing and other methods to simulate internal network attack behaviors to verify whether the configuration information or security strategy of related devices or components are effective.

If the test results show that the evaluated system has deployed intrusion detection devices or related components at key nodes, and the security protection strategy is properly configured and the strategy database is updated in a timely manner, which can detect, prevent or limit internal network attacks, the result is conforming; otherwise it is non-conforming or partially conforming.

【L3-ABS1-12/ L4-ABS1-14 Interpretation and Description】

The main evaluation targets of the evaluation metric that "Technical measures shall be taken to analyze network behaviors to achieve the analysis of cyberattacks, especially new types of cyberattacks" are anti-APT attack systems, network backtracking systems, threat intelligence detection systems or related components.

The key points of the evaluation implementation include: interviewing with the network administrator to obtain the network topology diagram, and confirming whether the network attack behavior detection system or related components are deployed in the network, such as anti-APT attack system, network backtracking system, threat intelligence detection system, etc.; checking whether relevant systems or components use machine learning, behavior analysis, dynamic sandboxing, known threat discovery and other technologies to analyze network traffic and form relevant analysis results, so as to realize the detection and analysis of network attacks, especially new types of network attacks; checking whether the rule base version or threat intelligence library of related systems or components is updated in a timely manner; checking whether the configuration information or security strategies of related devices or components are reasonable, and whether they can monitor network attacks (such as port scanning, Trojan horse backdoor attacks, buffer overflow attacks, network worm attacks, etc.), and verify them by viewing log records.

If the test results show that anti-APT device is deployed in the network and the attack incident detection strategy is enabled, the rule base or the threat intelligence library is updated in a timely manner, which can detect and analyze new cyberattack behaviors, the result is conforming; otherwise it is non-conforming or partially conforming.

[L3-ABS1-13/ L4-ABS1-15 Interpretation and Description]

The main evaluation targets of the evaluation metric that "When the attack behavior is detected, the attack source IP, attack type, attack target and attack time shall be recorded, and the alert shall be generated when any serious intrusion incident occurs" include the anti-APT attack system, network backtracking system, threat intelligence detection system, anti-DDoS attack system and intrusion protection system or related components.

The key points of the evaluation implementation include: checking whether the log information of the relevant system or component is complete, whether the content at least contains the attack source IP, attack type, attack target, attack time and other information; checking whether the relevant system or component is equipped with the alert mechanism, and when a serious intrusion incident occurs, whether the alert can be generated by means of sound and light, email, text message, etc.

If the test results show that the log records of the relevant devices meet the requirements, and the alert is generated in the event of any serious intrusion incident, the result is conforming; otherwise it is non-conforming or partially conforming.

3.3.4 Malicious Code and Spam Prevention

[Standard Requirements]

This control point for Level III includes the evaluation units of L3-ABS1-14, L3-ABS1-15, and Level IV includes that of L4-ABS1-16, L4-ABS1-17.

[L3-ABS1-14/ L4-ABS1-16 Interpretation and Description]

The main evaluation targets of the evaluation metric that "Malicious code shall be detected and removed at critical network nodes, and the upgrade and update of malicious code protection mechanism shall be maintained" include systems or related components that provide network layer anti-malware functions, such as anti-virus gateways, UTM, and firewalls with anti-virus function modules.

The key points of the evaluation implementation include: interviewing with the network administrator and viewing the network topology diagram to determine key network nodes; checking whether anti-malware code systems or related components are deployed at key network nodes, and logging into the device to check whether it is operating normally and whether malicious code is detection function is enabled; checking whether the malicious code library of the anti-malware code products or components is updated in time.

If the test results show that anti-malware code products are deployed at key network nodes, and anti-malware code products are operating normally, and the malicious code library is updated in time, the result is conforming; otherwise it is non-conforming or partially conforming.

[L3-ABS1-15/ L4-ABS1-17 Interpretation and Description]

The main evaluation targets of the evaluation metric that "Spam emails shall be detected and prevented at critical network nodes, and the maintenance and update of spam prevention mechanisms shall be maintained" are systems or related components that provide anti-spam functions such as anti-spam gateways. If the evaluated system does not involve email sending and receiving functions, this evaluation metric is not applicable.

The key points of the evaluation implementation include: checking whether the protection configuration strategy of spam prevention products or related components is turned on and operating normally; checking whether the anti-spam rule base of spam prevention products or related components is updated in a timely manner. In principle, the non-updated period shall not exceed one month; For the enterprise mailbox system provided by the service provider, that is, the devices and mailbox system are not deployed locally, the spam prevention strategy configuration system or related protection protocol provided by the service provider shall be checked to see if it can detect and prevent spam emails; sending spam emails to a specific mailbox to test and verify whether the security strategies of spam prevention products or related components are effective.

If the test results show that the Security email gateway is deployed at key network nodes, the Security mail gateway is operating normally, the security strategy is effective, and the anti-spam rule base can be updated in time, the result is conforming; otherwise it is non-conforming or partially conforming.

3.3.5 Security Audit

[Standard Requirements]

This control point for Level Ⅲ includes the evaluation units of L3-ABS1-16, L3-ABS1-17, L3-ABS1-18, L3-ABS1-19, and Level Ⅳ includes that of L4-ABS1-18, L4-ABS1-19, L4-ABS1-20.

[L3-ABS1-16/ L4-ABS1-18 Interpretation and Description]

The main evaluation targets of the evaluation metric that "Security audits shall be conducted at network boundaries and important network nodes. Auditing shall cover each user to audit important user behaviors and important security incidents" include the comprehensive security audit system and the network audit system and the system platform with similar functions.

The key points of the evaluation implementation include: interviewing with the network administrator and obtaining the network topology diagram, sorting out and analyzing network boundaries and important network nodes; checking whether the system platform with similar functions such as the comprehensive security audit system and the network audit system is deployed, and checking whether the device deployment location is reasonable; analyzing and determining whether the audit scope covers the network boundaries and important network nodes by checking the device network interface, configuration information, mirroring port configuration, etc. ; checking whether the audit of network traffic behavior is enabled (such as various illegal access protocols and their traffics, and the personnel behavior of accessing sensitive data, etc.); checking whether the security audit scope covers every user, and whether important user behaviors and important security incidents are audited.

If the test results show that devices with similar functions such as network behavior auditing and network traffic auditing are deployed at network boundaries and important network nodes, and the audit scope covers each user, the audit content includes important user behaviors and important security incidents, the result is conforming; otherwise it is non-conforming or partially conforming.

【L3-ABS1-17/ L4-ABS1-19 Interpretation and Description】

The main evaluation targets of the evaluation metric that "Audit records shall include the date and time of the incident, the user, the type of incident, whether the incident is successful, and other audit-related information" include the system platform with similar functions such as the comprehensive security audit system and the network audit system.

The key points of the evaluation implementation include: logging into the comprehensive security audit system and other system platforms with similar functions to check whether the audit record information includes the date and time of the incident, the user, the type of the incident, whether the incident is successful, and other audit-related information. Among them, the accuracy and clock synchronization issues of the date and time shall be concerned about.

If the test results show that the audit record of the relevant equipment includes the date and time of the incident, the user, the type of the incident, whether the incident is successful, and other audit-related information, the result is conforming; otherwise it is non-conforming or partially conforming.

【L3-ABS1-18/ L4-ABS1-20 Interpretation and Description】

The main evaluation target of the evaluation metric that "Audit records shall be protected and regularly backed up to avoid unexpected deletion, modification or coverage" is the system platform with similar functions such as the comprehensive security audit system.

The key points of the evaluation implementation include: checking whether a dedicated administrator is staffed to manage audit records (including local and backup logs of the device), unauthorized users have no right to delete, modify or overwrite audit records, and checking whether there are log storage rules and synchronization outgoing backup mechanism to prevent the original audit log from being accidentally lost or erased by attackers; checking whether technical measures are taken to regularly back up audit records, and checking whether the backup strategy is reasonable. The regular backup here generally refers to pushing logs to the log server and other devices for backup. If the device is configured with the log server or the IP address of centralized log analysis system, it is necessary to confirm that the assets corresponding to the IP are in normal working condition and to conform whether they have effectively stored historical backup logs. If it is the manual backup, checking the backup record and the validity of the backup. Regarding the retention time of log records, the requirements of the *Cybersecurity Law* shall be followed to keep records for at least 6 months.

If the test results show that the relevant system or device can normally collect various security incident audit logs, unauthorized users have no right to delete, modify or overwrite the audit records, and the audit records are regularly backed up, the result is conforming; otherwise it is non-conforming or partially conforming.

[L3-ABS1-19 Interpretation and Description]

The main evaluation targets of the evaluation metric that "It shall be able to conduct separate behavior audit and data analysis for the remote access users and Internet access users" include VPN devices, online behavior management systems or comprehensive security audit systems.

The key points of the evaluation implementation include: interviewing with the security administrator or the system administrator to understand what technical measures are taken to conduct separate behavior audits and data analysis on remote access user behaviors, and logging into relevant devices to check whether the audit strategy configuration is reasonable and whether the device is operating normally; interviewing with the security administrator or the system administrator and asking whether the operation and maintenance endpoints and business endpoints involved in the evaluated system need to access the Internet. If so, check what technical measures are taken to conduct independent behavior audit and data analysis on the behavior of users accessing the Internet, and logging into the relevant device to check whether the audit strategy configuration is reasonable and whether the device is operating normally.

If the test results show that relevant devices are deployed in the evaluated system to conduct separate audits and data analysis for remote access user behaviors and Internet user behaviors, the result is conforming; otherwise it is non-conforming or partially conforming.

Chapter 3　Application and Interpretation of the General Security Evaluation Requirements at Level Ⅲ and Level Ⅳ

3.3.6　Trusted Verification

[Standard Requirements]

This control point for Level Ⅲ includes the evaluation unit of L3-ABS1-20, and Level Ⅳ includes that of L4-ABS1-21.

[L3-ABS1-20 Interpretation and Description]

The main evaluation targets of the evaluation metric that "Trusted verification for the system boot program, system program, important configuration parameters and communication application of the communication device can be performed on the basis of trusted root, and dynamic trusted verification in the key execution link of the application shall be performed, after detecting that the credibility is damaged, the alarm shall be generated and the verification result shall be formed into an audit record and sent to the security management center" are devices or components that provide trusted verification, and systems that provide centralized audit functions.

The key points of the evaluation implementation include: interviewing with the network administrator and reviewing the technical white papers of communication device and other materials to understand whether the boundary device contains chip with trusted root; checking whether the system boot loader, the system program, important configuration parameters and communication application program can be measured and verified step by step through the algorithm and keys in the chip with trusted root and the integrated special micro-controller during the boot process of related boundary device; checking whether the boundary device can generate alarms when the damage to the credibility of relevant applications is detected; checking whether the trusted verification result is formed into an audit record and sent to the security management center.

If the test results show that the relevant devices can meet the requirements of the above evaluation metric, the result is conforming; otherwise it is non-conforming or partially conforming.

[L4-ABS1-21 Interpretation and Description]

The main evaluation targets of the evaluation metric that "The trusted verification of the system boot loader, the system program, important configuration parameters and communication application of the communication device can be performed on the basis of trusted root, and dynamic trusted verification shall be performed in all execution links of the application. After detecting that the credibility is damaged, an alarm shall be generated and the verification result shall be formed into an audit record and sent to the security management center for dynamic correlation and situation awareness" are devices or components that provide trusted verification, and systems that provide centralized audit functions.

The key points of the evaluation implementation include: checking whether the system boot loader, the system program, important configuration parameters and boundary protection application can be measured and verified step by step through the algorithm and keys in the chip with trusted root and the integrated special micro-controller during the boot process of related boundary device; checking whether boundary devices use trusted verification technologies such as the application behavior whitelist to dynamically obtain their operating characteristics and perform dynamic measurement and verification in the key execution links of the application, and judge whether the key execution links are operating normally according to rules or model analysis; checking whether boundary devices can detect the credibility of the system boot loader, system program, important configuration parameters and communication application program is damaged and then generate an alarm; checking whether the trusted verification result is formed into an audit record and sent to the security management center for dynamic correlation and situation awareness.

If the test results show that the relevant devices can meet the requirements of the above evaluation metric, the result is conforming; otherwise it is non-conforming or partially conforming.

3.4 Security Computing Environment

3.4.1 Network Device and Security Device

1. Identity Authentication

[Standard Requirements]

This control point for Level Ⅲ includes the evaluation units of L3-CES1-01, L3-CES1-02, L3-CES1-03, L3-CES1-04, and Level Ⅳ includes that of L4-CES1-01, L4-CES1-02, L4-CES1-03, L4-CES1-04.

[L3-CES1-01/ L4-CES1-01 Interpretation and Description]

The main evaluation targets of the evaluation metric that "Identification and authentication shall be implemented upon the log-in users. The identification shall be unique and the identity authentication information shall have complexity requirements and change regularly" are network devices (including virtual network devices), security devices (including virtual security devices) and other devices or related components that provide network communication functions.

The key points of the evaluation implementation include: logging into (locally or remotely) the relevant network or security devices to check user configuration information, and check the user list to confirm whether the user identification is unique, test and verify whether there exists users with an empty password; checking whether the user password is at least 8-digits long or longer; if the network device is managed remotely through a third-party device or system, and does not use the authentication function of the network device itself, ten checking whether the identity authentication measures of the third-party device or system meet the above requirements; checking whether the administrator changes the password regularly, and the replacement period is generally no longer than 90 days.

If the test results show that the relevant device has identified and authenticated the log-in users, and the identification is unique, the identification information meets the complexity requirements and the replacement period is reasonable, the result is conforming; otherwise it is non-conforming or partially conforming.

【L3-CES1-02/ L4-CES1-02 Interpretation and Description】

The main evaluation targets of the evaluation metric that "It shall have the function of handling login failure, shall configure and enable related measures such as session ending, limiting the number of illegal logins and automatically logging out when the login connection times out" are network devices (including virtual network devices), security devices (including virtual security devices) and other devices or related components that provide network communication functions.

The key points of the evaluation implementation include: logging into the device under test to review the configuration information and check whether different login methods (such as remote, local) have configured or enabled relevant measures such as session ending, limiting the number of illegal logins, and logging out automatically when the connection is timed out. The evaluation shall focus on checking whether the configuration is reasonable, such as the lock time is too short, the time-out time is too long, etc. ; if remote management is performed through a third-party operation and maintenance platform, the login failure lock function and connection time-out time of the third-party operation and maintenance platform shall be checked at the same time. The configuration requirements are the same as the network devices.

If the test results show that the relevant device has enabled the login failure processing function, and the number of illegal logins, lockout time and login connection timeout time are reasonably configured, the result is conforming; otherwise it is non-conforming or partially conforming.

【L3-CES1-03/ L4-CES1-03 Interpretation and Description】

The main evaluation targets of the evaluation metric that "When conducting remote management, necessary measures shall be taken to prevent the authentication information from being eavesdropped during network transmission" are network devices (including virtual network devices), security devices (including virtual security devices) and other devices or related components that provide network communication functions. If the evaluation target adopts local management, then the evaluation metric is not applicable.

The key points of the evaluation implementation include: interviewing with the network administrator to understand the remote management method; checking whether the device accessing to the third-party operation and maintenance platform uses SSH and HTTPS protocols .if devices are managed through bastions or other devices; checking whether the device is prohibited from using Telnet, HTTP and other in Security protocols for remote management, and trying to use these protocols to log into the evaluated device for testing and verification; if C/S method is used for remote management, checking whether security method such as encryption is used, and packet capture method can be used for analyzing and confirming whether the authentication information is encrypted and protected during network transmission.

If the test results show that security methods such as encryption are used for remote management, the result is conforming; otherwise it is non-conforming or partially conforming.

[L3-CES1-04/ L4-CES1-04 Interpretation and Description]

The main evaluation targets of the evaluation metric that "Two or more combinations of authentication technologies such as passwords, cryptography and biotechnology shall be used to authenticate users, and at least one of the authentication technologies shall be implemented by using cryptographic technology" are network devices (including virtual network devices), security devices (including virtual security devices) and other devices or related components that provide network communication functions.

The key points of the evaluation implementation include: verifying whether two or more combinations of authentication technologies such as dynamic passwords, digital certificates, biometric technology and device fingerprinting are used to authenticate user identities, and verifying whether one of the authentication technologies uses cryptographic technology; checking whether the cryptographic algorithm used in the identity authentication process is a Security cryptographic algorithm. Among them, DES, RSA1024 and below, MD5 and SHA-1 algorithms are cryptographic algorithms with potential security risks. If the above algorithms are used, special attention shall be paid in the risk assessment.

If the test results show that the relevant device uses passwords, cryptography, bioindex and other technologies to authenticate the user's identity, and at least one of the authentication technologies uses qualified and Security cryptographic technology, the result is conforming; otherwise it is non-conforming or partially conforming.

2. Access Control

[Standard Requirements]

This control pint for Level Ⅲ includes the evaluation units of L3-CES1-05, L3-CES1-06, L3-CES1-07, L3-CES1-08, L3-CES1-09, L3-CES1-10, L3-CES1-11, and Level Ⅳ includes that of L4-CES1-05, L4-CES1-06, L4-CES1-07, L4-CES1-08, L4-CES1-09, L4-CES1-10, L4-CES1-11.

[L3-CES1-05/ L4-CES1-05 Interpretation and Description]

The main evaluation targets of the evaluation metric that "Accounts and permissions shall be assigned to the logged-in users" are network devices (including virtual network devices), security devices (including virtual security devices) and other devices or related components that provide network communication functions.

The key points of the evaluation implementation include: checking whether each logged-in user is assigned with an account and permission, checking whether the account permission is manually assigned or created based on a permission template, and verifying the consistency of the permission scope and content of similar accounts; if 3A platform, bastions are used for the management, checking the state of account allocation and permissions in the 3A platform and bastions to confirm whether each logged-in user is assigned with the account and permission; checking whether access permissions of anonymous and default accounts are disabled or restricted. In principle, it is forbidden to log into the device with an anonymous or default account.

If the test results show that the evaluated system has assigned accounts and permissions to the logged-in users, the result is conforming; otherwise it is non-conforming or partially conforming.

[L3-CES1-06/ L4-CES1-06 Interpretation and Description]

The main evaluation targets of the evaluation metric that "The default account shall be renamed or deleted, and the default password of the default account shall be modified" are network devices (including virtual network devices), security devices (including virtual security devices) and other devices or related components that provide network communication functions.

The key points of the evaluation implementation include: checking whether the default account of the device (such as admin, etc.) has been renamed or deleted. If the device does not support renaming or deleting the default account, the default password of the default account shall be modified, and the login address and login method of the default user shall be restricted; the default account or default password shall be used for the device login trial, test and verify whether the default account has been renamed or deleted, and whether the default password of the default account has been modified.

If the test results show that devices or components under test have renamed or deleted the default accounts or modified the default passwords of the default accounts, the result is conforming; otherwise it is non-conforming or partially conforming.

[L3-CES1-07/ L4-CES1-07 Interpretation and Description]

The main evaluation targets of the evaluation metric that "The redundant, expired accounts shall be deleted or deactivated in time to avoid the existence of shared accounts" are network devices (including virtual network devices), security devices (including virtual security devices) and other devices or related components that provide network communication functions.

The key points of the evaluation implementation include: interviewing with the network administrator and reviewing the device configuration file, checking whether there is a one-to-one correspondence between administrator users and accounts, and analyzing whether there are redundant or expired accounts. For accounts whose purpose cannot be confirmed by the assessed party, attention shall be paid to them during risk analysis. If there are redundant and expired accounts, testing and verifying whether redundant and expired accounts are disabled; checking whether there exists the situation of multiple people sharing the same management account, which can be analyzed and judged by comparing the list of administrator users and device accounts. If the operation and maintenance model of the tested system is the multi-shift system or the job-rotation system, there may be situations where different operation and maintenance teams share the management account, and it shall be analyzed whether there are shared accounts in the same time period.

If the test results show that there are no redundant, expired, shared accounts for devices or components under test, and there is a one-to-one correspondence between the administrator user and the account, the result is conforming; otherwise it is non-conforming or partially conforming.

[L3-CES1-08/ L4-CES1-08 Interpretation and Description]

The main evaluation targets of the evaluation metric that "The administration user shall be granted with the needed least privilege to realize the privilege separation of the administration user" are network devices (including virtual network devices), security devices (including virtual security devices) and other devices or related components that provide network communication functions.

The key points of the evaluation implementation include: checking whether the roles are divided, such as setting up the role of system administrator, security administrator and audit administrator, etc.; checking whether the privileges of administration users are separated and understanding their corresponding scope of responsibility, analyzing whether there exists the possibility of unauthorized access; checking whether the privilege of the administration user is the minimum privilege required for the task.

If the test results show that the device or component under test is equipped with multiple administration accounts, and each administration account is granted with minimum privilege, the result is conforming; otherwise it is non-conforming or partially conforming.

[L3-CES1-09/ L4-CES1-09 Interpretation and Description]

The main evaluation targets of the evaluation metric that "The access control strategy shall be configured by the authorization subject, and the access control strategy stipulates the access rules of the subject to the object" are network devices (including virtual network devices), security devices (including virtual security devices) and other devices or related components that provide network communication functions.

The main points of the evaluation include: checking whether the authorization subject configures the access control strategy. Generally, the authorization subject in the network equipment refers to the administration account with overall operation permissions, and the access permission of the object shall be specified by the owner (that is, the subject), such as allowing the subject (administration user) to access what level of object (such as ACL strategy), and the classification of the object shall also be determined by the subject;

If the test results show that high-level administrators can configure the overall parameters, and different levels of administrators can configure configuration parameters under corresponding permissions, and low-privilege accounts cannot use high-privilege commands to configure, the result is conforming; otherwise it is non-conforming or partially conforming.

[L3-CES1-10/ L4-CES1-10 Interpretation and Description]

The main evaluation targets of the evaluation metric that "The granularity of access control shall achieve that the subject is at the user level or process level, and the object is at the file and database table level" are network devices (including virtual network devices), security devices (including virtual security devices) and other devices or related components that provide network communication functions.

The main points of the evaluation include: checking the configuration file of the device, and checking the control granularity of its access control strategy to see whether the subject reaches the user level or process level and the file, whether the object is at the file, database table, record or field level.

If the test results show that the access control granularity of the related devices meet the requirements of the evaluation metric, the result is conforming; otherwise it is non-conforming or partially conforming.

[L3-CES1-11 Interpretation and Description]

The main evaluation targets of the evaluation metric that "Security tags shall be set up for important subjects and objects, and the subject shall be restrained from accessing to information resources with security tags" are network devices (including virtual network devices), security devices (including virtual security devices) and other devices or related components that provide network communication functions.

The key points of the evaluation implementation include: interviewing with network administrators to understand what security model the device adopts to control user's access to device resources, checking whether the device supports the setting of security tags for important subjects and objects. If the device uses mandatory access control, checking whether the corresponding access control strategy is reasonable, testing and verifying users with low privilege can successfully access objects outside their permissions.

If the test results show that the evaluated device adopts mandatory access control, sets up security tags for important subjects and objects, and restrain the subject's access to information resources with security tags, the result is conforming; otherwise it is non-conforming or partially conforming.

[L4-CES1-11 Interpretation and Description]

The main evaluation targets of the evaluation metric that "Security tags shall be set for the subject and the object, the subject's access to the object shall be based on security tags and mandatory access control rules" are network devices (including virtual network devices), security devices (including virtual security devices) and other devices or related components that provide network communication functions.

The key points of the evaluation implementation include: interviewing with network administrators to understand what security model the device adopts to control user's access to device resources, checking whether the device supports the setting of security tags for important subjects and objects. If the device uses mandatory access control, checking whether the corresponding access control strategy is reasonable, testing and verifying users with low privilege can successfully access objects outside their permissions.

If the test results show that the evaluated device adopts mandatory access control, sets up security tags for important subjects and objects, and restrain the subject's access to information resources with security tags, the result is conforming; otherwise it is non-conforming or partially conforming.

3. Security Audit

[Standard Requirements]

This control point for Level III includes the evaluation units of L3-CES1-12, L3-CES1-13, L3-CES1-14, L3-CES1-15, and Level IV includes that of L4-CES1-12, L4-CES1-13, L4-CES1-14, L4-CES1-15.

[L3-CES1-12/ L4-CES1-12 Interpretation and Description]

The main evaluation targets of the evaluation metric that "The security audit function shall be enabled, covering each user and auditing important user behaviors and important security incidents" are devices or related components that provide network communication functions such as network devices (including virtual network devices), security devices (including virtual security devices).

The key points of the evaluation implementation include: checking the configuration file of related device, confirming whether the audit function is enabled, checking whether the audit scope covers each user, whether important user behaviors and important security incidents are audited. If the third-party security audit product is connected, checking whether the audit scope of third-party security audit products can cover each user, and whether important user behaviors and important security incidents are audited.

If the test results show that the device or system under test has the security audit function enabled, the audit covers every user, and important user behaviors and important security incidents are audited, the result is conforming; otherwise it is non-conforming or partially conforming.

Chapter 3 Application and Interpretation of the General Security Evaluation Requirements at Level Ⅲ and Level Ⅳ

【L3-CES1-13 Interpretation and Description】

The main evaluation targets of the evaluation metric that "The audit record shall include the date and time of the incident, the user, the type of incident, the success of the incident, and other audit-related information" are devices or related components that provide network communication functions such as network devices (including virtual network devices), security devices (including virtual security devices).

The key points of the evaluation implementation include: checking whether the audit record type is comprehensive, including at least the date and time of the incident, the user, the type of incident, the success of the incident, and other audit-related information.

If the test results show that the audit record includes the date and time of the incident, the user, the type of incident, whether the incident is successful or not, and other audit-related information, the result is conforming; otherwise it is non-conforming or partially conforming.

【L4-CES1-13 Interpretation and Description】

The main evaluation targets of the evaluation metric that "The audit record shall include the date and time of the incident, the type of incident, the subject identification, the object identification and results" are devices or related components that provide network communication functions such as network devices (including virtual network devices), security devices (including virtual security devices).

The key points of the evaluation implementation include: checking whether the data types of audit records are comprehensive, including the date and time of the incident, incident type, the subject identification, the object identification and results, and other audit-related information.

If the test results show that the audit record includes audit related information such as the date and time of the incident, incident type, the subject identification, the object identification and results, the result is conforming; otherwise it is non-conforming or partially conforming.

【L3-CES1-14/ L4-CES1-14 Interpretation and Description】

The main evaluation targets of the evaluation metric that "The audit record shall be protected and regularly backed up to avoid unexpected deletion, modification or overwriting" are devices or related components that provide network communication functions such as network devices (including virtual network devices), security devices (including virtual security devices).

The key points of the evaluation implementation include: checking whether the dedicated administrator is staffed to manage audit records, and unauthorized users have no right to delete, modify or overwrite audit records, and checking whether there are log storage rules and synchronization outgoing backup mechanism to prevent the original audit log from being accidentally lost or erased by attackers; checking whether technical measures are taken to regularly back up audit records and checking the backup strategy. The regular backup here generally refers to pushing logs to the log server and other devices for backup. Under this circumstance, if the device is configured with the log server or the IP address of centralized log analysis system, it is necessary to confirm that the assets corresponding to the IP are in normal working condition and to conform whether they have effectively stored historical backup logs. If it is a manual backup, check the backup record and its validity. Regarding the retention time of log records, the requirements of the *"Cybersecurity Law"* shall be followed to keep records for at least 6 months.

If the test results show that the audit record is forwarded to the log server in real-time, and the retention time of the audit record is more than six months, the result is conforming; otherwise it is non-conforming or partially conforming.

[L3-CES1-15/ L4-CES1-15 Interpretation and Description]

The main evaluation targets of the evaluation metric that "The audit process shall be protected against unauthorized interruptions" are devices or related components that provide network communication functions such as network devices (including virtual network devices), security devices (including virtual security devices).

The key points of the evaluation implementation include: interviewing with security administrators or network administrators to confirm whether the device is assigned with the audit management account; check the device configuration file to confirm whether the audit management account permission is limited to the use of audit instructions/commands by the method of verification testing; checking the network device configuration file to confirm whether other management accounts with non-audit permission have the permission to use audit instructions/commands by the method of verification testing, and verifying whether the audit process is protected by executing the instruction to interrupt the audit process.

If the test results show that the network device is configured with the audit account, and other accounts cannot operate the audit command, the result is conforming; otherwise it is non-conforming or partially conforming.

4. Intrusion Prevention

[Standard Requirements]

This control point for Level III includes the evaluation units of L3-CES1-18, L3-CES1-19, L3-CES1-21, and Level IV includes that of L4-CES1-17, L4-CES1-18, L4-CES1-20.

[L3-CES1-18/ L4-CES1-17 Interpretation and Description]

The main evaluation targets of the evaluation metric that "Unnecessary system services, default sharing and high-risk ports shall be closed" are devices or related components that provide network communication functions such as network devices (including virtual network devices), security devices (including virtual security devices).

The main points of the evaluation include: interviewing with network administrators, confirming the services, service ports and physical interfaces used by the device, checking whether unnecessary system services and default sharing are closed, such as checking whether default services such as finger, dns, tcpsmall, udpsmall are closed; checking whether there are unnecessary high-risk ports, such as ports 135, 139, and 445.

If the test results show that the device has closed unnecessary system services, default sharing and high-risk ports, the result is conforming; otherwise it is non-conforming or partially conforming.

[L3-CES1-19/ L4-CES1-18 Interpretation and Description]

The main evaluation targets of the evaluation metric that "The management terminal managed through the network shall be restricted by setting the endpoint access model or network address range" devices or related components that provide network communication functions such as network devices (including virtual network devices), security devices (including virtual security devices).

The main points of the evaluation include: checking the device configuration file to confirm whether the management terminal that manages through the network is restricted, understanding the allowed access method of the device or the network address range; checking whether the network address range limitation is reasonable, and it is recommended to limit the network address as the IP address of the management terminal. If the limited network address range is too large, the security requirements will not be met. If remote operation and maintenance management is performed through VPN, checking whether a specific virtual IP is set as the management host in the VPN configuration strategy.

If the test results show that the device has restricted the management terminal which is managed through the network, and the network address range is reasonably restricted, the result is conforming; otherwise it is non-conforming or partially conforming.

[L3-CES1-21/ L4-CES1-20 Interpretation and Description]

The main evaluation targets of the evaluation metric that "It shall be able to discover known vulnerabilities that may exist, and the vulnerabilities shall be fixed in a timely manner after sufficient evaluation" devices or related components that provide network communication functions such as network devices (including virtual network devices), security devices (including virtual security devices).

The key points of the evaluation implementation include: checking the configuration file of the device to confirm whether the system is the latest version; checking whether the device has security vulnerabilities through vulnerability scanning and other methods. For medium and high-risk vulnerabilities, penetration testing can be adopted to further verify whether the vulnerabilities can be exploited. Logging into the manufacturer's official website to check whether the above version is the latest version.

If the test results show that the device version is the latest version, the result is conforming; otherwise it is non-conforming or partially conforming.

5. Trusted Verification

[Standard Requirements]

This control point for Level III includes the evaluation unit of L3-CES1-24, and Level IV includes that of L4-CES1-23.

[L3-CES1-24 Interpretation and Description]

The main evaluation targets of the evaluation metric that "The trusted verification of the system boot loader, the system program, important configuration parameters and application of the device can be performed on the basis of trusted root, and after detecting that the credibility is damaged, an alarm shall be issued and the verification result shall be formed into an audit record and sent to the security management center" are devices or components that provide trusted verification, and systems that provide centralized audit function.

The key points of the evaluation implementation include: interview with the network administrator or the security administrator to ask whether the device adopts the technology of trusted verification based on the root of trust; checking whether the trusted verification for the system boot loader, the system program, important configuration parameters can be performed on the basis of the trusted root; checking whether the dynamic trusted verification is performed in the key execution link of the devices or components; it shall be tested to verify whether the alert is generated when any damage to the credibility of the network device is detected; it shall be tested to verify whether the alarm results are sent to the security management center in the form of audit records.

If the test results show that the relevant device has adopted the technique of trusted root to perform the trusted verification for the booting program, the system program, important configuration parameters and application programs, the dynamic trusted verification in the key execution link of the application program is performed, and the alarm is generated after detecting the credibility is damaged, and the verification result is formed into an audit record and sent to the security management center, the result is conforming; otherwise it is non-conforming or partially conforming.

[L4-CES1-23 Interpretation and Description]

The main evaluation targets of the evaluation metric that "The trusted verification of the system boot loader, the system program, important configuration parameters and communication application of the communication device can be performed on the basis of trusted root, and dynamic trusted verification shall be performed in all execution links of the application. After detecting that the credibility is damaged, an alarm shall be generated and the verification result shall be formed into an audit record and sent to the security management center for dynamic correlation and situation awareness" are devices or components that provide trusted verification, and systems that provide centralized audit functions.

The key points of the evaluation implementation include: interviewing with the network administrator to the security administrator to understand whether the technology of trusted verification based on the trusted root is adopted, whether the trusted verification for the system boot loader, the system program, important configuration parameters can be performed on the basis of the trusted root; checking whether the dynamic rusted verification is carried out in all links of devices or components; it shall be tested to verify whether the network device can detect the credibility of the system boot loader, system program, important configuration parameters and communication application program is damaged and then generate the alert; checking whether the trusted verification result is formed into an audit record and sent to the security management center for dynamic correlation and situation awareness.

If the test results show that the relevant device adopts the technique of trusted root to perform the trusted verification for the booting program, the system program, important configuration parameters and application programs, and the dynamic trusted verification is performed in all links of the application program, and the alarm is generated after detecting the credibility is damaged, and the verification result is formed into the audit record and sent to the security management center for dynamic correlation and situation awareness, the result is conforming; otherwise it is non-conforming or partially conforming.

3.4.2 Server and Terminal

1. Identity Authentication

[Standard Requirements]

This control point for Level III includes the evaluation of L3-CES1-01, L3-CES1-02, L3-CES1-03, L3-CES1-04, and Level IV includes that of L4-CES1-01, L4-CES1-02, L4-CES1-03, L4-CES1-04.

[L3-CES1-01/ L4-CES1-01 Interpretation and Description]

The main evaluation targets of the evaluation metric that "The identity of login users shall be identified and authenticated. The identification has the characteristic of uniqueness and the authentication information has the requirement of complexity and regular replacement" are server operation systems (including the operation system of host and virtual machine), terminal operation systems.

The key points of the evaluation implementation include: checking whether the remote and local logins have authenticated the logged-in user; reviewing the user list of the operation system to confirm whether the user identification is unique; checking whether the user password is at least 8 characters and meets the complexity requirements; checking whether the user of operating system adopts identity authentication measures when logging in, and testing to verify whether there is no empty password used in the existing accounts; and if servers are remotely managed through the third-party software, and the server's own authentication function is not used, checking the identity authentication measures of the third-party software; At the same time, the mapping relationship between the identity of the third-party software and the identity of the server shall be confirmed. For example, no matter which account is currently logged in on the server, the third-party software can directly take over the server desktop, then this operation does not meet the requirements.

If the test results show that the evaluated system has enabled the identity authentication mechanism, and the password complexity and regular replacement meet the requirements of the security strategy, the result is conforming; otherwise it is non-conforming or partially conforming.

[L3-CES1-02/ L4-CES1-02 Interpretation and Description]

The main evaluation targets of the evaluation metric that "It shall have a login failure processing function, shall configure and enable related measures such as ending the session, limiting the number of illegal logins, and automatically logging out when the login connection times out" are the server operation system (including host and virtual machine operation systems), Terminal operation system, etc.

The key points of the evaluation implementation include: checking whether the operating system is configured with a lock mode and lock period that conforms to the security policy, and checking specific actions taken after a certain number of consecutive login failures, such as account lockout. The number of consecutive login failures is generally 3—10 times, and there exists greater risk if failure exceeds 10 times, the lock time needs to be more than 30 minutes or continue to be locked until manual review and unlock. The evaluation shall focus on checking whether the configuration is reasonable, such as the lock time is too short, the timeout period is too long, etc. and checking whether the login connection timeout and automatic logout functions are configured and enabled. If the server is managed remotely through third-party software, checking whether the third-party software has enabled the login failure handling function and properly configured it.

If the test results show that the operation system has enabled the continuous login failure processing function and configured reasonable parameters, the result is conforming; otherwise it is non-conforming or partially conforming.

[L3-CES1-03/ L4-CES1-03 Interpretation and Description]

The main evaluation targets of the evaluation metric that "When performing remote management, necessary measures shall be taken to prevent the authentication information from being eavesdropped during network transmission" are server operation systems (including the operation system of host and virtual machine), terminal operation systems. If the tested system adopts local management, then this evaluation metric is not applicable.

The key points of the evaluation implementation include: checking the management protocol used by the administrator during the remote management, and use the packet capture tool to verify whether the authentication information is encrypted during the transmission process and whether replay attacks can be resisted. If the third-party software is used to remotely operate and maintain the management server, checking whether the third-party software is encrypted for transmission, and verify the function by means of packet capturing and replaying and reviewing the existing third-party software protocol attack cases.

If the test results show that security measures of encryption transmission, such as SSL or SSH, are used in the remote management, the result is conforming; otherwise it is non-conforming or partially conforming.

[L3-CES1-04/ L4-CES1-04 Interpretation and Description]

The main evaluation targets of the evaluation metric that "Two or more combinations of authentication technologies such as passwords, cryptography and biotechnology shall be used to authenticate users, and at least one of the authentication technologies shall be implemented by using cryptographic technology" are server operation systems (including the operation system of host and virtual machine), terminal operation systems.

The key points of the evaluation implementation include: verifying whether two or more combinations of authentication technologies such as dynamic passwords, digital certificates, biometric technology and device fingerprinting are used to authenticate user identities, and verifying whether one of the authentication technologies uses cryptographic technology; checking whether the cryptographic algorithm used in the identity authentication process is a Security cryptographic algorithm. Among them, DES, RSA1024 and below, MD5 and SHA-1 algorithms are cryptographic algorithms with potential security risks. If the above algorithms are used, special attention shall be paid in the risk assessment.

If the test results show that the operating system has enabled two-factor authentication measures to authenticate users, and at least one of the authentication technologies is implemented by using cryptography technology, the result is conforming; otherwise it is non-conforming or partially conforming.

2. Access Control

[Standard Requirements]

This control point for Level III includes the evaluation units of L3-CES1-05, L3-CES1-06, L3-CES1-07, L3-CES1-08, L3-CES1-09, L3-CES1-10, L3-CES1-11, and Level IV includes that of L4-CES1-05, L4-CES1-06, L4-CES1-07, L4-CES1-08, L4-CES1-09, L4-CES1-10, L4-CES1-11.

[L3-CES1-05/ L4-CES1-05 Interpretation and Description]

The main evaluation targets of the evaluation metric that "Accounts and permissions shall be assigned to logged-in users" are server operation systems (including host and virtual machine operation systems), terminal operation systems, etc.

The key points of the evaluation implementation include: interviewing with the system administrator and obtain the list of all users who need to log in to the operation system;

checking whether the operation system assigns accounts and permissions to each logged-in user, and for users that can locally or remotely log into the operation system; checking whether they are assigned into the existing user group; checking whether the operation system has disabled or restricted access to anonymous, default permission accounts.

If the test results show that the evaluated system assigns accounts and permissions to the logged-in user, and disables or restricts access to anonymous and default permission accounts, the result is conforming; otherwise it is non-conforming or partially conforming.

[L3-CES1-06/ L4-CES1-06 Interpretation and Description]

The main evaluation targets of the evaluation metric that "The default account shall be renamed or deleted, and the default password of the default account shall be modified" are server operation systems (including host and virtual machine operation systems), terminal operation systems, etc.

The key points of the evaluation implementation include: checking whether the default account has been renamed or the default account has been deleted; checking whether the default account of the operation system or the super administrator account has been renamed; checking whether the default password of the default account has been modified.

If the test results show that the evaluated system has renamed or deleted the default account, modified the default password of the default account, the result is conforming; otherwise it is non-conforming or partially conforming.

[L3-CES1-07/ L4-CES1-07 Interpretation and Description]

The main evaluation targets of the evaluation metric that "the redundant, expired accounts shall be deleted or deactivated in time to avoid the existence of shared accounts" are server operation systems (including host and virtual machine operation systems), terminal operation systems, etc.

The key points of the evaluation implementation include: asking the system administrator to obtain the list of all users who need to log into the operation system; checking one by one whether there are redundant, invalid, and long-term unused accounts in the operating system; for accounts whose purpose cannot be confirmed, special attention shall be paid during the risk analysis. In addition, checking whether there is a one-to-one correspondence between the administrator user and the account; checking whether there exists the situation of multiple people sharing a certain management account, which can be analyzed and judged by comparing the list of administrator users and device accounts. If the operation and maintenance model of the tested system is a multi-shift system or a duty-rotation system, there may be situations where different operation and maintenance teams share the same management account, and it shall be analyzed whether there is a shared account in the same time period. By the method of trying to log into account, testing and verifying whether redundant or expired accounts are effectively deleted or disabled.

If the test results show that the evaluated system has deleted or disabled redundant and expired accounts, and there are no shared accounts, the result is conforming; otherwise it is non-conforming or partially conforming.

【L3-CES1-08/ L4-CES1-08 Interpretation and Description】

The main evaluation targets of the evaluation metric that "the administration user shall be granted with the needed least privilege to realize the privilege separation of the administration user" are server operation systems (including host and virtual machine operation systems), terminal operation systems, etc.

The key points of the evaluation implementation include: checking the role division in the operating system; checking whether management user permissions of the operating system are separated; at least the accounts of the system administrator, security administrator and auditor shall be divided; auditor accounts only grant audit log query and read permissions; security administrator accounts shall not have system operation and maintenance permissions; the auditor account shall be granted with the audit log query and read permission only; the security administrator account shall not have permissions for system operation and maintenance;

If the test results show that the operating system has granted the minimum privilege required for the administration user, and the privilege separation of the administrator user is realized, the result is conforming; otherwise it is non-conforming or partially conforming.

【L3-CES1-09/ L4-CES1-09 Interpretation and Description】

The main evaluation targets of the evaluation metric that "The access control strategy shall be configured by the authorization subject, and the access control strategy stipulates the access rules of the subject to the object" are server operation systems (including host and virtual machine operation systems), terminal operation systems, etc.

The key points of the evaluation implementation include: checking whether the authorization subject (such as the security administrator) is responsible for configuring access control strategies in the operating system, and checking the specific operating system to confirm which authorization subject account can specify the access rules and control granularity of objects (files, services, programs) through configuration strategies. If there is a domain control server environment, the authorization subject shall be the security strategy administrator of the domain control server; checking whether the authorization subject has configured the subject's access rules to the object according to the security strategy. For the specifically applied server, It shall achieve the purpose of controlling access by setting the grant or denial of individual users and group users from accessing to a certain core resource (such as files, databases, etc.).

If the test results show that the operating system has been configured with the access control strategy by the authorized subject, the result is conforming; otherwise it is non-conforming or partially conforming.

[L3-CES1-10/ L4-CES1-10 Interpretation and Description]

The main evaluation targets of the evaluation metric that "The granularity of access control shall achieve that the subject is at the user level or process level, and the object is at the file and database table level" are server operation systems (including host and virtual machine operation systems), terminal operation systems, etc.

The main points of the evaluation include: checking the control granularity of its access control strategy to see whether the subject reaches the user level or process level and the file, whether the object is at the file, database table, record or field level. Performed operations include reading, writing, modifying or complete control, etc.

If the test results show that the granularity of operating system security strategy control has reached the user level or process level for the subject, and the file and database table level for the object, the result is conforming; otherwise it is non-conforming or partially conforming.

[L3-CES1-11 Interpretation and Description]

The main evaluation targets of the evaluation metric that "Security tags shall be set up for important subjects and objects, and the subject shall be restrained from accessing to information resources with security tags" are server operation systems (including host and virtual machine operation systems), terminal operation systems, etc.

The main points of the evaluation include: checking whether security tags are set for important subjects and objects, and determining the access control permission with reference to the strategy model according to the security level of the tags to ensure the one-way flow of data; checking whether the access control strategy controls the subject's access to the object in accordance with the subject and object security tags, such as trying to use a low-privilege account to perform the operation outside of its permission to see if the requirements are met.

If the test results show that the operating system has set up security tags for important subjects and objects, and restrained the subject's access to information resources with security tags, the result is conforming; otherwise it is non-conforming or partially conforming.

[L4-CES1-11 Interpretation and Description]

The main evaluation targets of the evaluation metric that "Security tags shall be set for the subject and the object, the subject's access to the object shall be based on security tags and mandatory access control rules" are server operation systems (including host and virtual machine operation systems), terminal operation systems, etc.

The main points of the evaluation include: checking whether security tags are set for important subjects and objects, and determining the access control permission with reference to the strategy model according to the security level of the tags to ensure the one-way flow of data; It shall be tested to verify whether the mandatory access control strategy of the subject to the object is controlled based on the subject and object security tags.

If the test results show that the operating system has set up security tags for important subjects and objects, and restrained the subject's access to information resources with security tags, the result is conforming; otherwise it is non-conforming or partially conforming.

3. Security Audit

[Standard Requirements]

This control point for Level Ⅲ includes the evaluation units of L3-CES1-12, L3-CES1-13, L3-CES1-14, L3-CES1-15, and Level Ⅳ includes that of L4-CES1-12, L4-CES1-13, L4-CES1-14, L4-CES1-15.

[L3-CES1-12/ L4-CES1-12 Interpretation and Description]

The main evaluation targets of the evaluation metric that "The security audit function shall be enabled, covering each user and auditing important user behaviors and important security incidents" are server operation systems (including host and virtual machine operation systems), terminal operation systems, etc.

The key points of the evaluation implementation include: checking whether the operating system has the security audit function enabled, whether the security audit rules is set, and whether historical logs is generated. Attention shall be paid to the difference between security audit and system log services; checking whether the security audit scope of the operating system covers every user. If the operating system has set security audit rules, it generally covers every user. If no security audit rules are set, and only the security audit service is activated, it is impossible to talk about covering every user; checking whether the operating system audits important user behaviors and important security incidents. The definition of "important user behaviors and important security incidents" is the key to evaluation. Important user behaviors shall at least include logging in, logging out, querying, modifying, deleting, and creating records. If no security audit rules are set, only the security audit service is activated, and some important user behaviors and important security incidents will be audited by default (for example, the Linux operating system audits the account modification, user login, configuration modification, etc. by default), but based on the evaluation experience, the default rules are not sufficient. Some security audit rules shall be extended and corresponding logs shall be generated. Asking and checking whether there are third-party audit tools or systems.

If the test results show that the operating system has enabled the security audit function, is configured with security audit rules that meet the characteristics of the system, there are historical logs that have been generated, or the third-party audit tool has been deployed, the audit covers every user, and important user behaviors and important security incidents are audited, the result is conforming; otherwise it is non-conforming or partially conforming.

【L3-CES1-13 Interpretation and Description】

The main evaluation targets of the evaluation metric that "The audit record shall include the date and time of the incident, the user, the type of incident, the success of the incident, and other audit-related information" are server operation systems (including host and virtual machine operation systems), terminal operation systems, etc.

The key points of the evaluation implementation implementation include: checking whether the audit record information includes the date and time of the incident, the user, the type of the incident, whether the incident is successful, and other audit-related information. Among them, the accuracy and clock synchronization issues of the date and time shall be concerned about. For some operating system logs, success or failure may not be clearly displayed, then need to find the corresponding content from the detailed description.

If the test results show that the operating system has enabled the audit and system logs, the elements can meet the standard requirements, the result is conforming; otherwise it is non-conforming or partially conforming.

【L4-CES1-13 Interpretation and Description】

The main evaluation targets of the evaluation metric that "The audit record shall include the date and time of the incident, the type of incident, the subject identification, the object identification and results" are server operation systems (including host and virtual machine operation systems), terminal operation systems, etc.

The key points of the evaluation implementation include: checking whether the audit record information includes the date and time of the incident, the type of incident, the subject identification, the object identification and results. Among them, the accuracy and clock synchronization issues of the date and time shall be concerned about. For some operating system logs, success or failure may not be clearly displayed, then need to find the corresponding content from the detailed description.

If the test results show that the operating system has enabled the audit andsyslog, which can meet the standard requirements in terms of elements, the result is conforming; otherwise it is non-conforming or partially conforming.

【L3-CES1-14/ L4-CES1-14 Interpretation and Description】

The main evaluation targets of the evaluation metric that "The audit records shall be protected and regularly backed up to avoid unexpected deletion, modification or coverage" are server operation systems (including host and virtual machine operation systems), terminal operation systems, etc.

The key points of the evaluation implementation include: checking whether the dedicated administrator is staffed to manage audit records; checking whether there are log storage rules and synchronization outgoing backup mechanism to prevent the original audit log from being accidentally lost or erased by attackers; checking whether technical measures are taken to regularly back up audit records, and checking whether the backup strategy is reasonable. The regular backup here generally refers to pushing logs to the log server and other devices for backup. If the device is configured with the log server or the IP address of centralized log analysis system, it is necessary to confirm that the assets corresponding to the IP are in normal working condition and to conform whether they have effectively stored historical backup logs. If it is the manual backup, checking the backup record and the validity of the backup. Regarding the retention time of log records, the requirements of the *"Cybersecurity Law"* shall be followed to keep records for at least 6 months.

If the test results show that the operating system security audit log permissions are reasonable and the log server is configured to collect logs synchronously, the result is conforming; otherwise it is non-conforming or partially conforming.

[L3-CES1-15/ L4-CES1-15 Interpretation and Description]

The main evaluation targets of the evaluation metric that "The audit process shall be protected against unauthorized interruptions" are server operation systems (including host and virtual machine operation systems), terminal operation systems, etc.

The key points of the evaluation implementation include: using the account of a non-auditing administrator to forcibly interrupt the audit process to verify whether the audit process is protected, and testing whether there is a protection process for the audit process, which can be terminated with non-ordinary permissions.

If the test results show that the operating system protects the audit process, the result is conforming; otherwise it is non-conforming or partially conforming.

4. Intrusion Prevention

[Standard Requirements]

This control point for Level III includes the evaluation units of L3-CES1-17, L3-CES1-18, L3-CES1-19, L3-CES1-21, L3-CES1-22, and Level IV includes that of L4-CES1-16, L4-CES1-17, L4-CES1-18, L4-CES1-20, L4-CES1-21.

[L3-CES1-17/ L4-CES1-16 Interpretation and Description]

The main evaluation targets of the evaluation metric that "the principle of minimum installation shall be followed, and only the necessary components and applications shall be installed" are server operation systems (including host and virtual machine operation systems), terminal operation systems, etc.

The key points of the evaluation implementation include: checking whether the operating system complies with the minimum installation principle based on the characteristics of the business system, evaluation experience, and the sorting results of services and ports.

If the test results show that the operating system has implemented the principle of minimum installation, only the required components and applications are installed, the result is conforming; otherwise it is non-conforming or partially conforming.

[L3-CES1-18/ L4-CES1-17 Interpretation and Description]

The main evaluation targets of the evaluation metric that "Unnecessary system services, default sharing and high-risk ports shall be closed" are server operation systems (including host and virtual machine operation systems), terminal operation systems, etc.

The key points of the evaluation implementation include: checking whether unnecessary system services and default sharing are closed in the operation system, recording the necessary system services and ports that need to be opened, and for services and ports that cannot be defined for their usage, they shall be treated as security issues and their security risks shall be analyzed; checking whether there are unnecessary high-risk ports in the operation system, such as 3389, 137, 139, 445 in Windows, 7001, 7002 in Weblogic, and address restrictions and other protection measures shall be taken for high-risk ports that must be opened; checking whether the operation system is able to block unnecessary high-risk ports by the built-in firewall or other effective measures.

If the test results show that unnecessary system services, default sharing and high-risk ports are closed in the operating system, the result is conforming; otherwise it is non-conforming or partially conforming.

[L3-CES1-19/ L4-CES1-18 Interpretation and Description]

The main evaluation targets of the evaluation metric that "The management terminal managed through the network shall be restricted by setting the endpoint access model or network address range" are server operation systems (including host and virtual machine operation systems), terminal operation systems, etc.

The key points of the evaluation implementation include: checking whether the configuration file of the operating system and the firewall settings of the host restrict the way of accessing the terminal and the range of network address, and checking whether the limitation of the network address range is reasonable.

If the test results show that the operating system restricts the endpoint access model or network address range, the result is conforming; otherwise it is non-conforming or partially conforming.

[L3-CES1-21/ L4-CES1-20 Interpretation and Description]

The main evaluation targets of the evaluation metric that "It shall be able to discover known vulnerabilities that may exist, and the vulnerabilities shall be fixed in a timely manner after sufficient evaluation" are server operation systems (including host and virtual machine operation systems), terminal operation systems, etc.

The key points of the evaluation implementation include: checking whether there are known high-risk vulnerabilities in the operation system through vulnerability scanning, penetration testing, etc.; checking whether there is a regular (generally at least once a month) vulnerability scanning system of using third-party tools for vulnerability scanning, and reviewing the history or report of each scanning; it's necessary to obtain the authorization of the evaluated organization and use the latest version of the vulnerability scanning tool for verification; checking whether the discovered vulnerabilities are fully tested and evaluated, and the vulnerabilities are patched in time, checking the vulnerability repair records and using the latest version of the vulnerability database scanning tool to verify the effectiveness of patching.

If the test results show that the evaluated organization has established the vulnerability scanning system, which can discover the known vulnerabilities that may exist in the operating system in a timely manner, and fix those vulnerabilities in time after sufficient evaluation, the result is conforming; otherwise it is non-conforming or partially conforming.

[L3-CES1-22/ L4-CES1-21 Interpretation and Description]

The main evaluation targets of the evaluation metric that "It shall be able to detect the behavior of intrusion on important nodes, and provide the alert in the event of a serious intrusion incident" are server operation systems (including host and virtual machine operation systems), terminal operation systems, etc.

The key points of the evaluation implementation include: interviewing with the system administrator to find out whether the relevant operating system has installed host intrusion detection software, checking the configuration to see whether it has alert function; obtaining the network topology diagram to check whether the intrusion detection device such as IDS is deployed in the network, and can detect the intrusion behavior on important nodes.

If the test results show that the operating system can detect the intrusion on important nodes and provide the alert when any serious intrusion incident occurs, the result is conforming; otherwise it is non-conforming or partially conforming.

5. Malicious Code Prevention

[Standard Requirements]

This control point for Level III includes the evaluation unit of L3-CES1-23, and Level IV includes that of L4-CES1-22.

[L3-CES1-23 Interpretation and Description]

The main evaluation targets of the evaluation metric that "Intrusion and virus behavior shall be identified in a timely manner and effectively blocked by adopting technical measures against malicious code attacks or the proactive immune trusted verification mechanism" are server operation systems (including host and virtual machine operation systems), terminal operation systems, etc.

The key points of the evaluation implementation include: checking whether the operating system is installed with anti-malware software or software with corresponding functions, checking whether the virus scanning engine in the protection software is turned on, and checking whether the anti-malware code library is regularly upgraded and updated; checking whether the operating system adopts proactive immune trusted verification technology to identify intrusions and virus behaviors in a timely manner; checking whether the operating system has effectively blocked the intrusion and virus behavior when it recognizes it; reviewing the blocking logs for verification.

If the test results show that the operating system has installed anti-malware software and regularly upgrades and updates the anti-malware code library, which can effectively block the intrusion and virus behavior, the result is conforming; otherwise it is non-conforming or partially conforming.

【L4-CES1-22 Interpretation and Description】

The main evaluation targets of the evaluation metric that "Intrusion and virus behavior shall be identified and effectively blocked in a timely manner by adopting the proactive immune trusted verification mechanism" are server operation systems (including host and virtual machine operation systems), terminal operation systems, etc.

The key points of the evaluation implementation include: checking whether the operating system adopts proactive immune trusted verification technology to identify intrusions and virus behaviors in a timely manner; checking whether the operating system has effectively blocked the intrusion and virus behavior when it recognizes it; reviewing the blocking logs for verification.

If the test results show that the operating system has installed anti-malware software and regularly upgrades and updates the anti-malware code library, which can effectively block the intrusion and virus behavior, the result is conforming; otherwise it is non-conforming or partially conforming.

6. Trusted Verification

【Standard Requirements】

This control point for Level III includes the evaluation unit of L3-CES1-24, and Level IV includes that of L4-CES1-24.

【L3-CES1-24 Interpretation and Description】

The main evaluation targets of the evaluation metric that "The trusted verification of the system boot loader, the system program, important configuration parameters and application of the device can be performed on the basis of trusted root, and after detecting that the credibility is damaged, an alarm shall be issued and the verification result shall be formed into an audit record and sent to the security management center" are server operation systems (including host and virtual machine operation systems), terminal operation systems, etc.

The key points of the evaluation implementation include: reviewing the technical white paper of the evaluated device to understand whether there is a root-of-trust chip on the motherboard; whether the device system boot loader, device system program, important configuration parameters and application programs are measured and verified step by step through the chip or the integrated dedicated micro-controller during the booting process; whether the alert is provided when it is found that the device system boot loader, the device system program, important configuration parameters and application programs are tampered with; the record information of trusted verification results can be viewed in the security management center.

If the test results show that the operating system has adopted the technique of trusted root to perform the trusted verification for the booting program, the system program, important configuration parameters and application programs, and whether the alert is generated after detecting the credibility is damaged, and the verification result is formed into an audit record and sent to the security management center, the result is conforming; otherwise it is non-conforming or partially conforming.

[L4-CES1-24 Interpretation and Description]

The main evaluation targets of the evaluation metric that "The trusted verification of the system boot loader, the system program, important configuration parameters and communication application of the communication device can be performed on the basis of trusted root, and dynamic trusted verification shall be performed in all execution links of the application. After detecting that the credibility is damaged, an alarm shall be generated and the verification result shall be formed into an audit record and sent to the security management center for dynamic correlation and situation awareness" are server operation systems (including host and virtual machine operation systems), terminal operation systems, etc.

The main points of the evaluation include: reviewing the technical white papers of related devices, and checking whether the related device has a root-of-trust chip; checking whether the operating system performs trusted verification for the system boot loader, system program, important configuration parameters and application programs of the computing device based on the trusted root; check whether the operating system performs the dynamic trusted verification in all execution links of the application program; it shall be tested to verify whether the operating system generates the alert when it detects that the credibility of the computing device is compromised; it shall be tested to verify whether the results are sent to the security management center in the form of audit records; checking whether the dynamic correlation and situation awareness can be performed.

If the test results show that the relevant operating system meets the requirements of the evaluation metric, the result is conforming; otherwise it is non-conforming or partially conforming.

3.4.3 Application System

1. Identity Authentication

[Standard Requirements]

This control point for Level III includes the evaluation units of L3-CES1-01, L3-CES1-02, L3-CES1-03, L3-CES1-04, and Level IV includes that of L4-CES1-01, L4-CES1-02, L4-

CES1-03, L4-CES1-04.

【L3-CES1-01/ L4-CES1-01 Interpretation and Description】

The main evaluation targets of the evaluation metric that "The identity of login users shall be identified and authenticated. The identification has the characteristic of uniqueness and the authentication information has the requirement of complexity and regular replacement" are business application systems, system management software, Client software, etc.

The key points of the evaluation implementation include: interviewing with the system administrator and checking system design documents to understand how users log in to the system and check whether the system provides a dedicated login module, and whether the identify of users are authenticated; checking how the system distinguishes the identification of each user, common ways of identity distinguishing are user ID, number, etc., and checking whether the identification is unique; if conditions permit, trying to create a new account with an existing user name and checking whether it can be created successfully; if user information is stored in the database, checking whether the user's identification in the database table is the primary key in the database table; when testing the user password module, checking whether the user configuration information allows a blank password, and if conditions permit, modifying the test user with a blank password to determine whether it can be successful; if user information is stored in the database, checking whether the attributes of the user's password field in the database table can be null; checking the user password module to determine whether there is a password complexity requirement and a regular replacement strategy. If conditions permit, modify the user's password to check whether there is a complexity requirement; when the user password is uniformly assigned by the background administrator, checking whether the user is mandatory to modify the password for the first login, and checking whether the modified password compulsorily meets the password complexity requirements; if conditions permit, trying to create a new user to check whether the above requirements are met. If there are other channels for users to modify passwords in the system, such as mini programs, mobile apps and registration interfaces provided to users, all of the above need to verify whether the password complexity of the password modification module is consistent and meets the security strategy requirements.

If the test results show that the system has identified the logged-in user, the system cannot create a blank password, and the password has the function of complexity and regular replacement, and the account assigned by the administrator is required to modify the default password for the first login, the result is conforming; otherwise it is non-conforming or partially conforming.

【L3-CES1-02/ L4-CES1-02 Interpretation and Description】

The main evaluation targets of the evaluation metric that "It shall have the function of handling login failure, shall configure and enable related measures such as session ending, limiting the number of illegal logins and automatically logging out when the login connection times out" are business application systems, system management software and Client software.

The main points of the evaluation implementation include: interviewing with the business administrator to understand whether there is the login failure handling function, and creating a new test account to verify whether the login failure handling function is enabled; creating a new test account to verify the number of login failure processing functions and specific actions after a certain number of login failures. If the account needs to be unlocked by the administrator or automatically unlocked during the lock period, special attention shall be paid on the error message after the login failure. The prompt information shall be obscured. For example, it shall not prompt overly obvious information such as "The user doesn't exist or the password is wrong" for cases of user name/password error.

If the test results show that the system restricts the user login failure measures, the account is locked when the login attempts fail for 6 times, and the administrator can unlock it from the administration platform after locking. The prompt information is obfuscated, and the login timeout period is set up, the result is conforming; otherwise it is non-conforming or partially conforming.

【L3-CES1-03/ L4-CES1-03 Interpretation and Description】

The main evaluation targets of the evaluation metric that "When performing remote management, necessary measures shall be taken to prevent the authentication information from being eavesdropped during network transmission" are business application systems, system management software, and Client software.

The main points of the evaluation implementation include: interviewing with the business administrator to understand whether the authentication information is transmitted in a Security manner such as encryption, and network sniffing tools can be used to capture the system's identity authentication process to check whether it is plaintext information; it shall be noted that there are two ways to prevent authentication information from being intercepted during network transmission. One is the encryption of the transmission protocol, such as HTTPS. When using the encryption protocol, attention shall be paid to whether the cryptographic algorithm suite used is a Security algorithm. Among them, DES, RSA1024 and below, MD5 and SHA-1 are cryptographic algorithms with hidden security risks. The other is to encrypt the user authentication information at the front end through cryptographic technology, such as the use of hash algorithms to encrypt the authentication information, or the use of symmetric cryptography. If symmetric cryptography is used, attention shall also be paid to the security of the symmetric key that encrypts the authentication information.

If the test results show that the relevant application system can meet the above metric requirements, the result is conforming; otherwise it is non-conforming or partially conforming.

【L3-CES1-04/ L4-CES1-04 Interpretation and Description】

The main evaluation targets of the evaluation metric that "Two or more combinations of authentication technologies such as passwords, cryptography and biotechnology shall be used to authenticate users, and at least one of the authentication technologies shall be implemented by using cryptographic technology" are business application systems, system management software, and Client software.

The key points of the evaluation implementation include: verifying whether two or more combinations of authentication technologies such as dynamic passwords, digital certificates, biometric technology and device fingerprinting are used to authenticate user identities, and verifying whether one of the authentication technologies uses cryptographic technology; checking whether the cryptographic algorithm used in the identity authentication process is a Security cryptographic algorithm. Among them, DES, RSA1024 and below, MD5 and SHA-1 algorithms are cryptographic algorithms with potential security risks. If the above algorithms are used, special attention shall be paid in the risk assessment.

If the test results show that the system uses two-factor authentication, and one of which uses cryptographic technology, the result is conforming; otherwise it is non-conforming or partially conforming.

2. Access Control

【Standard Requirements】

This control point for Level III includes the evaluation units of L3-CES1-05, L3-CES1-06, L3-CES1-07, L3-CES1-08, L3-CES1-09, L3-CES1-10, L3-CES1-11, and Level IV includes that of L4-CES1-05, L4-CES1-06, L4-CES1-07, L4-CES1-08, L4-CES1-09, L4-CES1-10, L4-CES1-11.

【L3-CES1-05/ L4-CES1-05 Interpretation and Description】

The main evaluation targets of the evaluation metric that "Accounts and permissions shall be assigned to logged-in users" are business application systems, system management software, and Client software.

The key points of the evaluation implementation include: understanding the types or roles of current users using the system, and it shall be able to create accounts for users who need to log in; checking whether the system has the function of assigning user permissions, and whether the assignment of user permissions for different types or roles is reasonable, only system administration users can assign minimum permissions for corresponding services to users of different types or roles; asking and checking whether there are anonymous accounts, test accounts or default accounts in the system, and confirming their permission settings and whether they are restricted or disabled.

If the test results show that the system has classified users, restricted the access right of anonymous accounts, the test account has been disabled or deleted, and the default account is controlled and managed by the security strategy, the result is conforming; otherwise it is

non-conforming or partially conforming.

【L3-CES1-06/ L4-CES1-06 Interpretation and Description】

The main evaluation targets of the evaluation metric that "The default account shall be renamed or deleted, and the default password of the default account shall be modified" are business application systems, system management software, and Client software.

The key points of the evaluation implementation include: It is necessary to understand whether the system is self-developed or commercial products/open source products sold in a unified manner. If it is a self-developed system, checking whether there exists commonly-used administration users such as admin, administrator, root, system, etc.; if it is commercial products/open source products sold in a unified manner, checking whether there is a default user that comes with the product; it is necessary to know whether the default account can be renamed. If it cannot be renamed, the default password of the default account shall be modified, and the terminal account login shall be restricted or the default account shall be deleted.

If the test results show that default accounts are disabled after the system acceptance, or default accounts are renamed and default passwords are modified, the result is conforming; otherwise it is non-conforming or partially conforming.

【L3-CES1-07/ L4-CES1-07 Interpretation and Description】

The main evaluation targets of the evaluation metric that "the redundant, expired accounts shall be deleted or deactivated in time to avoid the existence of shared accounts" are business application systems, system management software, and Client software.

The key points of the evaluation implementation implementation include: understanding the system's handling rules for expired accounts, such as the handling of users who haven't logged in for a long time; understanding the system's handling of temporary accounts, and deleting temporary accounts that are no longer used in the short term; understanding the use of A/B role accounts, the A/B posts with the same role shall use different accounts to avoid shared accounts.

If the test results show that there is a one-to-one correspondence between the system user and the user account, and there are no accounts shared by multiple people, and the account corresponding to the resigned staff is disabled, the result is conforming; otherwise it is non-conforming or partially conforming.

【L3-CES1-08/ L4-CES1-08 Interpretation and Description】

The main evaluation targets of the evaluation metric that "the administration user shall be granted with the needed least privilege to realize the privilege separation of the administration user" are business application systems, system management software, and Client software.

The key points of the evaluation implementation include: checking whether the system divides different roles for users, and understanding the rationale of the role division; it is necessary to understand the authorization rules and procedures of the system, whether there is a corresponding permission matrix to determine whether the permissions of different administration users are separated from each other; checking whether the administration users are granted with the minimum privilege required for the work task. For example, the super administrator shall not have operation permission for business data.

If the test results show that the system user has the minimum permissions required for the job responsibilities, the administrator grants the authorization to users according to the permission application and approval results after the change of job position, and the account of resigned user is disabled, the role of user administrator, system administrator and auditor are set separately, and the system does not have the super administrator or the password of the super administrator is managed in segments, and the alert is generated when the super administrator logs in, and all operations of the super administrator are recorded, the result is conforming; otherwise it is non-conforming or partially conforming.

【L3-CES1-09/ L4-CES1-09 Interpretation and Description】

The main evaluation targets of the evaluation metric that "The access control strategy shall be configured by the authorization subject, and the access control strategy stipulates the access rules of the subject to the object" are business application systems, system management software, and Client software.

The key points of the evaluation implementation include: It is necessary to understand whether the system permissions can be set by the administrator or the super administrator according to the access control strategy, for example, the system administrator sets the role permissions for the department system, and the user administrator assigns the corresponding role permissions according to the user's job responsibilities; checking whether unauthorized access is possible, such as directly accessing the ID card images collected by the system via the image address or directly accessing the controlled resources via the URL.

If the test results show that the system user is authorized by the administrator or super administrator to configure the permissions according to the job responsibilities, and the low-privileged users cannot access the system resources that only high-privileged users can access, the result is conforming; otherwise it is non-conforming or partially conforming.

【L3-CES1-10/ L4-CES1-10 Interpretation and Description】

The main evaluation targets of the evaluation metric that "The granularity of access controlshall achieve that the subject is at the user level or process level, and the object is at the file and database table level" are business application systems, system management software, and Client software.

The key points of the evaluation implementation include: it is necessary to understand the granularity of the system's access control to see whether the subject of the system authorizes a single user or a user group, and what degree of fine granularity can be achieved; among them, the granularity of the access subject (who will access) shall be at the user level or process level, and the granularity of the access object (accessed object) shall be at the file or database table level; it is necessary to understand the way the system controls the user access to resources, and whether each access requires the verification of user permissions.

If the test results show that the relevant business application system can meet the above mentioned evaluation metric requirements, the result is conforming; otherwise it is non-conforming or partially conforming.

[L3-CES1-11 Interpretation and Description]

The main evaluation targets of the evaluation metric that "Security tags shall be set up for important subjects and objects, and the subject shall be restrained from accessing to information resources with security tags" are business application systems, system management software, and Client software.

The key points of the evaluation implementation include: interviewing with the system administrators/developers to understand what security model the system adopts to control users' access to system resources; if mandatory access control is used, interviewing with the system administrator to determine whether the corresponding access control strategy is reasonable, and testing and verification shall be carried out.

If the test results show that the relevant business application system adopts mandatory access control (such as MAC, etc.) and it is proved to be effective after testing and verification, the result is conforming; otherwise it is non-conforming or partially conforming.

[L4-CES1-11 Interpretation and Description]

The main evaluation targets of the evaluation metric that "Security tags shall be set for the subject and the object, the subject's access to the object shall be based on security tags and mandatory access control rules" are business application systems, system management software, and Client software.

The key points of the evaluation implementation include: interviewing with the system administrators/developers to understand what security model the system adopts to control users' access to system resources; if mandatory access control is used, interviewing with the system administrator to determine whether the corresponding access control strategy is reasonable, and testing and verification shall be carried out.

If the test results show that the relevant business application system adopts mandatory access control (such as MAC, etc.) and it is proved to be effective after testing and verification, the result is conforming; otherwise it is non-conforming or partially conforming.

3. Security Audit

[Standard Requirements]

This control point for Level III includes the evaluation units of L3-CES1-12, L3-CES1-13, L3-CES1-14, L3-CES1-15, and Level IV includes that of L4-CES1-12, L4-CES1-13, L4-

CES1-14, L4-CES1-15.

[L3-CES1-12/ L4-CES1-12 Interpretation and Description]

The main evaluation targets of the evaluation metric that "The security audit function shall be enabled, covering each user and auditing important user behaviors and important security incidents" are business application systems, system management software, Client software and integrated audit system.

The key points of the evaluation implementation include: understanding how the relevant system conducts auditing; checking whether the audit function is enabled, and whether the audit can cover all users' important behaviors and important security incidents; checking whether the audit includes but is not limited to the user login/logout, the addition, deletion and modification of the system data, the query audit of sensitive information, the management of user's configuration changes to the system, and user permission changes, etc.

If the test results show that the relevant system audit logs can cover all operations of all users, and can audit important user behaviors and important security incidents, the result is conforming; otherwise it is non-conforming or partially conforming.

[L3-CES1-13 Interpretation and Description]

The main evaluation targets of the evaluation metric that "The audit record shall include the date and time of the incident, the user, the type of incident, the success of the incident, and other audit-related information" are business application systems, system management software, Client software and integrated audit system.

The key points of the evaluation implementation include: interviewing with the system administrator/business administrator, checking the system security audit function, querying the system audit log, and determining whether the evaluated system has a complete audit log; checking the audit record and interviewing with the system administrator/ business administrator and finding out whether the audit records contain credit queries initiated by users to confirm whether it contains the date and time of the operation, the user, the type of incident, whether the incident is successful, the query object, the details of the query object, the query IP, etc., and determining whether the audit record of the evaluated system is comprehensive.

If the test results show that the audit record of the relevant system contains all operation behaviors of users, operation incidents and important audit contents, the result is conforming; otherwise it is non-conforming or partially conforming.

[L4-CES1-13 Interpretation and Description]

The main evaluation targets of the evaluation metric that "The audit record shall include the date and time of the incident, the type of incident, the subject identification, the object identification and results" are business application systems, system management software, Client software and integrated audit system.

Chapter 3 Application and Interpretation of the General Security Evaluation Requirements at Level III and Level IV

The main points of the evaluation implementation include: checking the specific audit records to see whether they include but not limited to the date and time of the incident, incident type, subject identification, object identification and results, etc. According to the characteristics of each system, special attention shall be paid to the audit content of the special system. For example, the operation of the credit system shall also include the query target and query content of the incident.

If the test results show that the audit log of the relevant system contain all the operation behaviors of users, operation incidents and important audit contents, the result is conforming; otherwise it is non-conforming or partially conforming.

【L3-CES1-14/ L4-CES1-14 Interpretation and Description】

The main evaluation targets of the evaluation metric that "The audit record shall be protected and regularly backed up to avoid unexpected deletion, modification or overwriting" are business application systems, system management software, Client software and integrated audit system.

The key points of the evaluation implementation include: checking whether the dedicated administrator is staffed to manage audit records, and unauthorized users have no right to delete, modify or overwrite audit records, and checking whether there are log storage rules and synchronization outgoing backup mechanism to prevent the original audit log from being accidentally lost or erased by attackers; checking whether technical measures are taken to regularly back up audit records and checking the backup strategy. If it is the manual backup, checking the backup record and the validity of the backup. Regarding the retention time of log records, the requirements of the *Cybersecurity Law* shall be followed to keep records for at least 6 months. If the audit records are kept in the database, verifying the audit table in the database table to confirm whether the data is stored in the database, and verifying the database backup method and storage location. If the audit records are kept in the log of the program, the audit records cannot be protected and can be easily deleted, modified or overwritten.

If the test results show that the relevant system audit records are stored in a location that is not vulnerable to unexpected deletion, modification or overwriting, and audit records are regularly backed up, the result is conforming; otherwise it is non-conforming or partially conforming.

【L3-CES1-15/ L4-CES1-15 Interpretation and Description】

The main evaluation targets of the evaluation metric that "The audit process shall be protected against unauthorized interruptions" are business application systems, system management software, Client software and integrated audit system.

The key points of the evaluation implementation include: understanding the components that the system audit process depends on and how the audit process is deployed; checking whether the system audit process can be easily interrupted by non-audit administrator users, checking whether the process can be prevented from being interrupted maliciously, and whether the audit process is monitored; if a third-party program or component is relied on for auditing, checking the handling measures of the third-party program or component when the audit log is not received.

If the test results show that the relevant system audit process cannot be interrupted independently or there are protective measures in the audit process, the result is conforming; otherwise it is non-conforming or partially conforming.

4. Intrusion Prevention

【Standard Requirements】

This control point for Level III includes the evaluation units of L3-CES1-17, L3-CES1-20, L3-CES1-21, and Level IV includes that of L4-CES1-16, L4-CES1-19, L4-CES1-20.

【L3-CES1-17/ L4-CES1-16 Interpretation and Description】

The main evaluation targets of the evaluation metric that "the principle of minimum installation shall be followed, and only the necessary components and applications shall be installed" include business application systems, system management software, and Client software. If the evaluated system is the backbone network system or a self-developed system, then this metric is not applicable.

The key points of the evaluation implementation include: understanding whether the current system is the open source system/commercial system or self-developed system. If it is the self-developed system, then this item is not applicable; if the system is the open source software/commercial system, then it is necessary to understand the functional modules contained in the system. The system only deploys the modules required by the business system, and deploys the unsuitable or unauthorized modules in a stripping method.

If the test results show that the relevant system has only installed the necessary components and applications based on business requirements, the result is conforming; otherwise it is non-conforming or partially conforming.

【L3-CES1-20/ L4-CES1-19 Interpretation and Description】

The main evaluation targets of the evaluation metric that "The data validity check function shall be provided to ensure that the content input through the human-machine interface or through the communication interface meets the system setting requirements" include business application systems, system management software, and Client software.

The main points of the evaluation implementation include: obtaining the system design documents and checking the data input and human-machine interaction interface conditions, and checking whether there is a validity check; checking the validity check measures of the human-machine interaction interface in the system, and testing to see whether the data validity check is performed; if condition permits, penetration tools can be used to test the communication interface of the system in a test environment or with the authorization of the customer, trying to submit abnormal data and detecting the feedback information of the system; the evaluation process shall focus on whether there are vulnerabilities, such as SQL injection and XSS.

If the test results show that the relevant system performs the validity check for the content input through the human-machine interface or through the communication interface, the result is conforming; otherwise it is non-conforming or partially conforming.

[L3-CES1-21/ L4-CES1-20 Interpretation and Description]

The main evaluation targets of the evaluation metric that "It shall be able to identify known vulnerabilities that may exist, and the vulnerability shall be fixed in a timely manner after sufficient evaluation" include business application systems, system management software, and Client software.

The key points of the evaluation implementation include: interviewing with the system administrator to learn whether penetration testing and vulnerability scanning are regularly performed, and checking related reports and repairs; using the latest version of penetration testing, vulnerability scanning and other tools to check whether the system still has high-risk vulnerabilities, and checking the assessment and repair conditions of known vulnerabilities.

If the test results show that the system has regularly conducted penetration tests and vulnerability scans, and no high-risk vulnerabilities are found in this evaluation, the result is conforming; otherwise it is non-conforming or partially conforming.

5. Data Backup and Recovery

[Standard Requirements]

This control point for Level III includes the evaluation unit of L3-CES1-31, and Level IV includes that of L4-CES1-31.

[L3-CES1-31/ L4-CES1-31 Interpretation and Description]

The main evaluation targets of the evaluation metric that "Hot redundancy of important data processing system shall be provided to ensure high availability of the system" are business application system, system management software, etc.

The key points of the evaluation implementation include: understanding the requirements of the system's RTO and RPO, checking which data processing systems in the system have been deployed with hot redundancy, and whether the hot redundancy deployment meets the requirements of RTO and RPO; checking the effectiveness of the deployment of the hot redundancy system, such as database active-standby replication, checking the log application status of the backup database; checking whether the system's processing of dirty data during the hot redundancy switch meets business requirements, such as whether the system has undergone emergency switching, whether the system can operate normally during the switching, and whether there is any data loss, etc.

If the test results show that the important data processing system adopts hot redundancy deployment and high availability is achieved through configuration, system administrators regularly conduct high-availability system effectiveness drills, the result is

conforming; otherwise it is non-conforming or partially conforming.

6. Personal Information Protection

[Standard Requirements]

This control point for Level III includes the evaluation units of L3-CES1-34, L3-CES1-35, and Level IV includes that of L4-CES1-35, L4-CES1-36.

[L3-CES1-34/ L4-CES1-35 Interpretation and Description]

The main evaluation targets of the evaluation metric that "only user's personal information necessary for the business shall be collected and stored" are business application systems, system management software, and Client software.

The key points of the evaluation implementation include: understanding the current number and types of system users, understanding whether personal information needs to be collected during the system's business process, and understanding the amount of personal information collected; checking whether the process of collecting personal information by the system and the process of handling personal information is legal, and whether it complies with relevant standards, laws and regulations, such as GB/T 35273 *Information Security Technology—Personal Information Security Specification*; checking whether the user agreement and privacy statement is signed with users, and whether the agreement clearly specifies what kind of personal information is collected, and how to protect the collected personal information; the terms of the agreement shall be accurately and clearly stated.

If the test results show that the system only collects and stores the user's personal information necessary for the business, and the protection of personal information meets the requirements of GB/T 35273 *Information Security Technology—Personal Information Security Specification* and other standards, laws and regulations, the result is conforming; otherwise it is non-conforming or partially conforming.

[L3-CES1-35/ L4-CES1-36 Interpretation and Description]

The main evaluation targets of the evaluation metric that "Unauthorized access and illegal use of user's personal information shall be prohibited" include business application systems, system management software, and Client software.

The key points of the evaluation implementation include: understanding the storage methods and access methods of users' personal information collected by the system, verifying whether technical measures are adopted to restrict the access and use of users' personal information; checking whether the management system and procedures for the protection of users' personal information are formulated.

If the test results show that the relevant system adopts technical measures to restrict the access and use of users' personal information, and formulates the management system and procedures for the protection of users' personal information, the result is conforming; otherwise it is non-conforming or partially conforming.

3.4.4 Data Security

1. Data Integrity

[Standard Requirements]

This control point for Level Ⅲ include the evaluation units of L3-CES1-25, L3-CES1-26, and Level Ⅳ includes that of L4-CES1-24, L4-CES1-25, L4-CES1-26.

[L3-CES1-25 Interpretation and Description]

The main evaluation targets of the evaluation metric that "Verification techniques or cryptography techniques shall be used to ensure the integrity of important data in the transmission process, including but not limited to authentication data, important business data, important audit data, important configuration data, important video data and important personal information" include information system-related authentication data, important business data, important audit data, important configuration data, etc. These data may be stored in business application systems, database management systems, middleware, system management software, data security protection systems, servers, terminals, network devices and security devices.

The key points of the evaluation implementation include: reading the system design documents to understand whether verification technology or cryptographic technology is adopted in the transmission process of authentication data, important business data, important audit data, important configuration data, important video data and important personal information to ensure their integrity. Among them, commonly used verification techniques include cyclic redundancy check (CRC), parity check, and cryptography techniques including digital signature, message authentication code (MAC) and Hash-based Message Authentication Code (HMAC); for integrity verification by using cryptographic technology, the HTTPS protocol is usually used in the application system to achieve channel encryption. The communication parties implement handshake and key negotiation through the TLS/SSL protocol, and use MAC or HMAC to protect the integrity of the communication data; using network sniffing tools or other methods to capture network traffic, checking the Server Hello packet of HTTPS in the handshake phase, checking the cryptographic algorithm suite used, and paying attention to whether the cryptographic algorithm used is Security; for application systems that do not use the HTTPS protocol, checking whether source encryption is used instead of channel encryption. Fields with MAC, HMAC or digital signature results in the transmitted information can also protect data integrity; the network traffic shall be captured by network sniffing tools or other methods to check whether the data is ciphertext. By comparing system design documents or interviewing with technical developers to verify whether the length of the MAC or HMAC field meets expectations. Digital signatures can be verified to test the integrity protection; it shall be tested to verify whether the authentication data, important business data, important audit data, important configuration data, important video data, and important personal information can be tampered with during the transmission process, and verify whether the integrity of data during the transmission process can be restored in a timely manner in case of being damaged.

If the test results show that the data transmission protocol used to protect the integrity of data transmission is well-configured, Security cryptographic algorithm suite are used, and no relevant security issues are found in the on-site testing, the result is conforming; otherwise it is non-conforming or partially conforming.

[L4-CES1-24 Interpretation and Description]

The main evaluation targets of the evaluation metric that "cryptography techniques shall be used to ensure the integrity of important data in the transmission process, including but not limited to authentication data, important business data, important audit data, important configuration data, important video data and important personal information" include information system-related authentication data, important business data, important audit data, important configuration data, etc. These data may be stored in business application systems, database management systems, middleware, system management software, data security protection systems, servers, terminals, network devices and security devices.

The key points of the evaluation implementation include: reading the system design documents to understand whether cryptography techniques are adopted in the transmission process of authentication data, important business data, important audit data, important configuration data, important video data and important personal information to ensure their integrity. Commonly-used cryptography techniques including digital signature, message authentication code (MAC) and Hash-based Message Authentication Code (HMAC); for integrity verification by using cryptographic technology, the HTTPS protocol is usually used in the application system to achieve channel encryption. Both parties of the communications implement handshake and key negotiation through the TLS/SSL protocol, and use MAC or HMAC to protect the integrity of the communication data; using network sniffing tools or other methods to capture network traffic, checking the Server Hello packet of HTTPS in the handshake phase, checking the cryptographic algorithm suite used, and paying attention to whether the cryptographic algorithm used is Security; for application systems that do not use the HTTPS protocol, checking whether source encryption is used instead of channel encryption. Fields with MAC, HMAC or digital signature results in the transmitted information can also protect data integrity; the network traffic shall be captured by network sniffing tools or other methods to check whether the data is ciphertext. By comparing system design documents or interviewing with technical developers to verify whether the length of the MAC or HMAC field meets expectations. Digital signatures can be verified to test the integrity protection; it shall be tested to verify whether the authentication data, important business data, important audit data, important configuration data, important video data, and important personal information can be tampered with during the transmission process, and verify whether the integrity of data during the transmission process can be restored in a timely manner in case of being damaged.

If the test results show that the data transmission protocol used to protect the integrity of data transmission is well-configured, the Security cryptographic algorithm suite are used, and no relevant security issues are found in the field testing, the result is conforming; otherwise it is non-conforming or partially conforming.

[L3-CES1-26 Interpretation and Description]

The main evaluation targets of the evaluation metric that "Verification techniques or cryptography techniques shall be used to ensure the integrity of important data in the storage process, including but not limited to authentication data, important business data, important audit data, important configuration data, important video data and important personal information" include information system-related authentication data, important business data, important audit data, important configuration data, etc. These data may be stored in business application systems, database management systems, middleware, system management software, data security protection systems, servers, terminals, network devices and security devices.

The key points of the evaluation implementation include: reading the design documents, and understanding the integrity protection measures for the authentication data in the storage process; testing to verify whether the authentication data can be tampered with during the storage process, and testing to verify whether it can be restored in time when the data integrity in the storage process is damaged; checking whether the cryptographic algorithm used is Security. Among them, DES, RSA1024 and below, MD5 and SHA-1 algorithms are cryptographic algorithms with hidden security risks. If the above algorithms are used, special attention shall be paid in the risk assessment.

If the test results show that the important data of the relevant application system has been checked for integrity in the file system or database, it can realize timely alert feedback and data recovery after important files or data are maliciously tampered with, the result is conforming; otherwise it is non-conforming or partially conforming.

[L4-CES1-25 Interpretation and Description]

The main evaluation targets of the evaluation metric that "cryptography techniques shall be used to ensure the integrity of important data in the storage process, including but not limited to authentication data, important business data, important audit data, important configuration data, important video data and important personal information" include information system-related authentication data, important business data, important audit data, important configuration data, etc. These data may be stored in business application systems, database management systems, middleware, system management software, data security protection systems, servers, terminals, network devices and security devices.

The key points of the evaluation implementation include: reading the design documents, and understanding the integrity protection measures for the authentication data in the storage process; testing to verify whether the authentication data can be tampered with during the storage process, and testing to verify whether it can be restored in time when the data integrity in the storage process is damaged; checking whether the cryptographic algorithm used is a Security one. Among them, DES, RSA1024 and below, MD5 and SHA-1 algorithms are cryptographic algorithms with hidden security risks. If the above algorithms are used, special attention shall be paid in the risk assessment.

If the test results show that the important data of the relevant application system has been checked for integrity in the file system or database, it can realize timely alert feedback and data recovery after important files or data are maliciously tampered with, the result is conforming; otherwise it is non-conforming or partially conforming.

[L4-CES1-26 Interpretation and Description]

The main evaluation targets of the evaluation metric that "In applications that may involve legal liability determination, cryptography techniques shall be used to provide data originating evidence and data receiving evidence to achieve the non-repudiation of data originating behaviors and the non-repudiation of data receiving behavior" are business application system, system management software and Client software.

The key points of the evaluation implementation include: reviewing the design documents to see whether cryptographic technology is used to ensure the non-repudiation of data sending and data receiving operations. If the message digest and digital signature that adopt the SSL protocol can achieve non-repudiation, then the SSL configuration in the server shall be checked in detail to check whether the OPENSSL version is too low, whether there is a risk of SSL stripping and other security issues; it shall be tested to verify that the data cannot be tampered with during the process of transmission.

If the test results show that the relevant system adopts encryption technology to meet the non-repudiation requirements of the data originating behavior and the data receiving behavior, and no relevant security issues are found in the field test, the result is conforming; otherwise it is non-conforming or partially conforming.

2. Data Confidentiality

[Standard Requirements]

This control point for Level III include the evaluation units of L3-CES1-27, L3-CES1-28, and Level IV includes that of L4-CES1-27, L4-CES1-28.

[L3-CES1-27/L4-CES1-27 Interpretation and Description]

The main evaluation targets of the evaluation metric that "cryptography techniques shall be used to ensure the confidentiality of important data in the transmission process, including but not limited to authentication data, important business data, important audit data, important configuration data, important video data and important personal information" include information system-related authentication data, important business data, important audit data, important configuration data, etc. These data may be stored in business application systems, database management systems, middleware, system management software, data security protection systems, servers, terminals, network devices and security devices.

The key points of the evaluation implementation include: reviewing system design documents to check whether cryptography techniques are used to improve the confidentiality of authentication data, important business data, and important personal information; checking the cryptography techniques used, where channel encryption is adopted, the HTTPS protocol is usually used, and both parties of the communication implement handshake and key negotiation through the TLS/SSL protocol, and use symmetric cryptography to encrypt the communication data. The network sniffing tools or other methods shall be used to capture the network traffic, and checking the Server Hello packet of HTTPS in the handshake phase, checking the cryptographic algorithm suite used, and checking whether the cryptographic algorithm used is Security; for application systems that do not use the HTTPS protocol, checking whether source encryption is used, and the network sniffing tools or other methods shall be used to capture the network traffic, checking whether the authentication data is encrypted during transmission; If the block cipher algorithm is used, verifying whether the length of the ciphertext meets the block expectations by comparing the system design documents or interviewing with technical developers.

If the test results show that the data transmission protocol used to protect the confidentiality of data transmission is well-configured, the Security cryptographic algorithm suite are used, and no relevant security issues are found in the field testing, the result is conforming; otherwise it is non-conforming or partially conforming.

[L3-CES1-28/L4-CES1-28 Interpretation and Description]

The main evaluation targets of the evaluation metric that "Cryptography techniques shall be used to ensure the confidentiality of important data in the storage process, including but not limited to authentication data, important business data, important audit data, important configuration data, important video data and important personal information" include information system-related authentication data, important business data, important audit data, important configuration data, etc. These data may be stored in business application systems, database management systems, middleware, system management software, data security protection systems, servers, terminals, network devices and security devices.

The key points of the evaluation implementation include: checking whether technical measures (such as data security protection systems, etc.) are used to ensure the confidentiality of authentication data, important business data, and important personal information in the storage process; during the evaluation process, checking whether the various log information of the system contains important data, and checking whether these important data are encrypted and protected; checking whether the cryptographic algorithm used is a security one. Among them, DES, RSA1024 and below, MD5 and SHA-1 algorithms are cryptographic algorithms with hidden security risks. If the above algorithms are used, special attention shall be paid in the risk assessment. It shall be tested to verify whether the specific data is encrypted.

If the test results show that the system's identity authentication information and important business data are stored in an encrypted manner during the storage process, and the log information of the middleware does not contain sensitive business information and identity authentication information, the result is conforming; otherwise it is non-conforming or partially conforming.

3. Data Backup and Recovery

[Standard Requirements]

This control point for Level III includes the evaluation units of L3-CES1-29, L3-CES1-30, and Level IV includes that of L4-CES1-29, L4-CES1-30, L4-CES1-32.

[L3-CES1-29/L4-CES1-29 Interpretation and Description]

The main evaluation targets of the evaluation metric that "The local data backup and recovery functions for important data shall be provided" include information system-related authentication data, important business data, important audit data, important configuration data, etc. These data may be stored in business application systems, database management systems, middleware, system management software, data security protection systems, servers, terminals, network devices and security devices.

The key points of the evaluation implementation include: understanding the requirements of RTO and RPO of the evaluated system, verifying whether the backup strategy meets the requirements of the overall security strategy, and whether the backup strategy is performed according to the backup strategy; checking whether the backup strategy setting is reasonable and the configuration is correct, including whether the backup time is reasonable, whether the backup period is appropriate, whether the storage container and the transmission protocol are Security, etc. ; checking whether the backup result is consistent with the backup strategy, and checking whether the recent recovery test record can perform normal data recovery; checking whether the backup file contains important information such as user passwords. If so, checking whether the file viewing permissions are only restricted to the users who have the execution permission.

If the test results show that the application system data is regularly backed up according to the backup strategy, and the administrator regularly conducts data recovery tests, and at the same time restricts the file permissions of the backup script, the result is conforming; otherwise it is non-conforming or partially conforming.

Chapter 3 Application and Interpretation of the General Security Evaluation Requirements at Level III and Level IV

[L3-CES1-30/L4-CES1-30 Interpretation and Description]

The main evaluation targets of the evaluation metric that "The function of offsite backup in real time shall be provided, and the important data shall be backed up to the backup site in real time by using the communication network" include information system-related important configuration data, important business data, etc. These data may be stored in business application systems, database management systems, middleware, system management software, data security protection systems, servers, terminals, network devices and security devices.

The key points of the evaluation implementation include: check whether the function of offsite backup in real time is provided, checking the construction mode of the offsite backup, generally including the same-city disaster recovery center, offsite disaster recovery center, etc.; checking whether the important configuration data and important business data are backed up to the backup site in real time through the network; and checking the timeliness and effectiveness of offsite backup.

If the test results show that the communication network is used to back up important data to the backup site in real time, the result is conforming; otherwise it is non-conforming or partially conforming.

[L4-CES1-32 Interpretation and Description]

The main evaluation targets of the evaluation metric that "An offsite disaster recovery center shall be established to provide real-time switching of business applications" include information system related authentication data, business data, etc. These data may be stored in business application systems, database management systems, middleware, system management software, data security protection systems, servers, terminals, network devices and security devices.

The key points of the evaluation implementation include: checking whether an offsite disaster backup center is established, equipped with communication lines, network devices and data processing devices required for the disaster recovery; checking whether the real-time switching function of business applications is provided.

If the test results show that the important data processing system adopts hot redundancy deployment and high availability is achieved through configuration, system administrators regularly conduct high-availability system effectiveness drills and provide real-time switching functions for offsite disaster recovery-level business applications, the result is conforming; otherwise it is non-conforming or partially conforming.

4. Residual Information Protection

[Standard Requirements]

This control point for Level III includes the evaluation units of L3-CES1-32, L3-CES1-33, and Level IV includes that of L4-CES1-33, L4-CES1-34.

[L3-CES1-32/L4-CES1-33 Interpretation and Description]

The main evaluation targets of the evaluation metric that "It shall be ensured that the storage space where the authentication information is located is completely cleared before being released or reassigned" are authentication data related to the information system. The authentication data may be distributed in the operation systems, business application systems, database management systems, middleware, and system management software of devices such as terminals and servers.

The key points of the evaluation implementation include: understanding the processing technology used in the identity authentication of system users, checking whether the user authentication information is retained on the Client side or the server side after user authentication, checking what is the method of destroying user's authentication information, and whether it meet the requirement of complete clearance in the case of abnormal/normal log-out; checking whether the system has closed the client's automatic filling function for user authentication information, especially for important business systems that contain transaction information.

If the test results show that the system does not remember the user name and automatic login by default, does not retain the user name after the user logs out, and prohibits the browser from automatically filling the form, and at the same time, the cookie information or Server ID cannot be used to log in synchronously or asynchronously, the result is conforming; otherwise it is non-conforming or partially conforming.

[L3-CES1-33/L4-CES1-34 Interpretation and Description]

The main evaluation targets of the evaluation metric that "it shall be ensured that the storage space with sensitive data is completely cleared before being released or reassigned" include information system related authentication data, business data, etc. These data may be stored in business application systems, database management systems, middleware, system management software, data security protection systems, servers, terminals, network devices and security devices.

The key points of the evaluation implementation include: understanding the storage location of sensitive information of the system, the processing method of sensitive information when the system is shut down or offline, the processing method of server-side memory and storage, whether the Client will download the sensitive information of the system to the Client when using the system, if it is downloaded, checking whether it is cleared when shutting down or logging out of the system; understanding the handling methods and procedures (such as degaussing or physical damage) in case that the system storage medium is broken or damaged for a replacement, and whether a non-return agreement with the supplier has been signed for the storage of sensitive information.

If the test results show that the space where the sensitive data of the system exists has been completely cleared after being released or before reallocation, the result is conforming;

otherwise it is non-conforming or partially conforming.

3.5 Security Management Center

3.5.1 System Management

[Standard Requirements]

This control point for Level III includes the evaluation units of L3-SMC1-01, L3-SMC1-02, and Level IV includes that of L4-SMC1-01, L4-SMC1-02.

[L3-CNS1-01/L4-SMC1-01 Interpretation and Description]

The main evaluation targets of the evaluation metric that "The identity authentication for system administrators shall be conducted, only allowing them to perform system management operations through specific commands or operation interfaces, and these operations shall be audited" include operation and maintenance bastions, cloud management and control platforms, and virtualization management platforms (such as VMware Workstation, VMwarevSphere) and other equipment or related components that provide system management functions.

The key points of the evaluation implementation include: interviewing with the system administrator to understand whether the operation and maintenance bastion host and other devices that provide system management functions are deployed; verifying whether the operation and maintenance bastion host authenticates the logged-in system administrators and reviewing the operation and maintenance bastion host's auditing logs to check whether the system administrator's operations are audited; if the system administrator directly uses an account with system administrator permission to log in to the relevant device for management, checking whether the device authenticates the system administrator during the login process, and checking whether the login protocol of the system administrator is restricted, and checking whether the device has the log audit function enabled.

If the test results show that the system administrator can only perform security management operations through specific commands or operation interfaces, and the system or the device audits the system administrator's operations, the result is conforming; otherwise it is non-conforming or partially conforming.

[L3-SMC1-02/L4-SMC1-02 Interpretation and Description]

The main evaluation targets of the evaluation metric that "The configuration, control and management of the resources and operations of the system shall be performed through the system administrator, including user's identity, system resource configuration, system loading and booting, system operation exception handling, backup and recovery of the data and device, etc." include operation and maintenance bastion hosts, cloud management and control platforms, and virtualization management platforms (such as VMware Workstation, VMwarevSphere) and other equipment or related components that provide system management functions.

The key points of the evaluation implementation include: checking whether the system administrator's permission configuration is reasonable, generally speaking, the permissions of the system administrator include user management, resource configuration, system loading and booting, abnormal handling of system operation, the backup and recovery of data and device, etc. ; checking whether non-system administrators can perform operations that only system administrators can perform, such as attempting to perform system resource configuration operations after logging in as ordinary users.

If the test results show that only the system administrator can configure, control and manage the resources and operation of the system, and the non-system administrator cannot perform operations that only the system administrator can perform, the result is conforming; otherwise it is non-conforming or partially conforming.

3.5.2 Audit Management

[Standard Requirements]

This control point for Level Ⅲ includes the evaluation units of L3-SMC1-03, L3-SMC1-04, and Level Ⅳ includes that of L4-SMC1-03, L4-SMC1-04.

[L3-SMC1-03/L4-SMC1-03 Interpretation and Description]

The main evaluation targets of the evaluation metric that "The identity of the audit administrator shall be authenticated, and they are allowed to perform security audit operations only through specific commands or operation interfaces, and these operations shall be audited later" are systems that provide centralized audit functions such as integrated security audit systems and database audit systems.

The key points of the evaluation implementation include: interviewing with the audit administrator to understand how they conduct audit management operations, and whether they have deployed the device that provides audit management functions such as a comprehensive security audit system; if the audit administrator conducts audit management operations through the integrated security audit system, checking whether the integrated security audit system authenticates the logged-in audit administrator and reviewing the audit log of the integrated security audit system to verify whether the operations of the audit administrator; are audited; if the audit administrator directly uses an account with the audit administrator permission to log in to the relevant device for management, checking whether the device authenticates the audit administrator during the login process, checking whether the login protocol of the audit administrator is limited, and checking whether the log audit function is enabled on the device.

If the test results show that the audit administrator can only perform security audit operations through specific commands or operation interfaces, and the relevant system or device audits the operations of the audit administrator, the result is conforming; otherwise it is non-conforming or partially conforming.

[L3-SMC1-04/L4-SMC1-04 Interpretation and Description]

The main evaluation targets of the evaluation metric that "The audit records shall be analyzed by the audit administrator and processed according to the analysis results, including the storage, management and query of audit records according to the security audit strategy" are systems that provide centralized audit functions such as integrated security audit systems and database audit systems.

The key points of the evaluation implementation include: checking whether the audit management of the system can only be operated by the audit administrator; checking whether the audit administrator can analyze the audit records, and what technical means are used for analysis; checking whether the audit administrator processes the analysis result of the audit record, including storing, managing, and querying audit records according to security audit strategies; checking whether non-audit administrator users can perform operations that only audit administrators can perform.

If the test results show that the audit records can only be analyzed by the audit administrator and processed according to the analysis results, and the non-audit administrator cannot perform operations that only the audit administrator can perform, the result is conforming; otherwise it is non-conforming or partially conforming.

3.5.3　Security Management

[Standard Requirements]

This control point for Level Ⅲ includes the evaluation units of L3-SMC1-05, L3-SMC1-06, and Level Ⅳ includes that of L4-SMC1-05, L4-SMC1-06.

[L3-SMC1-05/L4-SMC1-05 Interpretation and Description]

The main evaluation targets of the evaluation metric that "The identity of the system administrator shall be authenticated, and they are allowed to perform system management operations only through specific commands or operation interfaces, and these operations shall be audited later" are systems that provide centralized security management functions such as the operation and maintenance bastion hosts, the security management and control platform, and the cloud security center.

The key points of the evaluation implementation include: interviewing with security administrators to understand how they conduct security management operations, and whether they have deployed systems that provide centralized security management functions such as the operation and maintenance bastion hosts. If the security administrator performs security management operations through the operation and maintenance bastion hosts, checking whether the operation and maintenance bastion hosts authenticate the logged-in security administrator, and checking the audit log of the operation and maintenance bastion host to verify whether it audits the operations of the security administrator. If the security administrator directly uses an account with security administrator permissions to log in to the relevant device for management, checking whether the device authenticates the security administrator during the login process, checking whether the login protocol of the security administrator is restricted, and checking whether the device has the log audit function enabled.

If the test results show that the security administrator can only perform security management operations through specific commands or operating interfaces, and the relevant system or equipment audits the security administrator's operations, the result is conforming; otherwise it is non-conforming or partially conforming.

【L3-SMC1-06/L4-SMC1-06 Interpretation and Description】

The main evaluation targets of the evaluation metric that "The security strategy in the system shall be configured by the security administrators, including the setting of security parameters, the unified security tags for the subjects and the objects, authorization to the subjects, configuration of trusted verification strategies, etc." are systems, devices or related components that provide centralized security management functions. Typical evaluation targets include bastion hosts, the security management platform, the security management and control center, the cloud security center, and situational awareness, etc.

The key points of the evaluation implementation include: interviewing with security administrators to understand whether their responsibilities include the setting of security parameters, performing the unified security tags for the subject and object, authorization of the subject, configuration of trusted verification strategies, etc.; logging into other users who do not have the corresponding permissions to see whether they can perform operations that only security administrators can perform.

If the test results show that the security strategy in the system can only be configured by the security administrator, and non-security administrators cannot perform operations that only security administrators can perform, the result is conforming; otherwise it is non-conforming or partially conforming.

3.5.4 Centralized Management and Control

【Standard Requirements】

This control point for Level Ⅲ includes the evaluation unit of L3-SMC1-07, L3-SMC1-08, L3-SMC1-09, L3-SMC1-10, L3-SMC1-11, L3-SMC1-12, and Level Ⅳ includes that of L4-SMC1-07, L4-SMC1-08, L4-SMC1-09, L4-SMC1-10, L4-SMC1-11, L4-SMC1-12, L4-SMC1-13.

【L3-SMC1-07/L4-SMC1-07 Interpretation and Description】

The main evaluation targets of the evaluation metric that "A specific management area

Chapter 3 Application and Interpretation of the General Security Evaluation Requirements at Level Ⅲ and Level Ⅳ

shall be allocated to manage and control the security devices or security components distributed in the network" are the overall network architecture and the network topology diagram, etc.

The key points of the evaluation implementation include: checking whether a separate network area is used to deploy and manage security devices or security components in the network topology. A separate network area can be a separate network segment or a dedicated VPC; for security devices deployed at the network boundary, such as gatekeepers, firewalls, IPS/IDS, etc., checking whether it is controlled by the management console deployed in the management area. For devices or systems such as bastion hosts, centralized management and control platforms, and comprehensive log audit platforms, checking whether they are deployed in the management area; if an independent out-of-band management network is established, checking whether the management addresses of each security device or security component are correctly assigned in the out-of-band network segment.

If the test results show that the network topology divides a separate network area for the deployment of security devices or security components, and each security device or security component is deployed in a separate network area, the result is conforming; otherwise it is non-conforming or partially conforming.

[L3-SMC1-08/L4-SMC1-08 Interpretation and Description]

The main evaluation targets of the evaluation metric that "A Security information transmission path shall be established to manage security devices or security components in the network" are routers, switches, firewalls and other devices or related components.

The key points of the evaluation implementation include: checking whether encryption protocols (such as SSH, HTTPS, IPSec, SSL/TLS protocols, etc.) are used to manage security devices or security components, and the corresponding network traffic can be captured through network packet sniffing tools, and analyze the communication protocol to check whether the encryption protocol is used; if the encryption protocol is used for management, checking whether the cryptographic algorithm suite in the encryption protocol is Security; if the encryption channel is not used for management, checking whether the independent out-of-band management network is used for management to ensure the security of the information transmission path. Out-of-band management network includes the network device management and maintenance system, the KVM, the power manager, and the network centralized and integrated management system.

If the test results show that the encryption protocol of the Security cryptographic algorithm suite is used when managing the security device or security component, or the management and control are performed through the out-of-band management network (either one meets the requirement shall suffice), the result is conforming; otherwise it is non-conforming or partially conforming.

[L3-SMC1-09/L4-SMC1-09 Interpretation and Description]

The main evaluation targets of the evaluation metric that "The operation of network links, security devices, network devices and servers shall be monitored in a centralized manner" are systems that provide operating status monitoring functions such as integrated network management systems.

The key points of the evaluation implementation include: interviewing with the system administrator or operation and maintenance administrator to understand whether a centralized monitoring system or device, such as an integrated network management system, is deployed; if the corresponding system or device is deployed, logging into the system to verify whether it is really capable of real-time monitoring of the network devices, security devices, servers, and network links, and whether the monitoring content includes CPU, memory, load, inflow and outflow bandwidth, and disk usage, etc. ; checking whether the system can set threshold value according to the working status, and provide real-time alert function if abnormalities occur (exceeding the threshold value). Reviewing history records to check alerting conditions, or testing to verify whether the real-time alert function is effective.

If the test results show that the operating conditions of network links, security devices, network devices and servers are monitored centrally, and the alarm occurs in case of abnormalities, the result is conforming; otherwise it is non-conforming or partially conforming.

[L3-SMC1-10/L4-SMC1-10 Interpretation and Description]

The main evaluation targets of the evaluation metric that are "The audit data distributed on each device shall be collected, aggregated and analyzed in a centralized manner, and the retention time of the audit records shall be in compliance with requirements of relevant laws and regulations" are systems that provide centralized audit functions such as the integrated security audit system and the database audit system.

The key points of the evaluation implementation include: check whether each device (server, database, network device, security device, etc. important objects) is configured and enabled with related logs or audit policies, and whether the audit data is sent to an independent comprehensive security audit system; log in to comprehensive security Audit system, check the list of logs collected by the system, check whether the received audit data meets audit requirements; check whether the comprehensive security audit system has functions such as centralized analysis; check whether the retention time of audit records meets the requirements of laws and regulations.

If the test results show that a comprehensive security auditing system is deployed to collect and protect equipment logs, and the log retention time is more than 6 months, the unit judges the result to be compliant, the result is conforming; otherwise it is non-conforming or partially conforming.

[L3-SMC1-11/L4-SMC1-11 Interpretation and Description]

The main evaluation targets of the evaluation metric that "Security-related issues such

Chapter 3 Application and Interpretation of the General Security Evaluation Requirements at Level Ⅲ and Level Ⅳ

as security strategies, malicious code, and patch upgrades shall be centrally managed" include the security management center, the cloud security center, the domain strategy controller, the host management and control system, the network antivirus system, WSUS and other systems that provide centralized security management and control functions.

The key points of the evaluation implementation include: interviewing with the network administrator and checking the network topology diagram to confirm whether the centralized security management and control systems are deployed, such as the security management center, the cloud security center, the domain strategy controller, the host management and control system, the network antivirus system, WSUS, etc. ; checking whether the security strategy (such as firewall access control strategy, IPS protection strategy, WAF security protection strategy, etc.) can be centrally managed; checking whether the operating system's anti-malicious code system and the network malicious code protection device can be centrally managed; checking whether the patch upgrade of each system or device can be centrally managed, and checking the recent patch update status of devices.

If the test results show that a centralized security management and control system is deployed, which can centrally manage security-related issues such as security strategy, malicious code, patch upgrade, etc., the result is conforming; otherwise it is non-conforming or partially conforming.

[L3-SMC1-12/L4-SMC1-12 Interpretation and Description]

The main evaluation targets of the evaluation metric that "It shall be able to identify, alarm and analyze various types of security incidents occurring in the network" are systems that provide centralized security management and control functions, such as the security management platform, the security management and control center, the cloud security center, and situational awareness.

The key points of the evaluation implementation include: understanding the deployment of relevant system platforms, checking the coverage of security incidents, and whether security incidents such as security vulnerabilities, websites with Trojan code, vulnerability intelligence, website with malicious links, phishing websites, illegal content, and spam emails can be reported and alerted; checking whether the detection range of relevant systems covers all critical network paths and characterized network traffic security services, and has the function of security incident analysis, such as network traffic level, intrusion prevention level, anti-malware code level, email service level, unauthorized device management level.

If the test results show that the system with centralized security management and control function is deployed, which can analyze various security incidents and generate alerts in real time through sound, light, email, etc. , and the monitoring range can cover all critical paths of the network, the result is conforming; otherwise it is non-conforming or partially conforming.

[L4-SMC1-13 Interpretation and Description]

The main evaluation targets of the evaluation metric that "It shall be ensured that the time within the system is generated by the uniquely determined clock to ensure the consistency of time in the management and analysis of various data" are devices that provide the unified time function, such as NTP servers.

The key points of the evaluation implementation include: checking whether a unique clock source (NTP service) is used uniformly within the system, checking the service component version of the NTP service, and analyzing whether there are security vulnerabilities in the time synchronization service caused by the lower version of the NTP service; checking whether the time of each device and system is consistent with the NTP service time.

If the test results show that the time of the evaluated object is generated by a uniquely determined clock, and the time of the device is consistent with the NTP service time, the result is conforming; otherwise it is non-conforming or partially conforming.

3.6 Security Management Systems

3.6.1 Security Strategy

[Standard Requirements]

This control point for Level III includes the evaluation unit of L3-PSS1-01, and Level IV includes that of L4-PSS1-01.

[L3-PSS1-01/L4-PSS1-01 Interpretation and Description]

The main evaluation target of the evaluation metric that "The overall policy and security strategy for cybersecurity work shall be developed to elaborate the overall objective, scope, principles and security framework of the organization's security work" is the overall policy and strategy document.

The key points of the evaluation implementation include: checking whether there is a document containing the overall policy and security strategy of the cybersecurity work; checking whether the document clarifies the overall objective, scope, principles of the organization's security work, and whether various security strategies are clearly defined. The security management framework includes organizational structure and job responsibilities, personnel security management, environment and asset security management, system construction management, system security operation management, incident handling and emergency response.

If the test results show that the evaluated organization has formulated the overall policy and security strategy for cybersecurity work, including the overall objective, scope, principles, and security framework of the security work, etc., the result is conforming; otherwise it is non-conforming or partially conforming.

3.6.2 Management Systems

[Standard Requirements]

This control point for Level Ⅲ includes the evaluation units of L3-PSS1-02, L3-PSS1-03, L3-PSS1-04, and Level Ⅳ includes that of L4-PSS1-02, L4-PSS1-03, L4-PSS1-04.

[L3-PSS1-02/L4-PSS1-02 Interpretation and Description]

The main evaluation targets of the evaluation metric that "The security management system for various management contents in security management activities shall be established" are security management system documents containing various security management activities.

The key points of the evaluation implementation include: in combination with the overall policy and security strategy of the cybersecurity work of the evaluated organization, checking the establishment of various security management systems, and whether they cover organizational structuring, personnel management (recruitment, dismissal, external personnel management, security education and training, etc.), construction management (classified protection management, software development, project implementation, launch acceptance, etc.), operation and maintenance management (work environment management, computer device management, network operation management, system security management, security incident management, media management, antivirus management, asset management, computer room management, backup and recovery management, etc.).

If the test results show that the evaluated organization has established a complete security management system, the result is conforming; otherwise it is non-conforming or partially conforming.

[L3-PSS1-03/L4-PSS1-03 Interpretation and Description]

The main evaluation targets of the evaluation metric that "The operational procedures for daily management operations performed by administrators or operators shall be established" are operating procedures documents.

The key points of the evaluation implementation include: checking the operating procedures documents, confirming whether security operating procedures and configuration specifications for the daily management operations of system administrators, security administrators, security auditors, operation and maintenance personnel and other related personnel are established respectively; checking whether the scope of security operating procedures and the configuration specification is comprehensive, and common operating procedures include: the network device operating procedure, the security device operating procedure, the operating system operating procedure, the database management system operating procedure, the data backup and recovery security operating procedures, etc. Common configuration specifications include: the account and password configuration specification, the remote login configuration specification, the access control strategy configuration specification, the security audit configuration specification, the closed ports and service configuration specification, the permission allocation management specification, the patch upgrade configuration specification, etc.

If the test results show that the evaluated organization has established the comprehensive and sound security operating procedures for daily management operations, the result is conforming; otherwise it is non-conforming or partially conforming.

[L3-PSS1-04/L4-PSS1-04 Interpretation and Description]

The main evaluation targets of the evaluation metric that "A comprehensive security management system consisting of security strategies, management systems, operating procedures, record forms shall be established" are overall policy and strategies, various security management systems, operation procedures and record form documents, etc.

The key points of the assessment implementation include: checking all strategies and systems documents, and confirming whether the evaluated organization has the overall policy document, management system documents, operating procedures documents, and record form documents. In general, a comprehensive set of security management system can be composed of 4 levels, including the "overall policy and strategy for cybersecurity work", various "management systems for managing security management activities", "operating procedures and management configuration specifications for daily system operating behavior" and various "recording forms", etc.

If the test results show that the evaluated organization has established a comprehensive security management system consisting of security strategies, management systems, operating procedures, record forms, etc. , and the daily record forms are comprehensive, the result is conforming; otherwise it is non-conforming or partially conforming.

3.6.3　Development and Release

[Standard Requirements]

This control point for Level Ⅲ includes the evaluation units of L3-PSS1-05, L3-PSS1-06, and Level Ⅳ includes that of L4-PSS1-05, L4-PSS1-06.

[L3-PSS1-05/L4-PSS1-05 Interpretation and Description]

The main evaluation targets of the evaluation metric that "The designated department or personnel shall be appointed or authorized to be responsible for the development of the security management system" are security supervisors and other personnel.

The key points of the evaluation implementation include: interviewing with information or network security supervisors to understand whether the designated department or personnel is appointed or authorized to be responsible for the development of the security management system; checking the drafting and review records of the relevant management systems, as well as the approval and implementation documents of the relevant management systems.

If the test results show that the evaluated organization has appointed the designated department or personnel to be responsible for the development of the security management system, the result is conforming; otherwise it is non-conforming or partially conforming.

[L3-PSS1-06/L4-PSS1-06 Interpretation and Description]

The main evaluation targets of the evaluation metric that "The security management system shall be released in a formal and effective manner, and version control shall be performed" include the system release process, management system and security management system release, version control and other record form documents.

The key points for the implementation of the evaluation include: reviewing management system documents, confirming whether regulations for system release, revision, maintenance and management are formulated, and whether the regulations include the release process, release method, release scope, format, and version control of the security management system; checking whether the published safety management system has undergone administrative or management procedures such as examination and issuance by the person in charge; reviewing the document revision and update records, and checking whether there are version control measures.

If the test results show that the evaluated organization has established the security management system release and process-related system, and the system release process is complete and sound, and the version control is effective, the result is conforming; otherwise it is non-conforming or partially conforming.

3.6.4 Review and Revision

[Standard Requirements]

This control point for Level III includes the evaluation unit of L3-PSS1-07, and Level IV includes that of L4-PSS1-07.

[L3-PSS1-07/L4-PSS1-07 Interpretation and Description]

The main evaluation targets of the evaluation metric that "The rationality and applicability of the security management system shall be regularly demonstrated and examined, and the security management system that has deficiencies or in need of improvement shall be revised" are form documents such as security management system review and revision records.

The key points of the evaluation implementation include: interviewing with information or network security supervisors to determine whether relevant departments and personnel regularly demonstrate and examine the rationality and applicability of the security management system; checking record forms documents to determine whether there is an examination and approval record of the security management system, the examination and approval record includes the time of examination, the purpose of examination, the departments/personnel involved in the examination, the content of examination, revision opinions, etc.; checking the examination and approval record of the security management system to determine whether the security management system that needs to be modified are revised.

If the test results show that the evaluated organization has regularly examined the rationality and applicability of the security management system and revised the security management system that needs to be improved, and the revision records are standardized and complete, the result is conforming; otherwise it is non-conforming or partially conforming.

3.7 Security Management Organization

3.7.1 Post Setting

[Standard Requirements]

This control point for Level Ⅲ includes the evaluation units of L3-ORS1-01, L3-ORS1-02, L3-ORS1-03, and Level Ⅳ includes that of L4-ORS1-01, L4-ORS1-02, L4-ORS1-03.

[L3-ORS1-01/L4-ORS1-01 Interpretation and Description]

The main evaluation targets of the evaluation metric that "The working committee or leading group that guides and manages the cybersecurity work shall be established, with the highest leadership being headed or authorized by the competent executives of the organization" are the files and record forms of the committees or leading groups for the establishment of cybersecurity work.

The key points of the evaluation implementation include: making an interview with the information or network security supervisor to understand whether a "committee or leading group of guiding and managing cybersecurity work" is established; checking the composition and related responsibilities of the "committee or leading group guiding and managing network security work", and confirm whether the top leadership of the "committee or leading group guiding and managing cybersecurity work" is held or authorized by the competent executives of the organization.

If the test results show that the evaluated organization has specified the committee or leading group that guides and manages the cybersecurity work in the form of an official document, and the highest leadership position of the committee or leading group of cybersecurity work is held or authorized by the unit's top leader, the result is conforming; otherwise it is non-conforming or partially conforming.

[L3-ORS1-02/L4-ORS1-02 Interpretation and Description]

The main evaluation targets of the evaluation metric that "The functional department for cybersecurity management shall be established, including various leader positions in charge of security supervision and security management, and the responsibilities of each person in charge shall be defined" are roles and responsibilities documents.

The key points of the evaluation implementation include: making an interview with the information or cybersecurity supervisor to confirm whether a functional department for cybersecurity management is established; reviewing the organization management structure and job responsibilities documents, post responsibility statements and other documents; checking whether there are positions such as security supervisor and head of security management in the department, and whether the roles and responsibilities of relevant positions are clarified.

If the test results show that the evaluated organization has established a functional department responsible for security management and explicitly defined the positions and responsibilities of each responsible person, the result is conforming; otherwise it is non-conforming or partially conforming.

[L3-ORS1-03/L4-ORS1-03 Interpretation and Description]

The main evaluation targets of the evaluation metric that "Posts such as the system administrator, audit administrator and security administrator shall be established, and the responsibilities of each department and job post shall be defined" are roles and responsibilities documents.

The key points of the evaluation implementation include: interviewing with the information or cyberecurity supervisor to understand whether posts such as system administrators, audit administrators and security administrators are set up; obtaining relevant documents to check the establishment and scope of responsibilities of the system administrator, audit administrator and security administrator. Generally speaking, the system administrator is responsible for the operation and maintenance of the system, including operations such as strategy configuration; the audit administrator is responsible for the management of various logs; the security administrator is responsible for managing user access permissions and maintaining the network to ensure its security and normal operation.

If the test results show that the evaluated organization has set up roles such as system administrators, audit administrators and security administrators, and clearly defined the role and responsibility of each post, the result is conforming; otherwise it is non-conforming or partially conforming.

3.7.2 Staffing

[Standard Requirements]

This control point for Level III includes the evaluation units of L3-ORS1-04, L3-ORS1-05, and Level IV includes that of L4-ORS1-04, L4-ORS1-05, L4-ORS1-06.

[L3-ORS1-04/L4-ORS1-04 Interpretation and Description]

The main evaluation targets of the evaluation metric that "A certain number of system administrators, audit administrators and security administrators shall be manned" are record form documents, such as the staff list of job responsibilities.

The key points of the evaluation implementation include: interviewing information or network security supervisors, obtaining relevant documents, checking the establishment of system administrators, auditing administrators, and security administrators and the number of staffing; reviewing related management documents, checking system administrators, auditing whether the number of administrators and security administrators meets the security management requirements.

If the test results show that the evaluated organization has been equipped with a certain number of system administrators, audit administrators and security administrators, etc., and can meet the security management requirements, the result is conforming; otherwise it is non-conforming or partially conforming.

[L3-ORS1-05/L4-ORS1-05 Interpretation and Description]

The main evaluation targets of the evaluation metric that "Full-time rather than part-time security administrators shall be staffed" are record form documents such as the staff list of job responsibilities.

The key points of the evaluation implementation include: interviewing with the information or network security supervisor to confirm the staffing of security administrators; reviewing name lists and job responsibilities management regulations and other system documents to check whether the security administrator role is full-time and whether he or she assumes other administrator positions.

If the test results show that the evaluated organization is staffed with the security administrator who does not assume other administrator role, the result is conforming; otherwise it is non-conforming or partially conforming.

[L4-ORS1-06 Interpretation and Description]

The main evaluation targets of the evaluation metric that "Key business positions shall be managed by multiple people" are record form documents such as the staff list of job responsibilities.

The key points of the evaluation implementation include: interview with the information or network security supervisor and other personnel to confirm which positions are critical business positions. Generally speaking, network administrators, security administrators, system administrators are key positions; checking the name list of personnel responsible for related positions to confirm whether the network administrators, host administrators, system administrators and other positions are staffed with multiple people for joint management.

If the test results show that the key business positions of the evaluated organization are all staffed with multiple people to manage together, the result is conforming; otherwise it is non-conforming or partially conforming.

3.7.3 Authorization and Approval

[Standard Requirements]

This control point for Level Ⅲ includes the evaluation units of L3-ORS1-06, L3-ORS1-07, L3-ORS1-08, and Level Ⅳ includes that of L4-ORS1-07, L4-ORS1-08, L4-ORS1-09.

[L3-ORS1-06/L4-ORS1-07 Interpretation and Description]

The main evaluation targets of the evaluation metric that "The authorized approval item, approval department and approvers shall be clearly defined according to the responsibilities of each department and position." are the management system that includes authorized approval content and related record form documents.

The key points of the evaluation implementation include: understanding the existing organizational structure and the scope of responsibilities of each department and position; checking the relevant management regulations (such as approval management system, change management system, computer room safety management system, cybersecurity management system) in combination with the responsibilities of each department and position to confirm whether the relevant approval information such as the authorized approval item, approval department and approver are clearly defined; checking relevant approval records (such as system changes, physical access, system access, etc.) to confirm whether the existing authorized approval item, approval department and approver comply with the management regulation, and whether the approval records are regularized.

If the test results show that the authorized approval item, the approval department and approver of the evaluated organization are reasonably and effectively set, and consistent with the responsibilities of each department and position, the result is conforming; otherwise it is non-conforming or partially conforming.

[L3-ORS1-07/L4-ORS1-08 Interpretation and Description]

The main evaluation targets of the evaluation metric that "The approval procedures shall be established for issues such as system changes, important operations, physical access and system access, and the approval process shall be carried out in accordance with the approval procedures, a level-by-level approval system for important activities shall be established" are the management system that includes authorized approval content and related record form documents.

The key points of the evaluation implementation include: interviewing with the information or network security supervisor and other personnel to understand the specific scope of important items, such as system changes, important operations, physical access and system access, etc. ; reviewing the corresponding approval system and the approval record document; checking the approval procedure, the approval department and approver are consistent with the approval system.

If the test results show that the evaluated organization has established relevant approval procedures for system changes, important operations, physical access, and system access, there are tiered approval nodes in the approval process, the result is conforming; otherwise it is non-conforming or partially conforming.

[L3-ORS1-08/L4-ORS1-09 Interpretation and Description]

The main evaluation targets of the evaluation metric that "The approval items shall be examined and reviewed regularly, and the information about the items that need to be authorized and approved, the approval department and the approver shall be updated in a timely manner" are the management system that includes authorized approval content and related record form documents.

The key points of the evaluation implementation include: interviewing with the information or network security supervisor and other personnel to confirm whether various authorized approval items are reviewed and updated regularly (quarterly, every six months or every year); checking relevant meeting minutes or document review forms and other document records to confirm the update and review status of the authorized approval items; checking the list of approval items to see if the approval items, the approval department involved, the approver are clarified in the list, and whether the list of approval items is required to be updated and maintained on a regular basis.

If the test results show that the evaluated organization has regularly reviewed the authorized approval items, the approval department and the approver, and there are review records for updating the approval process, the result is conforming; otherwise it is non-conforming or partially conforming.

3.7.4 Communication and Cooperation

[Standard Requirements]

This control point for Level Ⅲ includes the evaluation units of L3-ORS1-09, L3-ORS1-10, L3-ORS1-11, and Level Ⅳ includes that of L4-ORS1-10, L4-ORS1-11, L4-ORS1-12.

[L3-ORS1-09/L4-ORS1-10 Interpretation and Description]

The main evaluation targets of the evaluation metric that "The cooperation and communication among various management personnel, internal departments andcybersecurity management department shall be strengthened, and coordination meetings shall be held regularly to jointly deal with cybersecurity issues" are record form documents containing the communication and cooperation content among various management personnel, various departments and cybersecurity management department.

The key points for the implementation of the evaluation include: interviewing with the information or network security supervisor to understand whether various management personnel, internal organizations andcybersecurity management department have established the cooperation and communication mechanism, and to understand the scope and main responsibilities of various management personnel, internal organizations and cybersecurity management department; reviewing relevant working meeting documents or meeting minutes to check whether the cooperation and communication mechanism is feasible and effective.

If the test results show that the evaluated organization has established a communication and cooperation mechanism among various departments on cyber security issues, and security work coordination meetings are regularly held, and complete meeting records (such as meeting minutes, cybersecurity work decision documents) are kept properly, the result is conforming; otherwise it is non-conforming or partially conforming.

[L3-ORS1-10/L4-ORS1-11 Interpretation and Description]

The main evaluation targets of the evaluation metric that "The cooperation and communication with cybersecurity administration authorities, various suppliers, industry experts and security organizations shall be strengthened" are record form documents that contain the content of communication and cooperation with external organizations.

The key points of the evaluation implementation include: interviewing with the information orcybersecurity supervisor and other personnel to understand whether the cooperation and communication mechanism has been established with the cybersecurity administration authorities, various suppliers, industry experts and security organizations; reviewing relevant communication records or documents, such as: the cybersecurity status report submitted to the cybersecurity authorities, the security planning/security design scheme provided by various security vendors, and the relevant records of security reviewing/ security review activities conducted by industry security experts, etc.

If the test results show that the evaluated organization has established a cooperation and communication mechanism with cybersecurity authorities, various suppliers, industry experts and security organizations, and relevant records of daily communication and cooperation are properly kept, the result is conforming; otherwise it is non-conforming or partially conforming.

[L3-ORS1-11/L4-ORS1-12 Interpretation and Description]

The main evaluation targets of the evaluation metric that "A contact list of the outreach units shall be established, including the name of the outreach unit, cooperation content, point of contacts and contact information" are record form documents containing the contact information of the outreach unit.

The key points of the evaluation implementation include: checking whether a contact list of outreach units is established. Outreach units may include: suppliers, industry experts, professional security vendors, security organizations (professional associations), superior authorities in charge, partners, security service agencies, telecommunications operations departments, law enforcement agencies, etc.; checking whether the content of the contact list of outreach units includes information such as the name of the outreach unit, cooperation content, contact person and contact details; the contact list of outreach units shall be maintained and updated according to the actual situation.

If the test results show that the evaluated organization has established a contact list of outreach units, and the content of the outreach unit contact list includes information such as the name of the outreach unit, cooperation content, contact person and contact information, the result is conforming; otherwise it is non-conforming or partially conforming.

3.7.5 Review and Inspection

[Standard Requirements]

This control point for Level Ⅲ includes the evaluation units of L3-ORS1-12, L3-ORS1-13, L3-ORS1-14, and Level Ⅳ includes that of L4-ORS1-13, L4-ORS1-14, L4-ORS1-15.

[L3-ORS1-12/L4-ORS1-13 Interpretation and Description]

The main evaluation targets of the evaluation metric that "the routine security inspections shall be carried out on a regular basis, including the daily operation of the system, system vulnerabilities and data backup" are management system documents and record form documents that include routine security inspections.

The key points of the evaluation implementation include: interviewing with the information or network security supervisor to understand whether a regular routine security inspection mechanism is established, reviewing related management documents, checking routine security inspection cycles, security inspection methods, security inspection procedures, security inspection content, etc.; routine security inspections are generally carried out monthly, quarterly, half a year or once a year; the inspection content may not be limited to the daily operation of the system, system vulnerabilities and data backup, etc., but can also include the consistency of the security configuration and security strategy, the implementation of the security management system, the implementation of technical protection measures and the development of classified security protection work, etc.; checking whether the records and documents of routine security inspections are regularized and complete.

If the test results show that the evaluated organization has established a regular routine security inspection mechanism and security inspection activities are performed regularly as required, and the security inspection records include the daily operation of the system, system vulnerabilities and data backup, etc., the result is conforming; otherwise it is non-conforming or partially conforming.

[L3-ORS1-13/L4-ORS1-14 Interpretation and Description]

The main evaluation targets of the evaluation metric that "The comprehensive security inspections shall be carried out on a regular basis, including the effectiveness of existing security technical measures, the consistency of security configurations and security strategies, and the implementation status of the security management system" are management system documents and record form documents that include routine security inspections.

The key points of the evaluation implementation include: interviewing with the information or cybersecurity supervisor to understand whether a comprehensive security inspection system is established on a regular basis (every six months or every year), obtaining relevant documents or records to check the time of comprehensive security inspections, security inspection methods, security inspection process, security inspection content, etc.; determining whether the comprehensive security inspection is carried out regularly in accordance with the security inspection requirements by checking the comprehensive security inspection documents and security inspection records; checking whether relevant records include the effectiveness of existing security technical measures, the consistency of security configuration and security strategy, the implementation status of the security management system, etc.; the comprehensive security inspections can be conducted by the organization itself or through a third-party agency. Either way, the inspection content shall cover the implementation status of technical and managerial security measures.

If the test results show that the evaluated organization has established a regular and comprehensive security inspection mechanism, and security inspection activities are conducted regularly as required, and the security inspection record includes the effectiveness of the existing security technology measures, the consistency of the security configuration and the security strategy, the implementation status of the security management system, etc., the result is conforming; otherwise it is non-conforming or partially conforming.

[L3-ORS1-14/L4-ORS1-15 Interpretation and Description]

The main evaluation targets of the evaluation metric that "The security inspection forms shall be developed to implement security inspections, summarizing the security inspection data, forming the security inspection report and notifying security inspection results" are security inspection forms, security inspection records, security inspection reports and other record form documents.

The key points of the evaluation implementation include: interviewing with the security administrator to understand whether specific requirements are formulated during the security inspection process, obtaining relevant documents to review whether the security inspection forms are developed, reviewing whether there are clear regulations on the collection of security inspection data, the generation of security inspection reports, and the notification of security inspection results; checking whether the security inspection forms, security inspection records, security inspection reports and other record forms meet the specified requirements; checking the notification record of security inspection results; checking whether the content of the notification is consistent with the security inspection report; checking whether the scope of the notification includes the relevant responsible departments and responsible personnel, especially information management department and administrators of each position.

If the test results show that the evaluated organization has record form documents such as security inspection forms, security inspection records, security inspection reports, etc., and the security inspection results are notified, the result is conforming; otherwise it is non-conforming or partially conforming.

3.8　Security Management Personnel

3.8.1　Staff Recruitment

[Standard Requirements]

This control point for Level III includes the evaluation units of L3-HRS1-01, L3-HRS1-02, L3-HRS1-03, and Level IV includes that of L4-HRS1-01, L4-HRS1-02, L4-HRS1-03, L4-HRS1-04.

[L3-HRS1-01/L4-HRS1-01 Interpretation and Description]

The main evaluation targets of the evaluation metric that "The designated department or personnel shall be appointed or authorized for employee hiring" are the cybersecurity supervisor and human resource management related personnel of the evaluated organization, including the management system containing staff recruitment and relevant record form documents.

The key points of the evaluation implementation include: interviewing with the cybersecurity supervisor and human resources management supervisor to understand whether the designated department or personnel is appointed or authorized to be responsible for employee recruitment; obtaining and reviewing the employee recruitment system related documents to check whether the designated department or personnel is appointed to be responsible for staff hiring, checking the work specifications and procedures of staff recruitment and hiring.

If the test results show that the employee recruitment management system of the evaluated organization is sound, and there exists dedicated department or personnel responsible for staff hiring, the result is conforming; otherwise it is non-conforming or partially conforming.

[L3-HRS1-02/L4-HRS1-02 Interpretation and Description]

The main evaluation targets of the evaluation metric that "The identity, security background, professional certifications or qualifications of the recruited personnel shall be examined and reviewed" are the management system for staff hiring and skill assessment related record form documents.

The key points of the evaluation implementation include: interviewing with the cybersecurity supervisor and human resources supervisor, obtaining personnel recruitment

Chapter 3　Application and Interpretation of the General Security Evaluation Requirements at Level III and Level IV

related system to check whether there are clear regulations on the review of the recruited personnel's identity, security background, professional certifications or qualifications, etc. ; checking personnel recruitment history audit records; checking the retained skill assessment documents or records when they were hired.

If the test results show that the evaluated organization examines and reviews the identity, security background, professional certifications or qualifications of the recruited personnel, assesses their technical skills, and keeps relevant records and documents, the result is conforming; otherwise it is non-conforming or partially conforming.

[L3-HRS1-03/L4-HRS1-03 Interpretation and Description]

The main evaluation targets of the evaluation metric that "A confidentiality agreement shall be signed with the recruited personnel and a post responsibility agreement with personnel at critical positions shall be signed" are record form documents such as the management system for staff hiring, the confidentiality agreement, the post responsibility agreement.

The key points of the evaluation implementation include: interviewing with the cybersecurity supervisor and human resources supervisor, reviewing the management systems related to personnel recruitment and hiring, checking whether there are relevant provisions such as "signing confidentiality agreements and key post responsibility agreements"; checking the signed confidentiality agreements, key post responsibility agreements; the confidentiality agreement shall include the scope of confidentiality, confidentiality responsibility, liability for breach of contract, as well as the validity period of the agreement and the signature of the responsible person; the key post responsibility agreement shall include the scope of work, security responsibilities, penalties, and the validity period of the agreement and the signature of the responsible person and other content.

If the test results show that the evaluated organization and the hired personnel have signed a complete confidentiality agreement, and a sound post responsibility agreement with key personnel are signed, the result is conforming; otherwise it is non-conforming or partially conforming.

[L4-HRS1-04 Interpretation and Description]

The main evaluation targets of the evaluation metric that "Personnel engaged in key positions shall be selected from within the organization" are the person in charge of the personnel recruitment and hiring in the evaluated organization, the system documents and the record form documents of personnel management.

The key points of the evaluation implementation include: interviewing with the cybersecurity supervisor and human resource supervisor to understand the personnel selection system for key positions; checking the relevant human resource management system, checking whether there is a provision of "Personnel in key positions shall be selected from within", and checking the selection files, documents or records of personnel currently engaged in key positions.

If the test results show that the evaluated organization stipulates the selection of key positions from within, the relevant selection and appointment documents are standardized and complete, the result is conforming; otherwise it is non-conforming or partially conforming.

3.8.2 Staff Dismissal

[Standard Requirements]

This control point for Level Ⅲ includes the evaluation units of L3-HRS1-04, L3-HRS1-05, and Level Ⅳ includes that of L4-HRS1-05, L4-HRS1-06.

[L3-HRS1-04/L4-HRS1-05 Interpretation and Description]

The main evaluation targets of the evaluation metric that "All access permissions of the resigned staff shall be terminated in time, and various identification documents, keys, badges as well as software and hardware devices provided by the organization shall be retrieved" are the human resource management related personnel in the evaluated organization, as well as the management system and related record form documents containing employee dismissal.

The key points of the evaluation implementation include: interviewing with the cybersecurity supervisor or human resource supervisor to understand the employee dismissal process and procedures, and check whether the relevant content or requirements such as termination of access permissions, retrieval of various identification documents, software and hardware devices are included; checking the personnel dismissal record form documents, checking the registration records of the resigned employee's terminating their access permissions, retrieving their identification documents, hardware and software devices, etc.

If the test results show that the evaluated organization has terminated all access permissions of the resigned personnel in a timely manner, retrieved various identification documents, keys, badges as well as the software and hardware devices provided by the organization, and a complete registration record is properly kept, the result is conforming; otherwise it is non-conforming or partially conforming.

[L3-HRS1-05/L4-HRS1-06 Interpretation and Description]

The main evaluation targets of the evaluation metric that "A strict resignation procedure shall be implemented, and the departing employees shall make the commitment to continue their confidentiality obligation after resignation" are the human resource management related personnel in the evaluated organization, as well as the management system and related record form documents containing employee dismissal.

The key points of the evaluation implementation include: interviewing with the cybersecurity and human resources supervisor to understand the employee dismissing process and procedure; reviewing the human resources management system to check whether there are management provisions, such as employee dismissing procedures, employees signing confidentiality agreement or letter of confidentiality commitment; checking the records of dismissal procedures, the confidentiality agreement or the letter of commitment signed by the resigned employees.

If the test results show that the departing personnel in the evaluated organization need to go through strict resignation procedures, signing a confidentiality agreement or a letter of confidentiality commitment, and the resignation record and effective letter of confidentiality commitment is properly kept, the result is conforming; otherwise it is non-conforming or partially conforming.

3.8.3 Security Awareness Education and Training

[Standard Requirements]

This control point for Level III includes the evaluation units of L3-HRS1-06, L3-HRS1-07, L3-HRS1-08, and Level IV includes that of L4-HRS1-07, L4-HRS1-08, L4-HRS1-09.

[L3-HRS1-06/L4-HRS1-07 Interpretation and Description]

The main evaluation targets of the evaluation metric that "Security awareness education and job skills training shall be conducted for all types of personnel, and relevant security responsibilities and disciplinary measures shall be informed" include the cybersecurity supervisor and personnel security education and training related management system documents, record form documents of the evaluated organization.

The key points of the evaluation implementation include: interviewing with the cybersecurity supervisor to understand the situation of security awareness education and job skill training; obtaining and reviewing the related management system to check whether it contains training content, training methods, training period, assessment methods and other relevant regulations, and whether it contains the security responsibilities and disciplinary measures for various personnel; checking security education and job skill training and other related record form documents.

If the test results show that the personnel security training system of the evaluated organization is well-established, the employee security awareness education and skills training are organized, and the relevant personnel understand the relevant regulations on security responsibilities and punishment, etc., the result is conforming; otherwise it is non-conforming or partially conforming.

[L3-HRS1-07/L4-HRS1-08 Interpretation and Description]

The main evaluation targets of the evaluation metric that "Different training plans shall be developed for different roles, and basic security knowledge and job operation procedures shall be trained" include the cybersecurity supervisor, employee security training related management system documents and record form documents.

The key points of the evaluation implementation include: interviewing with the cybersecurity supervisor to understand the development of training plans and the implementation of training plans, etc.; checking whether different training plans are developed for different positions, and checking whether the training content includes courses such as basic security knowledge and job operation procedures; checking whether the meeting minutes, training notes, assessment results and other records related to security education and job skill training are clear, standardized and complete.

If the test results show that the evaluated organization has formulated different training plans for different positions, and carried out training on basic security knowledge and job operating procedures, the result is conforming; otherwise it is non-conforming or partially conforming.

[L3-HRS1-08/L4-HRS1-09 Interpretation and Description]

The main evaluation targets of the evaluation metric that "The skill assessment shall be conducted regularly for employees in different positions" include the cybersecurity supervisor and related management system documents, record documents, etc.

The key points of the assessment implementation include: interviewing with the cybersecurity supervisor to understand the skill assessment situation. Reviewing relevant human resource management documents to check whether requirements or regulations are put forward for skill assessment for each position, whether the content of the assessment, assessment method, result evaluation are clarified; reviewing the skill assessment record form documents, and checking whether the assessment record documents include the appraisee information, assessment content, assessment result and other information.

If the test results show that the evaluated organization has established a regular skill assessment system for different positions, and the assessment records are standardized, complete and effective, the result is conforming; otherwise it is non-conforming or partially conforming.

3.8.4 External Visitor Access Management

[Standard Requirements]

This control point for Level III includes the evaluation units of L3-HRS1-09, L3-HRS1-10, L3-HRS1-11, L3-HRS1-12, and Level IV includes that of L4-HRS1-10, L4-HRS1-11, L4-HRS1-12, L4-HRS1-13, L4-HRS1-14.

[L3-HRS1-09/L4-HRS1-10 Interpretation and Description]

The main evaluation targets of the evaluation metric that "The written application shall be submitted before the external personnel physically access the controlled area. After approval, the visitors shall be accompanied by the designated employee all the time and the visiting shall be registered and filed for the record" include the main person in charge of the information department in the organization, the management system for external personnel's physical access to the controlled area, the written application document or application process record file for external personnel's physical access, and the registration record form for external personnel's access to important areas.

The key points of the evaluation implementation include: interviewing with the cybersecurity supervisor or the person in charge of the information department to understand the management system and procedures for external personnel to access the controlled area; checking whether the scope of external personnel's access, the conditions for external personnel's entry, and the access control measures for external personnel's entry are clearly defined; check on-site the written application document for external personnel to visit important areas, whether there is the approval signature from the head of the competent department, etc.; conducing the on-site checking of the registration records of external personnel visiting important areas, and whether the contents, such as the entry time of external personnel's visiting important areas, departure time, visiting area and accompanying person are recorded.

If the test results show that the evaluated organization has the external personnel visiting management system, the complete approval records and registration records for visiting important areas, the result is conforming; otherwise it is non-conforming or partially conforming.

[L3-HRS1-10/L4-HRS1-11 Interpretation and Description]

The main evaluation targets of the evaluation metric that "A written application shall be submitted before the external personnel accessing to the controlled network access system. After approval, the account creation and permission allocation is done by the designated person, and the access shall be registered and filed for the record" include the main responsible person of information department in the organization, the security service management system for external personnel, the application record of external personnel accessing the network, and the form of external personnel accessing the network and the registration record form of external personnel accessing the network.

The key points of the evaluation implementation include: interviewing with the relevant person in charge to determine whether a security service management system for external personnel are established; checking whether the management system includes the definition of external personnel, the application approval process before external personnel accessing to the controlled network, and measures to restrict the permissions of external personnel, and the division of related responsibilities, etc.; checking whether the external personnel's access application approval form contains the approval signature from the head of the competent department; checking whether the registration form for external personnel accessing important areas or information systems includes the recording of personnel information, time, reason, account access permission, valid time, etc.

If the test results show that the evaluated organization has developed rules and regulations for external personnel to access the controlled area, and approval records and related registration records for external personnel to access the controlled area are properly kept, the result is conforming; otherwise it is non-conforming or partially conforming.

【L3-HRS1-11/L4-HRS1-12 Interpretation and Description】

The main evaluation targets of the evaluation metric that "All external visitors shall be cleared of all access permissions in time after leaving the area" include the management system of external personnel accessing the controlled area, the registration record of external personnel accessing the system, etc.

The key points of the evaluation implementation include: reviewing the external personnel accessing management system to check whether it contains the relevant regulations of "clearing all access permissions of external personnel after leaving the are"; logging into the relevant device to check whether there are redundant or temporary accounts; checking the registration records of external personnel accessing the system, whether the account name, activation time, effective time limit, access permission clearance time are recorded, and checking whether the corresponding account is deleted or disabled in time.

If the test results show that the evaluated organization has relevant documents that clarify the requirements for clearing the permissions of external personnel after they leave the area, and has records of clearing access permissions, the result is conforming; otherwise it is non-conforming or partially conforming.

【L3-HRS1-12/L4-HRS1-13 Interpretation and Description】

The main evaluation targets of the evaluation metric that "External personnel who have obtained system access authorization shall sign a confidentiality agreement, may not perform unauthorized operations, and may not copy or disclose any sensitive information" are confidentiality agreements or record form documents of visiting external personnel.

The key points of the evaluation implementation include: reviewing the related security management system to check whether there are clear terms requiring external personnel who are authorized to access the system to sign a confidentiality agreement; checking whether the content of the confidentiality agreement contains the regulations of no unauthorized operations, no copying and disclosure of any sensitive information, etc.; checking the actual implementation of the management system, and checking the history records of external personnel visiting and signing of confidentiality agreements.

If the test results show that the evaluated organization has established a relevant system and requires signing a confidentiality agreement with external personnel, and the confidentiality agreement specifies the confidentiality obligation, the result is conforming; otherwise it is non-conforming or partially conforming.

【L4-HRS1-14 Interpretation and Description】

The main evaluation targets of the evaluation metric that "External personnel are not allowed to visit critical areas or critical systems" are the management system for external personnel physically accessing the controlled area.

The key points of the evaluation implementation include: checking the accessing management system for external personnel, and checking whether it is clear that external personnel are not allowed to access critical areas or critical business systems.

If the external personnel access management system clarifies that external personnel are not allowed to access critical areas or critical business systems, the result is conforming; otherwise it is non-conforming or partially conforming.

3.9   Security Development Management

3.9.1   Grading and Filing

[Standard Requirements]

This control point for Level Ⅲ includes the evaluation units of L3-CMS1-01, L3-CMS1-02, L3-CMS1-03, L3-CMS1-04, and Level Ⅳ includes that of L4-CMS1-01, L4-CMS1-02, L4-CMS1-03, L4-CMS1-04.

[L3-CMS1-01/L4-CMS1-01 Interpretation and Description]

The main evaluation targets of the evaluation metric that "The security protection level of the protection object, the method and reason for determining the level shall be explained in written form" are the filing form of the classified security protection for the information system, the grading report of the classified security protection for information system

The key points of the evaluation implementation include: checking the record form and the grading report to confirm whether the security protection level of the grading object is clearly defined; with reference to relevant standards such as the classification guide for classified protection of cybersecurity or industry classification guide, checking whether the methods and reasons for determining the system level in the grading report are reasonable, and whether the grading method meets the requirements of relevant standards.

If the test results show that the grading report provided by the evaluated organization clearly defines the security protection level of the system, and explains the method and reason for the grading of the system, the result is conforming; otherwise it is non-conforming or partially conforming.

[L3-CMS1-02/L4-CMS1-02 Interpretation and Description]

The main evaluation targets of the evaluation metric that "Relevant departments and security technical experts shall be organized to demonstrate and validate the rationality and correctness of the grading results" are the expert review opinions of the information system grading results, meeting sign-in forms and other record forms.

The key points of the evaluation implementation include: it's necessary to know whether relevant departments or security technical experts are organized to demonstrate and validate the grading results; checking whether there are evaluation and demonstration record documents for the grading results, such as expert review opinions.

If the test results show that the evaluated organization has organized relevant departments and relevant security technical experts to demonstrate and validate the rationality and correctness of the grading results, and relevant records are properly kept, the result is conforming; otherwise it is non-conforming or partially conforming.

[L3-CMS1-03/L4-CMS1-03 Interpretation and Description]

The main evaluation targets of the evaluation metric that "It shall be ensured that the grading results is approved by the relevant departments" are record form documents such as the approval opinions of the superior competent department or the relevant department of the organization.

The key points of the evaluation implementation include: checking whether there is a higher-level competent authority or relevant department of the organization to approve the grading results; checking the approval records of the determination of grading results.

If the test results show that the evaluated organization has the approval opinion of the competent authority or the confirmation document of the relevant departments, the result is conforming; otherwise it is non-conforming or partially conforming.

[L3-CMS1-04/L4-CMS1-04 Interpretation and Description]

The main evaluation targets of the evaluation metric that "The filing materials shall be reported to the competent department and the public security organ for the record" is the filing certificate issued by the public security organ.

The key points for the implementation of the evaluation include: checking whether it is filed with the competent department and the public security organ; checking whether there is a filing certificate issued by the public security organ.

If the test results show that the filing materials of the evaluated organization have been reported to the competent department and the public security organ for the record, and there is a filing certificate issued by the public security organ, the result is conforming; otherwise it is non-conforming or partially conforming.

3.9.2 Security Scheme Design

[Standard Requirements]

This control point for Level III includes the evaluation units of L3-CMS1-05, L3-CMS1-06, L3-CMS1-07, and Level IV includes that of L4-CMS1-05, L4-CMS1-06, L4-CMS1-07.

[L3-CMS1-05/L4-CMS1-05 Interpretation and Description]

The main evaluation targets of the evaluation metric that "Basic security measures shall be selected according to the level of security protection, and security measures shall be supplemented and adjusted according to the results of risk analysis" are security planning and design documents such as security requirements analysis documents and overall system security design documents.

The main points of the evaluation and implementation include: reviewing the security level protection design, implementation scheme and other documents, checking whether the basic security protection requirements are proposed and the basic security measures are selected according to the requirements of the corresponding national classified protection specifications to form a design scheme and an implementation (construction) plan; After the completion of the system construction, whether analysis is made according to the classified evaluation results to find hidden security risks, further supplement and adjust the security measures; check the records and documents formed in the process of relevant analysis, demonstration and rectification for verification.

If the test results show that the security planning and design documents provided by the evaluated organization clearly describe the system security protection level, and the basic security measures are selected according to the security protection level, and the security measures can be adjusted in time according to the risk analysis results, the result is conforming; otherwise it is non-conforming or partially conforming.

[L3-CMS1-06/L4-CMS1-06 Interpretation and Description]

The main evaluation targets of the evaluation metric that "The overall security plan and security scheme design shall be carried out according to the security protection level of the protection object and the relationship with other protection objects of various levels. And the design content shall include the relevant content of the cryptography technology and its supporting documents shall be developed" are security planning and design documents such as the overall security planning scheme for the classified protection objects and security design scheme.

The main points of the evaluation include: checking whether the overall security plan and security scheme design are made according to the security protection level of the protection object and the relationship with other levels of protection objects; the overall security plan shall focus on whether it includes but not limited to the overall objective, overall structure, and Security Development tasks (such as security infrastructure construction, network Security Development, Security Development of system platform and application platform, Security Development of data system, system construction of security standard, personnel training system construction, security management system construction), short-term or long-term construction plans, etc. The security scheme design shall focus on whether it includes, but is not limited to, the detailed security design scheme, the implementation plan of technical measures, and the implementation plan of management measures. The security technology framework shall include content related to cryptography, the level of security protection, the use of network security products, the division of network subsystems, IP address planning, etc.; checking whether the overall security plan and Security Development plan documents are consistent. In general, the supporting documents include security strategies, security system architecture, Security Development plan, security design scheme, etc.

If the test results show that the security planning and design documents provided by the evaluated organization have carried out the overall security planning and security scheme design based on the security protection level of the classified protection object and the relationship with other levels of protection objects, the design content includes cryptography technology related content, and supporting documents are formed, the result is conforming; otherwise it is non-conforming or partially conforming.

[L3-CMS1-07/L4-CMS1-07 Interpretation and Description]

The main evaluation targets of the evaluation metric that "Relevant departments and relevant security experts shall be organized to demonstrate and validate the rationality and correctness of the overall security plan and its supporting documents, and the formal implementation can be carried out only after approval" are record form documents such as approval opinions and demonstration comments of the overall security plan, security design scheme and related supporting documents.

The key points of the evaluation implementation include: checking whether the overall security plan and its supporting documents have the approval opinions of relevant departments and relevant security technical experts; checking whether the review form of the overall security plan and its supporting documents has the signatures and seals of relevant departments and relevant security technical experts; checking whether the relevant departments and relevant technical experts have the qualifications and capabilities for the demonstration and review of the overall security planning and the design of supporting documents.

If the test results show that the evaluated organization has organized relevant departments and relevant security experts to demonstrate and validate the rationality and correctness of the overall security plan and its supporting documents, and which have been approved, the result is conforming; otherwise it is non-conforming or partially conforming.

3.9.3 Product Procurement and Usage

[Standard Requirements]

This control point for Level Ⅲ includes the evaluation units of L3-CMS1-08, L3-CMS1-09, L3-CMS1-10, and Level Ⅳ includes that of L4-CMS1-08, L4-CMS1-09, L4-CMS1-10, L4-CMS1-11.

[L3-CMS1-08/L4-CMS1-08 Interpretation and Description]

The main evaluation targets of the evaluation metric that "It shall be ensured that the procurement and usage of network security products comply with relevant national regulations" are product procurement management system documents, product procurement lists, product usage management system documents, and computer security product sales licenses.

The key points of the evaluation implementation include: interviewing with the person in charge of system construction or procurement to understand whether the network security product procurement management system and procurement specifications and procedures are established; checking whether a network security product procurement list are formulated; checking whether the installed network security products have obtained the sales licenses.

If the test results show that the product procurement and usage management system of the evaluated organization is well-established, and all the network security products purchased and used have obtained the sales licenses, the result is conforming; otherwise it is non-conforming or partially conforming.

[L3-CMS1-09/L4-CMS1-09 Interpretation and Description]

The main evaluation targets of the evaluation metric that "It shall be ensured that the procurement and usage of cryptographic products and services meet the requirements of the national cryptography authority" include the person in charge of system construction, product procurement management system document, product procurement lists, product usage management and other system documents, etc.

The key points of the evaluation implementation include: interviewing with the person in charge of system construction, asking whether the system uses cryptographic products and related services; checking whether the cryptographic products used have obtained the commercial cryptographic product model certificate issued by the State Cryptography Administration, and checking whether the cryptographic products used and the model certificates are consistent; checking whether the cryptographic services used has obtained the permit issued by the the State Cryptography Administration, such as the permit for use of cryptography for E-authentication services.

If the test results show that the cryptographic products purchased and used by the evaluated organization have obtained the commercial cryptographic product model certificate issued by the State Cryptography Administration, and the cryptographic services used has obtained the permit of the State Cryptography Administration, the result is conforming; otherwise it is non-conforming or partially conforming.

[L3-CMS1-10/L4-CMS1-10 Interpretation and Description]

The main evaluation targets of the evaluation metric that "The product selection test shall be conducted in advance to determine the range of alternative products, and the list of candidate products shall be reviewed and updated on a regular basis" are product selection test result documents, alternative product purchase list, validation or update records, etc.

The main points of the evaluation include: interviewing with the person in charge of system construction to understand the product procurement process and related regulations; checking the management system of product procurement to confirm whether the control method of the procurement process (such as product selection test before procurement, clarifying the required product performance indicators, determining the range of alternative products, etc.) and personnel code of conduct are clearly stated; checking whether there are product selection test results records, the validation record of the alternative products list or the updated alternative product list.

If the test results show that the product procurement management system of the evaluated organization is well-established, there is a standardized product selection test report, the alternative product list review record or the updated candidate product list, etc., the result is conforming; otherwise it is non-conforming or partially conforming.

[L4-CMS1-11 Interpretation and Description]

The main evaluation targets of the evaluation metric that "The products of important parts shall be commissioned to the professional test and evaluation agency for specialized testing, and products shall be then selected according to the testing result" include the person in charge of system construction, and the product testing record form documents.

The key points of the evaluation implementation include: interviewing with the person in charge of system construction, reviewing the product procurement system, checking whether there is a requirement that "products in important parts must undergo specialized tests before they can be used"; checking procurement records, and checking whether there are specialized test reports for related products.

If the test results show that the procurement management system of the evaluated organization is well-established, and the products in important parts retain the evaluation report issued by the professional test and evaluation agency, the result is conforming; otherwise it is non-conforming or partially conforming.

3.9.4 Software Self-Development

[Standard Requirements]

This control point for Level III includes the evaluation units of L3-CMS1-11, L3-CMS1-12, L3-CMS1-13, L3-CMS1-14, L3-CMS1-15, L3-CMS1-16, L3-CMS1-17, and Level IV includes that of L4-CMS1-12, L4-CMS1-13, L4-CMS1-14, L4-CMS1-15, L4-CMS1-16, L4-CMS1-17, L4-CMS1-18.

[L3-CMS1-11/L4-CMS1-12 Interpretation and Description]

The main evaluation targets of the evaluation metric that "The development environment shall be physically separated from the actual operation environment, and the testing data and testing results shall be controlled" include the person in charge of system construction and software development management system documents. If the evaluated organization does not involve self-developed software, the evaluation metric is not applicable.

The key points of the evaluation implementation include: reviewing the software development management system and other documents to checking whether there exists the specification that the development environment and the actual operating environment of the evaluated organization must be operated separately, whether there exists the requirement that the testing data and testing results are controlled; interviewing the person in charge of system construction to check whether the development environment is physically separated from the actual operating environment; checking whether the implementation of testing data and testing results control measures is sound and effective.

If the test results show that the development environment of the evaluated organization is physically separated from the actual operating environment, the testing data and testing results are under control, the result is conforming; otherwise it is non-conforming or partially conforming.

[L3-CMS1-12/L4-CMS1-13 Interpretation and Description]

The main evaluation targets of theevaluation metric that "A software development management system shall be established to clearly define the control methods and personnel code of conduct during the development process" are the software development management system, system development manuals and the code of conduct record list for the software development process. If the evaluated organization does not have its own software development, the evaluation metric is not applicable.

The key points of the evaluation implementation include: interviewing the person in charge of system construction to understand whether a software development management system is established; checking whether the software development system includes the control methods of the development process and the developer's code of conduct; checking relevant documents formed during the software development process, such as meetings minutes, developer education and training, and process (software design, development, testing, acceptance, etc.) control records and other documents.

If the test results show that the evaluated organization has formulated a sound software development system document, and there are clear terms that stipulate the control methods and personnel code of conduct for the software design, development, testing, and acceptance process, and the management system is fully implemented, the result is conforming; otherwise it is non-conforming or partially conforming.

[L3-CMS1-13/L4-CMS1-14 Interpretation and Description]

The main evaluation targets of the evaluation metric that "A Security coding specifications shall be developed, requiring developers to write code according to the specifications" include the cybersecurity supervisor of the evaluated organization, system documents for self-developed software and record form documents. If the system does not involve self-developed software, the evaluation metric is not applicable.

The key points of the evaluation implementation include: interviewing the cybersecurity supervisor to see whether a management system such as code writing security specifications are formulated; checking whether the source code of the relevant software meets the specification requirements as appropriate; checking the documents and records formed during the source code writing process.

If the test results show that the evaluated organization has formulated a unified code writing specification document and implemented it effectively, the result is conforming; otherwise it is non-conforming or partially conforming.

[L3-CMS1-14/L4-CMS1-15 Interpretation and Description]

The main evaluation targets of the evaluation metric that "Relevant documents and instructions for software design shall be provided and the use of documents shall be controlled" include the cybersecurity supervisor of the evaluated organization, system documents for self-developed software and record form documents. If the system does not involve self-developed software, the evaluation metric is not applicable.

The key points of the evaluation implementation include: interviewing the cybersecurity supervisor to understand the formulation of the self-developed software development management system, to see whether the system includes software design related documents and usage guidelines, whether the use of the documents is controlled; checking whether the preparation, usage and storage of related documents are managed and control; checking the compiled software design documents, software usage guides, and the content of the documents shall meet the relevant requirements of the *National Standard for Software Development Documentation GB 8567—88*".

If the test results show that the evaluated organization has formulated a software development management system, the relevant documents and usage guidelines for software design are standardized and complete, and the use of documents is controlled, the result is conforming; otherwise it is non-conforming or partially conforming.

[L3-CMS1-15/L4-CMS1-16 Interpretation and Description]

The main evaluation targets of the evaluation metric that "Security testing shall be performed during software development and potential malicious code shall be detected prior to software installation" include the cybersecurity supervisor of the evaluated organization, system documents for self-developed software and record form documents. If the system does not involve self-developed software, the evaluation metric is not applicable.

The key points of the evaluation implementation include: interviewing the cybersecurity supervisor to understand the relevant self-developed software development management system, and checking whether it contains the specification such as "Security testing shall be performed during software development and potential malicious code shall be detected prior to software installation"; checking whether security testing is carried out during the software development process; checking relevant documents or records for verification; checking whether malicious code detection has been carried out before software installation, and checking relevant test reports or records for verification.

If the test results show that the evaluated organization has established a corresponding system that stipulates that the security of the software system needs to be tested during the development of the software system, and there are process documents to prove that the necessary tests have been carried out, the result is conforming; otherwise it is non-conforming or partially conforming.

[L3-CMS1-16/L4-CMS1-17 Interpretation and Description]

The main evaluation targets of the evaluation metric that "The modification, update and release of the application resource repository shall be authorized and approved, and the version control shall be strictly implemented" include the cybersecurity supervisor of the evaluated organization, system documents for self-developed software and record form documents. If self-developed software is not involved, then the evaluation metric is not applicable.

The key points of the evaluation implementation include: interviewing with the cybersecurity supervisor to understand the management regulations for the application resource repository, checking whether there are authorization and approval requirements for the modification of the application resource repository, and whether there are provisions for version control; checking the authorization and approval documents or related records formed in the process of modification, update, and release of application resource repository; reviewing the version control documents or records.

If the test results show that the evaluated organization has established rules and regulations for the application resource repository, and clearly stipulated that the modification, update, and release of the application resource repository need to be authorized and approved, and there are records of version control, authorization, modification and update, the result is conforming; otherwise it is non-conforming or partially conforming.

[L3-CMS1-17/L4-CMS1-18 Interpretation and Description]

The main evaluation targets of the evaluation metric that "It shall be ensured that the developers are full-time employees, and that the development activities of the developers are controlled, monitored and reviewed" include the cybersecurity supervisor of the evaluated organization, system documents for self-developed software and record form documents. If self-developed software is not involved, then the evaluation metric is not applicable.

The key points of the evaluation implementation include: interviewing with the cybersecurity supervisor to understand the relevant management system of software development, checking whether there are clear requirements for software developers and their activities, that is, the developers are full-time and the development activities are controlled; interviewing with the person in charge of software development, checking the developer education, training record, resumes or relevant professional qualification certificates, checking their job responsibilities documents to determine whether they are full-time developers; interviewing with the person in charge of software development to understand whether software development control procedures or related specifications have been formulated for software development activities (generally divided into five stages:

outline design, detailed design, coding and self-testing, assembly and system testing, trial and improvement) to implement control, monitoring and review of development activities; checking relevant documents and records, such as software development process records, requirements review forms, design review forms and case review forms, etc.

If the test results show that the evaluated organization has established the relevant system, stipulating that the developer shall be full-time, and has specific measures in place to control, monitor and review the development activities, the result is conforming; otherwise it is non-conforming or partially conforming.

3.9.5 Outsourcing Software Development

[Standard Requirements]

This control point for Level III includes the evaluation units of L3-CMS1-18, L3-CMS1-19, L3-CMS1-20, and Level IV includes that of L4-CMS1-19, L4-CMS1-20, L4-CMS1-21.

[L3-CMS1-18/L4-CMS1-19 Interpretation and Description]

The main evaluation targets of the evaluation metric that "the potential malicious code shall be detected before the software is delivered" are source code audit reports, malicious code detection reports, etc. If there is no outsourcing software development, the evaluation metric is not applicable.

The key points of the evaluation implementation include: interviewing with the system administrator to understand the project implementation management process, and reviewing the relevant management system in the system construction and management process, checking whether the delivered software has been tested for malicious code; checking the malicious code detection report.

If the test results show that the outsourcing development software is tested for malicious code and a test report is issued before delivery, the result is conforming; otherwise it is non-conforming or partially conforming.

[L3-CMS1-19/L4-CMS1-20 Interpretation and Description]

The main evaluation targets of the evaluation metric that "It shall be ensured that the development organization provides the relevant software design document and user guide" are needs analysis statement, software design specifications and other operating procedure documents and record form documents. If there is no outsourcing software development, then this evaluation metric is not applicable.

The key points of the evaluation implementation include: interviewing with the security supervisor or project leader, reviewing related outsourcing software development management system to check whether there are provisions for signing outsourcing software development agreements or contracts, and checking whether development agreements or contracts require software development organizations to provide corresponding software design documents and user guide; obtaining all the documents of the outsourcing software that has been put into operation, checking whether the software design document and user guide are provided; checking whether the content of the document is comprehensive and meets the application requirements; when there are multiple outsourcing development software in the information system, all software design documents and user guide shall be provided.

If the test results show that the evaluated organization has established a relevant system, put forward clear requirements on the software design documents and user guide, and the documentation of the software that has been put into operation is complete, the result is conforming; otherwise it is non-conforming or partially conforming.

[L3-CMS1-20/L4-CMS1-21 Interpretation and Description]

The main evaluation targets of the evaluation metric that "The software source code shall be provided by the development party and potential backdoors and covert channels in the software shall be examined and reviewed" include the person in charge of construction, software source code, software code audit reports and other record forms. If there is no outsourcing software development, then this evaluation metric is not applicable.

The key points of the evaluation implementation include: obtaining the outsourcing software development management system or software development agreement to check whether there is a specification of "providing software source code, examining and reviewing potential backdoors and hidden channels in the software"; checking the relevant documents or records of the submitted software source code, checking whether there is a code audit report issued by a third-party review agency.

If the test results show that the evaluated organization has established an outsourcing software development management system, the source code of the outsourcing software put into operation is submitted, and there is a test report issued by a third party, the result is conforming; otherwise it is non-conforming or partially conforming.

3.9.6 Security Engineering Implementation

[Standard Requirements]

This control point for Level III includes the evaluation units of L3-CMS1-21, L3-CMS1-22, L3-CMS1-23, and Level IV includes that of L4-CMS1-22, L4-CMS1-23, L4-CMS1-24.

[L3-CMS1-21/L4-CMS1-22 Interpretation and Description]

The main evaluation targets of the evaluation metric that "The designated department or personnel shall be appointed or authorized to be responsible for the management of the security engineering implementation process" are the person in charge of system construction, the system documents and record forms of security engineering implementation, etc..

The main points of the evaluation implementation include: interviewing with the person in charge of system construction, reviewing the project implementation management system or related documents to check whether a designated department or person is appointed or authorized to be responsible for the project implementation management; checking the department or personnel authorization documents, checking the scope of responsibility; checking the documents or records formed in the process of project implementation, checking they are fully implemented in accordance with management requirements.

If the test results show that the relevant system documents of the evaluated organization specify that a designated person or department is responsible for the management of the protect implementation process, and the person in charge of implementation knows the job responsibilities, the result is conforming; otherwise it is non-conforming or partially conforming.

【L3-CMS1-22/L4-CMS1-23 Interpretation and Description】

The main evaluation targets of the evaluation metric that "The security project implementation scheme shall be developed to control the implementation process of the project" are the person in charge of system construction, the security project implementation scheme, etc.

The key points of the evaluation implementation include: interviewing with the cybersecurity supervisor or the person in charge of system construction to understand whether the project implementation scheme has been formulated, checking whether the project plan, schedule control and quality control are included in the scheme; checking the phased documents generated during the project implementation process, such as schedule adjustment, rectification of quality problems and other documents or records, checking whether the requirements of the project implementation scheme are met.

If the test results show that the evaluated organization has formulated a security project implementation scheme, the content of the scheme includes the project time control, schedule control and quality control, etc. , and it is implemented in strict accordance with the scheme, and phased documents are produced at different stages of the project, the result is conforming; otherwise it is non-conforming or partially conforming.

【L3-CMS1-23/L4-CMS1-24 Interpretation and Description】

The main evaluation targets of the evaluation metric that "The implementation process of the project shall be controlled by third-party project supervision agency" are the person in charge of system construction, project supervision reports, etc. If the evaluated system is self-built by the evaluated organization, and a designated department or person is appointed to perform third-party supervision responsibilities and control the implementation process of the project, then this evaluation metric is not applicable.

The key points of the evaluation implementation include: interviewing with the person in charge of system construction to understand whether the relevant project has entrusted a third-party engineering supervision to control the implementation of the project; reviewing the contract or agreement entrusted to the third-party supervision; checking the project

Chapter 3 Application and Interpretation of the General Security Evaluation Requirements at Level Ⅲ and Level Ⅳ

supervision report to see whether the project progress, time schedule and control measures are clearly specified; For the construction or major transformation of important information system projects related to the national economy and people's livelihood, a third-party project supervision shall be entrusted.

If the test results show that the project construction of the evaluated organization has entrusted a third-party project supervision, and the content of the project supervision report meets the supervision requirements, the result is conforming; otherwise it is non-conforming or partially conforming.

3.9.7 Test and Acceptance

[Standard Requirements]

This control point for Level Ⅲ includes the evaluation units of L3-CMS1-24, L3-CMS1-25, and Level Ⅳ includes that of L4-CMS1-25, L4-CMS1-26.

[L3-CMS1-24/L4-CMS1-25 Interpretation and Description]

The main evaluation targets of the evaluation metric that "A acceptance testing plan shall be formulated, and the acceptance testing shall be implemented according to the acceptance testing plan, and a acceptance testing report shall be formed" are the acceptance testing plan, the acceptance testing report.

The key points of the evaluation implementation include: interviewing the cybersecurity supervisor or the person in charge of project construction to understand whether a special or general acceptance testing plan is formulated; checking whether the acceptance testing plan specifies the contents such as the departments and personnel involved in the test, acceptance details, processes, operating procedures; checking the relevant documents and records of the acceptance testing, reviewing the acceptance report, and checking whether the acceptance testing process meets the requirements of the acceptance testing plan.

If the test results show that the evaluated organization has formulated a acceptance testing plan, and the relevant records of the acceptance process are consistent with the requirements in the acceptance plan, the acceptance links and results of the acceptance report are complete and the relevant person in charge of the acceptance has signed and confirmed the report, the result is conforming; otherwise it is non-conforming or partially conforming.

[L3-CMS1-25/L4-CMS1-26 Interpretation and Description]

The main evaluation targets of the evaluation metric that "The security test before launch shall be performed, and a security test report shall be issued, which shall contain the content related to the cryptography application security test" are security test report and other record form documents.

The key points of the evaluation implementation include: checking whether the system has been tested for security before launch, such as penetration testing, vulnerability scanning, risk assessment, cryptography application security test, etc.; whether the security test report is obtained; checking whether the used cryptographic algorithms, products, technologies, and services have been evaluated for compliance, correctness, and effectiveness, the security of cryptography application is tested and a security test report is obtained.

If the test results show that the evaluated system has completed the pre-launch security test and the cryptography application security test, the result is conforming; otherwise it is non-conforming or partially conforming.

3.9.8 System Delivery

[Standard Requirements]

This control point for Level III includes the evaluation units of L3-CMS1-26, L3-CMS1-27, L3-CMS1-28, and Level IV includes that of L4-CMS1-27, L4-CMS1-28, L4-CMS1-29.

[L3-CMS1-26/L4-CMS1-27 Interpretation and Description]

The main evaluation targets of the evaluation metric that "A delivery list shall be developed and the devices, software and documentations handed over shall be counted according to the delivery list" are the delivery list and record form documents.

The key points of the evaluation implementation include: understanding the development of the delivery list, reviewing the actual delivery list; checking whether the delivery list contains all the devices and software of the system, as well as documentations such as system operation, deployment, operation and maintenance, and user account information.

If the test results show that the evaluated organization has formulated a delivery list, the relevant information in the list is complete and consistent with the real situation of the information system, the result is conforming; otherwise it is non-conforming or partially conforming.

[L3-CMS1-27/L4-CMS1-28 Interpretation and Description]

The main evaluation targets of the evaluation metric that "The corresponding skill training shall be carried out for the technical staff responsible for operation and maintenance" are technical training delivery records and other record forms.

The key points of the evaluation implementation include: checking whether the technical personnel responsible for operation and maintenance are properly trained after the delivery of the system; checking whether the training content is comprehensive and whether the training effect meets the operation and maintenance requirements; checking whether the training plan, course setting, training result assessment and other document records are authentic.

If the test results show that the project developer or construction party has trained relevant personnel on the system operation, maintenance, and business skills, and corresponding training records are properly retained, the result is conforming; otherwise it is non-conforming or partially conforming.

Chapter 3　Application and Interpretation of the General Security Evaluation Requirements at Level III and Level IV

[L3-CMS1-28/L4-CMS1-29 Interpretation and Description]

The main evaluation targets of the evaluation metric that "Construction process documents, operation and maintenance documents shall be provided" are construction process documents and operation and maintenance documents.

The main points of the evaluation implementation include: interviewing with the system operation and maintenance personnel, reviewing system construction process documents and operation and maintenance documents to check whether the system operation and maintenance documents are consistent with the current system operation and maintenance situation; checking whether the construction process documents, operation and maintenance documents conform to the requirements of the management system related to the construction of the information system.

If the test results show that the evaluated organization has provided construction process documents, operation and maintenance documents, and the documents meet the requirements of management regulations, the result is conforming; otherwise it is non-conforming or partially conforming.

3.9.9　Classified Security Evaluation

[Standard Requirements]

This control point for Level III includes the evaluation units of L3-CMS1-29, L3-CMS1-30, L3-CMS1-31, and Level IV includes that of L4-CMS1-30, L4-CMS1-31, L4-CMS1-32.

[L3-CMS1-29/L4-CMS1-30 Interpretation and Description]

The main evaluation targets of the evaluation metric that "Classified security evaluation shall be carried out regularly, and timely rectification shall be implemented if the corresponding level of protection requirements are not met" are classified evaluation reports, security rectification plans, etc. If it is the first evaluation, then this evaluation metric is not applicable.

The key points of the evaluation implementation include: interviewing with the cybersecurity supervisor and reviewing the previous classified evaluation report and security rectification plan; reviewing the relevant rectification documents or records to check the rectification implementation of the security issues; checking the time of the previous classified evaluation report to confirm whether the system classified evaluation cycle is in compliance with the management requirements, such as information systems graded as Level III shall be evaluated at least once a year.

If the test results show that the system evaluation cycle of the evaluated organization meets the management requirements, and security problems found are rectified in a timely manner, the result is conforming; otherwise it is non-conforming or partially conforming.

[L3-CMS1-30/L4-CMS1-31 Interpretation and Description]

The main evaluation targets of the evaluation metric that "Classified security evaluation shall be carried out when major changes or level changes occur" are the person in charge of operation and maintenance, classified evaluation reports, etc. If this is the first evaluation, then the evaluation metric is not applicable.

The key points of the evaluation implementation include: interviewing with the person in charge of operation and maintenance, and checking whether the grading of the system has changed or whether the system has undergone major changes, such as major business adjustments, equipment room relocation, and important application systems replacement, etc.; whether the classified evaluation has been re-conducted after a major change, and checking the classified evaluation report.

If the test results show that the system of the evaluated organization has not undergone major changes or level changes since the last classified evaluation, or the classified evaluation work has been carried out in a timely manner after corresponding changes and grading changes occurred, the result is conforming; otherwise it is non-conforming or partially conforming.

[L3-CMS1-31/L4-CMS1-32 Interpretation and Description]

The main evaluation targets of the evaluation metric that "It shall be ensured that the selection of the evaluation agency complies with the relevant national regulations" are classified evaluation report and relevant qualification documents.

The key points for the implementation of the evaluation include: checking whether the evaluation agencies that have implemented the classified evaluation in the past are in compliance with the relevant national regulations. Qualified test and evaluation agencies can be found on the official website of *China Classified Protection for Cybersecurity* (http://www.djbh.net/).

If the selection of the evaluation agency meets the relevant national regulations, the result is conforming; otherwise it is non-conforming or partially conforming.

3.9.10 Service Provider Management

[Standard Requirements]

This control point for Level III includes the evaluation units of L3-CMS1-32, L3-CMS1-33, L3-CMS1-34, and Level IV includes that of L4-CMS1-33, L4-CMS1-34, L4-CMS1-35.

[L3-CMS1-32/L4-CMS1-33 Interpretation and Description]

The main evaluation targets of the evaluation metric that "It shall be ensured that the selection of service providers conforms to the relevant national regulations" are the person in charge of construction, the list of safety service providers, etc.

The main points of the evaluation implementation include: interviewing with the person in charge of construction and asking whether the information system involves the third-party security consulting, supervision, training, planning, involvement, implementation, evaluation and other security services; if the above-mentioned security services are involved, checking whether the security service provider has relevant qualifications, such as security integration service qualification certification, information system classified security protection certification, CNAS (China National Accreditation Service for Conformity Assessment) certification, etc.

If the test results show that the security service provider selected by the evaluated organization meets the relevant national regulations, the result is conforming; otherwise it is non-conforming or partially conforming.

【L3-CMS1-33/L4-CMS1-34 Interpretation and Description】

The main evaluation targets of the evaluation metric that "Relevant agreements shall be signed with the selected service providers to clarify the cyber security related obligations that all parties to the entire service supply chain need to perform" are record form documents such as service contracts or security responsibility documents.

The key points of the evaluation implementation include: checking whether a service contract or security responsibility letter has been signed with the service provider; checking whether the service contract or security responsibility letter clearly specifies the content of subsequent technical support and service commitments.

If the test results show that the evaluated organization has signed a relevant agreement with the selected service provider to clarify the cybersecurity related obligations that all parties to the entire service supply chain need to perform, the result is conforming; otherwise it is non-conforming or partially conforming.

【L3-CMS1-34/L4-CMS1-35 Interpretation and Description】

The main evaluation targets of the evaluation metric that "The services provided by service providers shall be regularly monitored, reviewed and audited, and their changes to the service content shall be controlled" are security service reports, service audit reports and other record form documents.

The key points of the evaluation implementation include: checking whether the service provider evaluation and audit management system is established, and whether the corresponding system includes provisions for supervision, review, audit, and control service content; checking the security service report submitted by the service provider regularly; reviewing the documents or records such as the service audit report of the service provider for regular review and evaluation.

If the test results show that the evaluated organization has developed a service provider evaluation audit management system, the result is conforming; otherwise it is non-conforming or partially conforming.

3.10 Security Operation and Maintenance Management

3.10.1 Environment Management

[Standard Requirements]

This control point for Level Ⅲ includes the evaluation units of L3-MMS1-01, L3-MMS1-02, L3-MMS1-03, and Level Ⅳ includes that of L4-MMS1-01, L4-MMS1-02, L4-MMS1-03, L4-MMS1-04.

[L3-MMS1-01/L4-MMS1-01 Interpretation and Description]

The main evaluation targets of the evaluation metric that "The designated department or personnel shall be assigned to be responsible for the safety of the equipment room, managing the access to the equipment room and regularly maintaining and managing the power supply and distribution, air conditioning, temperature and humidity control, fire protection and other facilities in the equipment room" are department or personnel's job responsibilities documents and record form documents.

The main points of the evaluation include: reviewing the human resource management system or related documents, checking whether the designated department or personnel is responsible for the safety of the equipment room, and checking whether their scope of responsibility includes the equipment room accessing management, maintenance and management of the equipment room environment, fire protection facilities, etc. ; checking the access control system, personnel accessing approval and registration, checking the maintenance and management records of the equipment room power supply and distribution, air conditioning, temperature and humidity control, fire protection and other facilities, and evaluating whether the safety of the equipment room meets the safety management requirements.

If the test results show that the evaluated organization has appointed the designated department or personnel to be responsible for the safety management of the equipment room, and there are complete records of facility maintenance such as equipment room accessing, power supply and distribution equipment, air-conditioning equipment, temperature and humidity control equipment, fire-fighting equipment, etc. , the result is conforming; otherwise it is non-conforming or partially conforming.

[L3-MMS1-02/L4-MMS1-02 Interpretation and Description]

The main evaluation targets of the evaluation metric that "The equipment room safety management system shall be developed to put forward the regulations of physical access, entry and exit of articles, and environmental safety" are the system document for equipment room management and record form documents.

Chapter 3 Application and Interpretation of the General Security Evaluation Requirements at Level III and Level IV

The key points of implementation of the evaluation include: reviewing the equipment room safety management system, checking whether there are provisions on physical access, item entry and exit, and environmental safety; checking the relevant approval documents, registration forms, activity records of the equipment room physical access, items brought in and out, and equipment room environmental safety, and evaluating the implementation of the equipment room safety management system.

If the test results show that the evaluated organization has established a equipment room safety management system, and put forward the regulations on the management of physical access, entry and exit of items, and environmental safety, and corresponding records are kept as the proof, the result is conforming; otherwise it is non-conforming or partially conforming.

[L3-MMS1-03/L4-MMS1-03 Interpretation and Description]

The main evaluation targets of the evaluation metric that "Visitors shall not be received in important areas, and paper files and mobile media containing sensitive information shall not be placed randomly" are the management system and office environment.

The main points of the evaluation include: reviewing the visitor management system, checking whether the reception area is limited, whether there are requirements for the storage of written documents and mobile media containing sensitive information; checking the office environment, and checking whether there are randomly placed paper files and removable media containing sensitive information in the office area.

If the test results show that the evaluated organization has formulated a management system for visitor reception, clearly defined the reception area for visitors, and has taken control measures for paper documents and mobile media containing sensitive information, the result is conforming; otherwise it is non-conforming or partially conforming.

[L4-MMS1-04 Interpretation and Description]

The main evaluation targets of the evaluation metric that "The entering and exiting personnel shall be granted with corresponding level of authorization, and the personnel and their activities in the important safety areas shall be monitored in real time" are record form documents and video materials containing personnel activities in important safety areas.

The main points of the evaluation include: checking whether there are authorization regulations for personnel entering and exiting; checking whether there are real-time monitoring requirements for personnel entering and exiting and activities in important safety areas; reviewing the authorized approval records of entering and exiting personnel to check whether they are authorized to enter and exit according to the corresponding level; checking the personnel access and activity monitoring measures in important safety areas, such as the installed monitor system, and the provision of personnel being accompanied throughout the process, etc.

If the test results show that the evaluated organization has established a relevant management system to authorize personnel entering and exiting according to their levels, and the activities of personnel in important areas are monitored, the result is conforming; otherwise it is non-conforming or partially conforming.

3.10.2 Asset Management

[Standard Requirements]

This control point for Level III includes the evaluation units of L3-MMS1-04, L3-MMS1-05, L3-MMS1-06, and Level IV includes that of L4-MMS1-05, L4-MMS1-06, L4-MMS1-07.

[L3-MMS1-04/L4-MMS1-05 Interpretation and Description]

The main evaluation targets of the evaluation metric that "A list of assets related to the objects of protection shall be compiled and kept, including the assets responsible department, degree of importance and their location, etc." are the asset list and asset management system of the evaluated organization.

The key points of the evaluation implementation include: checking whether a list of assets has been compiled for the protection objects, whether the list contains various assets such as device assets, facility assets, software assets, document assets, etc.; checking whether the asset list clearly indicates the assets responsible department and the physical location of the assets, asset information, asset importance, etc.

If the test results show that the evaluated organization has developed a comprehensive and complete list of assets, the result is conforming; otherwise it is non-conforming or partially conforming.

[L3-MMS1-05/L4-MMS1-06 Interpretation and Description]

The main evaluation targets of the evaluation metric that "Identification management of assets shall be carried out according to the importance of the assets, and corresponding management measures shall be selected according to the value of the assets" are the relevant personnel of the asset management in the evaluated organization, as well as the management system and related record form documents containing the asset management.

The key points of the evaluation implementation include: checking whether the asset classification management system has been established, understanding the principles and methods of assets management; checking whether assets are identified and managed according to the importance of the assets, and checking whether the asset identification method is consistent with the provisions of the assets security management system; checking whether different management measures have been taken for assets of different values, checking whether the relevant management documents and records are complete, whether the relevant measures are feasible and effective, etc.

If the test results show that the evaluated organization has established a security management system that includes assets classification and identification, and the asset identification method is consistent with the system, corresponding management measures can be taken according to the importance of the assets, the result is conforming; otherwise it is non-conforming or partially conforming.

[L3-MMS1-06/L4-MMS1-07 Interpretation and Description]

The main evaluation targets of the evaluation metric that "Information classification and identification methods shall be stipulated, and the use, transmission and storage of information shall be regularized in management" are information classification and identification management system and related record form documents.

The key points of the evaluation implementation include: understanding whether information classification and identification management regulations have been formulated, checking the principles and methods of information classification and identification, and checking whether regularized requirements are put forward for the use, transmission, and storage of information; checking relevant documents or management records to evaluate whether the management effect meets the security management requirements.

If the test results show that the evaluated organization has formulated the regulations of information classification and identification management, information are classified and identified accordingly, and the use, transmission and storage of information are regularized in management, the result is conforming; otherwise it is non-conforming or partially conforming.

3.10.3 Media Management

[Standard Requirements]

This control point for Level III includes the evaluation units of L3-MMS1-07, L3-MMS1-08, and Level IV includes that of L4-MMS1-08, L4-MMS1-09.

[L3-MMS1-07/L4-MMS1-08 Interpretation and Description]

The main evaluation targets of the evaluation metric that "The media shall be stored in a safe environment, and various types of media shall be controlled and protected. The storage of media shall be managed by the designated personnel, and the inventory shall be regularly counted according to the catalogue of the archived media" are the media management related personnel in the evaluated organization and the management system including media management and related record form documents.

The main points of the evaluation include: understanding whether a media management system has been established, checking whether the storage environment of the media, media management are clearly defined; checking whether the media is managed by the designated personnel, and checking whether a media catalog list is established; checking the daily media management documents to see whether there are records of use, return, filing and destruction, etc.; checking the records of regular inventory counting of storage media; checking whether the storage environment of the media is safe, such as stored in a safe or file cabinet.

If the test results show that the media of the evaluated organization are stored in a safe environment, there is a complete record of media control and protection, and the media are managed by the designated personnel and inventory counting is carried out on a regular basis, the result is conforming; otherwise it is non-conforming or partially conforming.

[L3-MMS1-08/L4-MMS1-09 Interpretation and Description]

The main evaluation targets of the evaluation metric that "The personnel selection, packaging, delivery of the media shall be controlled during the physical transmission process, and the archiving and querying of the media shall be registered and recorded" are the media management related personnel in the evaluated organization and the management system including media management and related record form documents.

The key points of the evaluation implementation include: understanding the relevant media management system, and checking whether there are regulations on physical media transmission; whether the regulations have clear requirements for each link in the transmission process, such as requirements for personnel selection, packaging, delivery, etc.; checking physical media transmission management record, checking whether the recorded content includes information such as performer, storage media information, storage media packaging, storage media delivery, storage media archiving, storage media query and other information.

If the test results show that the relevant security management system in the evaluated organization contains provisions on physical transmission, archiving and querying of the media, and the relevant record forms are complete, the result is conforming; otherwise it is non-conforming or partially conforming.

3.10.4　Device Maintenance Management

[Standard Requirements]

This control point for Level Ⅲ includes the evaluation units of L3-MMS1-09, L3-MMS1-10, L3-MMS1-11, L3-MMS1-12, and Level Ⅳ includes that of L4-MMS1-10, L4-MMS1-11, L4-MMS1-12, L4-MMS1-13.

[L3-MMS1-09/L4-MMS1-10 Interpretation and Description]

The main evaluation targets of the evaluation metric that "The designated department or personnel shall be appointed for the regular maintenance and management of various devices (including backup and redundant devices) and lines" are the device maintenance and management related personnel in the evaluated organization and the job responsibility documents, including device maintenance and management system and related record form documents.

The key points of the evaluation implementation include: reviewing the related management system to check whether management regulations have been formulated for the

Chapter 3　Application and Interpretation of the General Security Evaluation Requirements at Level Ⅲ and Level Ⅳ

maintenance and management of various devices and lines; reviewing the department and job responsibilities documents to check whether the designated department or personnel are appointed to be responsible, and whether the maintenance scope, maintenance requirements and maintenance procedures are clearly defined. Requirements, maintenance procedures, etc.; checking the relevant maintenance records, and evaluating whether the maintenance management system has been strictly implemented.

If the test results show that the evaluated organization has appointed the designated department or personnel to perform regular maintenance and management of various devices and lines and relevant record forms are retained, the result is conforming; otherwise it is non-conforming or partially conforming.

[L3-MMS1-10/L4-MMS1-11 Interpretation and Description]

The main evaluation targets of the evaluation metric that "A management system for supporting facilities, hardware and software maintenance shall be established to effectively manage their maintenance, including clarifying the responsibilities of maintenance personnel, the approval of maintenance and services, and the supervision and control of the maintenance process" are device maintenance management system and related record form documents.

The main points of the evaluation implementation include: reviewing the device maintenance management system to check whether it contains the responsibilities of the maintenance personnel, the approval of repairs and services, and the supervision and control of the repairing process; checking the records of approval, supervision and repairing generated during the maintenance process, and checking whether the approval and record content meets the requirements of the system.

If the test results show that the evaluated organization has formulated a management system for supporting facilities, software and hardware maintenance, and has retained relevant approval documents and record forms, the result is conforming; otherwise it is non-conforming or partially conforming.

[L3-MMS1-11/L4-MMS1-12 Interpretation and Description]

The main evaluation targets of the evaluation metric that "Information processing device shall be approved before being taken out of the equipment room or office, and important data shall be encrypted when the device containing storage media is taken out of the working environment" are the device maintenance and management related personnel in the evaluated organization, the management system including device maintenance and management and the related record form documents.

The key points of the evaluation implementation include: reviewing the related security management systems to check whether relevant regulations have been formulated for the information processing device being taken away from the equipment room. The regulations shall include measures such as the approval process and encryption protection of important data in the device media; checking the approval form for the device to be taken away from the equipment room, confirming whether the operation is carried out in accordance with the established approval process, and focusing on whether the form contains the content of such as the whereabouts of the device, the time of being taken away, and the signature of the reviewer; checking the records of the device containing the storage media that are taken away from the equipment room, whether there are encryption measures for important data in the records, and focusing on the examination and approval record documents for the encryption measures.

If the test results show that the evaluated organization has established an approval process for the device to be taken away from the equipment room or office location, the important data in the storage media has been encrypted, and the relevant record forms are retained properly, the result is conforming; otherwise it is non-conforming or partially conforming.

[L3-MMS1-12/L4-MMS1-13 Interpretation and Description]

The main evaluation targets of the evaluation metric that "Equipment containing storage media shall be completely erased orSecurityly covered before being scrapped or reused to ensure that the sensitive data and authorized software on the device cannot be restored and reused" are the device maintenance and management related personnel in the evaluated organization, the management system including device maintenance and management and the related record form documents.

The key points of the evaluation implementation include: reviewing related security management systems to check whether there are corresponding provisions for scrapping or reusing devices containing storage media, and whether specific measures are specified for the removal of sensitive data on the device; checking related scrap records, sensitive data and authorized software removal records; checking whether the removal measures and methods are effective to make sure that they cannot be restored for reuse.

If the test results show that the evaluated organization has formulated a management system for the scrap or reuse of device containing storage media, making sure that the sensitive data and authorized software in the scrapped device cannot be recovered, and the relevant record forms are retained properly, the result is conforming; otherwise it is non-conforming or partially conforming.

3.10.5 Vulnerability and Risk Management

[Standard Requirements]

This control point for Level Ⅲ includes the evaluation units of L3-MMS1-13, L3-MMS1-14, and Level Ⅳ includes that of L4-MMS1-14, L4-MMS1-15.

[L3-MMS1-13/L4-MMS1-14 Interpretation and Description]

The main evaluation targets of the evaluation metric that "Necessary measures shall be taken to identify security vulnerabilities and hidden risks, and the discovered security vulnerabilities and hidden risks shall be fixed in a timely manner or after the possible impact is assessed" are security management related personnel, as well as related security management systems, vulnerabilities identification and repair record form documents.

The key points of the evaluation implementation include: interviewing with the security administrators to understand whether there are regulations for regularly conducting security vulnerabilities and hidden risks detection; checking the detection methods or tools used for security vulnerabilities and hidden risks; checking the detection reports, such as vulnerability scanning reports, penetration test reports and security test report, etc.; checking the vulnerability repair report or the operation record of eliminating hidden risks to review whether it contains the vulnerability repairing method, repairing time, repairing result, etc.

If the test results show that the evaluated organization is equipped with security vulnerabilities and hidden risks detection devices or tools, which can assess and fix related vulnerabilities in a timely manner, and related vulnerability detection and vulnerability fix record forms are retained properly, the result is conforming; otherwise it is non-conforming or partially conforming.

[L3-MMS1-14/L4-MMS1-15 Interpretation and Description]

The main evaluation targets of the evaluation metric that "Security test and evaluation shall be carried out on a regular basis, a security evaluation report shall be formed, and measures shall be taken to deal with identified security issues" are documents, such as filing proof documents and evaluation reports over the years.

The key points of the evaluation implementation include: reviewing the filing evidence and evaluation reports over the past years, checking whether the security evaluation period meets the relevant national regulations; reviewing the rectification scheme, rectification acceptance report, etc., checking the rectification results on the spot, and checking whether effective protection measures are taken against the security problems found in the evaluation.

If the test results show that the evaluated organization regularly conducts security evaluations and takes effective rectification measures against the security problems found in the evaluations, the result is conforming; otherwise it is non-conforming or partially conforming.

3.10.6 Network and System Security Management

[Standard Requirements]

This control point for Level III includes the evaluation units of L3-MMS1-15, L3-MMS1-16, L3-MMS1-17, L3-MMS1-18, L3-MMS1-19, L3-MMS1-20, L3-MMS1-21, L3-MMS1-22, L3-MMS1-23, L3-MMS1-24, and Level IV includes that of L4-MMS1-16, L4-MMS1-17, L4-MMS1-18, L4-MMS1-19, L4-MMS1-20, L4-MMS1-21, L4-MMS1-22, L4-MMS1-23, L4-MMS1-24, L4-MMS1-25.

[L3-MMS1-15/L4-MMS1-16 Interpretation and Description]

The main evaluation targets of the evaluation metric that "Different administrator roles shall be divided to manage the network and system operation and maintenance, and the responsibilities and permissions of each role shall be specified" are network and system operation and maintenance management systems and job responsibility documents.

The key points of the evaluation implementation include: reviewing the network and system operation and maintenance management system and personnel job responsibility documents to check whether different administrator roles (such as network administrator, system administrator, etc.) are divided; checking the network and system operation and maintenance management system and personnel job responsibility document to confirm whether the responsibilities and permissions of each role are clearly defined; checking whether the roles, responsibilities and permissions of the network and system operation and maintenance administrators match.

If the test results show that the evaluated organization is staffed with different operation and maintenance administrators for the network and system management, and the responsibilities of each role are explicitly defined, the result is conforming; otherwise it is non-conforming or partially conforming.

[L3-MMS1-16/L4-MMS1-17 Interpretation and Description]

The main evaluation targets of the evaluation metric that "The designated department or personnel shall be appointed for account management, controlling the account application, account creation and account deletion" are security related personnel, as well as security operation and maintenance management systems and record form documents.

The key points of the evaluation implementation include: interviewing with the cybersecurity supervisor to find out whether the designated department or personnel is responsible for account management and whether it is specified in the management system; checking whether the relevant approval process or operation specifications are established, and the activities such as account application, account creation and account deletion shall be controlled effectively.

If the test results show that the evaluated organization has appointed the designated department or personnel to be responsible for account management, and the account application, creation and deletion are strictly controlled, the result is conforming; otherwise it is non-conforming or partially conforming.

[L3-MMS1-17/L4-MMS1-18 Interpretation and Description]

The main evaluation target of the evaluation metric that "The network and system security management system shall be established to put forward provisions for security strategies, account management, configuration management, log management, daily operations, upgrades and patches, password update cycle, etc." is the network and system security operation and maintenance management System.

Chapter 3 Application and Interpretation of the General Security Evaluation Requirements at Level III and Level IV

The key points of the evaluation implementation include: reviewing the network and system security management system to check whether security strategies, account management, configuration management, log management, daily operations, upgrades and patches, password update cycle are included.

If the test results show that the evaluated organization has established the network and system security management system to provide provisions on security strategies, account management, configuration management, log management, daily operations, upgrades and patches, and password update cycle, etc., the result is conforming; otherwise it is non-conforming or partially conforming.

【L3-MMS1-18/L4-MMS1-19 Interpretation and Description】

The main evaluation targets of the evaluation metric that "The configuration and operation manuals of important devices shall be developed, and the security configuration and configuration optimization for devices shall be conducted according to the manuals" are important device configuration and operation manual documents.

The key points of the evaluation implementation include: checking the configuration and operation manuals of important devices; checking the content integrity of the device configuration and operation manuals, and checking whether the operating procedures, operating specifications, parameter configuration of the operating system, database, network devices (core switches, routers, etc.), security devices (firewalls, WAF, etc.), applications and components are included.

If the test results show that the evaluated organization has developed the configuration and operation manuals for important devices, and operations are performed according to the manuals, the result is conforming; otherwise it is non-conforming or partially conforming.

【L3-MMS1-19/L4-MMS1-20 Interpretation and Description】

The main evaluation targets of the evaluation metric that "The operation logs of operation and maintenance shall be recorded in detail, including daily inspection work, operation and maintenance records, parameter setting and modification, etc." are operation record form documents for operation and maintenance.

The key points of the evaluation implementation include: reviewing the operation and maintenance operation logs, and checking whether the contents of the log records include inspection work, operation and maintenance records, etc.

If the test results show that the operation log is recorded in detail during the device operation and maintenance, including information such as daily inspection, operation and maintenance records, parameter setting, modification, the result is conforming; otherwise it is non-conforming or partially conforming.

【L3-MMS1-20/L4-MMS1-21 Interpretation and Description】

The main evaluation targets of the evaluation metric that "The designated department or personnel shall be appointed to make analysis and statistics of logs, monitoring and alert data, and suspicious behaviors shall be discovered in time" are job responsibility documents and operation record documents for operation and maintenance.

The key points of the evaluation implementation include: interviewing with system administrators to understand whether the designated department or personnel are assigned to perform statistics and analysis on logs, monitoring and alert data, checking whether there are specific system or responsibility provisions; reviewing the previous statistics and analysis reports of logs, monitoring and alert data to check whether suspicious behaviors or hidden risks can be found in time.

If the test results show that the evaluated organization has appointed the designated department or personnel to perform analysis and statistics on the log, monitoring and alert data, and the corresponding disposal report are retained properly, the result is conforming; otherwise it is non-conforming or partially conforming.

[L3-MMS1-21/L4-MMS1-22 Interpretation and Description]

The main evaluation targets of the evaluation metric that "he change operation and maintenance shall be strictly controlled. Only after approval can the connection, installation of system components or adjustment of configuration parameters be changed. The unchangeable audit logs shall be retained during the operation process. The configuration information database shall be updated synchronously after the operation" are network and system security operation and maintenance management system and record documents.

The main points of the evaluation implementation include: checking whether the approval procedure for configuration changes are established, such as the approval process for changing connections, installing system components or adjusting configuration parameters; checking the approval form for the relevant change process, and making sure that all change operations and maintenance are approved in advance before being implemented; checking the audit records of configuration changes, making sure that unchangeable audit records are retained during the operation; checking whether there are update records of the configuration information database.

If the test results show that the evaluated organization has developed the application and approval process for changeable operation and maintenance, retained the corresponding audit logs, and updated the configuration database in time, the result is conforming; otherwise it is non-conforming or partially conforming.

[L3-MMS1-22/L4-MMS1-23 Interpretation and Description]

The main evaluation targets of the evaluation metric that "The use of operation and maintenance tools shall be strictly controlled. Tools can be accessed for operations only after approval. The unchangeable audit log shall be retained during the operation, and the sensitive data in the tool shall be deleted after the operation" are security management personnel in the evaluated organization and related record form documents.

Chapter 3   Application and Interpretation of the General Security Evaluation Requirements at Level Ⅲ and Level Ⅳ

The key points of the evaluation implementation include: interviewing with information or network security supervisors to ask about the operation and maintenance tools used, and whether the corresponding approval process are established for the use of operation and maintenance tools; checking the approval records of the operation and maintenance tools of the management system, and reviewing the approval record to see whether it contains relevant information such as the applicant, approval time, the approver, etc.; checking the usage record of the operation and maintenance tools to confirm whether it contains the record of sensitive data deletion.

If the test results show that the evaluated organization has established the approval process for the use of operation and maintenance tools, and has approval records and the unchangeable audit log, and the sensitive data is cleared after the operation is completed, the result is conforming; otherwise it is non-conforming or partially conforming.

[L3-MMS1-23/L4-MMS1-24 Interpretation and Description]

The main evaluation targets of the evaluation metric that "The permission granting of remote operation and maintenance shall be strictly controlled, and the remote operation and maintenance interface or channel can be granted after approval, and the unchangeable audit log shall be kept during the operation, and the interface or channel shall be closed immediately after the operation" are security administrators in the evaluated organization, and security record form documents. If there is no remote operation and maintenance management, then this evaluation metric is not applicable.

The key points of the evaluation implementation include: interviewing with system administrators to understand whether remote operation and maintenance approval procedures are established; reviewing the approval records to check whether the operation and maintenance items, the start time of remote operation and maintenance, the applicant, approval time and the approver's signature are included; checking whether the audit records of remote operation and maintenance operations can be changed, and whether the corresponding interfaces or channels are closed in time after the operation and maintenance operations are completed.

If the test results show that the evaluated organization has developed the remote operation and maintenance approval procedure, the operation and maintenance interface or channel are closed in time after the operation and maintenance is completed, and there exists the complete audit record of the operation and maintenance operation, and the audit log cannot be changed, the result is conforming; otherwise it is non-conforming or partially conforming.

[L3-MMS1-24/L4-MMS1-25 Interpretation and Description]

The main evaluation targets of the evaluation metric that "It shall be ensured that all connections with the outside are authorized and approved, and violations of wireless Internet access and other violations of network security strategies shall be regularly inspected" are the security management personnel in the evaluated organization, as well as record form documents for violation behavior ofcybersecurity strategy.

The key points of the evaluation implementation include: interviewing with system administrators to understand the approval procedures for external connections, checking the approval process; checking historical approval records, whether there are items such as the reason, time, approver, performer for external connection, etc.; checking whether there are regular inspections of violations of external connection behavior or activities; checking the corresponding inspection records.

If the test results show that all connections to the outside are authorized and approved, and regular inspections of wireless Internet access violations and other violations of network security policies are performed, the result is conforming; otherwise it is non-conforming or partially conforming.

3.10.7 Malicious Code Prevention Management

[Standard Requirements]

This control point for Level III includes the evaluation units of L3-MMS1-25, L3-MMS1-26, and Level IV includes that of L4-MMS1-26, L4-MMS1-27.

[L3-MMS1-25/L4-MMS1-26 Interpretation and Description]

The main evaluation targets of the evaluation metric that "All users' awareness of anti-malware codes shall be raised, and malicious code inspections shall be conducted before any external computers or storage devices are connected to the system" are security management personnel in the evaluated organization, the malicious code prevention management system and related records.

The key points of the evaluation implementation include: reviewing training documents or records to check whether it contains the training content of raising the awareness of malicious code prevention; reviewing the related security management system to check whether there are requirements for malicious code inspection before external devices are connected to the system; checking the record of malicious code inspection.

If the test results show that the evaluated organization can regularly carry out activities training to improve the awareness of malicious code prevention, and require the inspection of malicious code before the external devices are connected to the system, and the inspection record are properly kept, the result is conforming; otherwise it is non-conforming or partially conforming.

[L3-MMS1-26/L4-MMS1-27 Interpretation and Description]

The main evaluation targets of the evaluation metric that "The effectiveness of technical measures to prevent against malicious code attacks shall be regularly verified" are the security management personnel in the evaluated organization, as well as the technical measures and related records of prevention against malicious code attacks.

Chapter 3 Application and Interpretation of the General Security Evaluation Requirements at Level Ⅲ and Level Ⅳ

The key points of the evaluation implementation include: checking the implementation of malicious code prevention measures, checking the technical measures against the malicious code attacks, such as anti-virus software, IDS, IPS, anti-virus firewall, WAF, etc.; reviewing related operation logs or records to check the implementation of technical measures (or software operation); checking the records of the regular update of the malicious code prevention measures signature database.

If the test results show that the evaluated organization regularly upgrades the malicious code database and analyzes the detected malicious code, the result is conforming; otherwise it is non-conforming or partially conforming.

3.10.8 Configuration Management

[Standard Requirements]

This control point for Level Ⅲ includes the evaluation units of L3-MMS1-27, L3-MMS1-28, and Level Ⅳ includes that of L4-MMS1-28, L4-MMS1-29.

[L3-MMS1-27/L4-MMS1-28 Interpretation and Description]

The main evaluation targets of the evaluation metric that "The basic configuration information shall be recorded and saved, including the network topology, software components installed in each device, version and patch information of software component, configuration parameters of each device or software component" are the configuration information record form.

The key points of the evaluation implementation include: interviewing with the personnel in charge of operation and maintenance to understand which configuration information of the evaluated system needs to be recorded and saved, and checking whether there is a configuration information storage record sheet and other documents; checking whether the configuration information storage record sheet includes the network topology, software components installed in each device, software component version and patch information, and configuration parameters of each device or software component.

If the test results show that the evaluated organization has recorded and saved the basic configuration information in a timely manner, and the configuration information content is recorded comprehensively, the result is conforming; otherwise it is non-conforming or partially conforming.

[L3-MMS1-28/L4-MMS1-29 Interpretation and Description]

The main evaluation targets of the evaluation metric that "The basic configuration information changes shall be included in the scope of change, the control of configuration information changes shall be implemented, and the basic configuration information database shall be updated in time" are configuration change management documents and configuration information change record sheets.

The main points of the evaluation implementation include: checking the configuration information change document, checking whether the configuration information change process has corresponding control procedures or measures; checking the configuration information change record, checking whether it contains the change applicant, the change approver, change time, change content, change result and other items.

If the test results show that the evaluated organization has developed the control procedure or measures for the change of configuration information, after the configuration is changed, the basic configuration information database can be updated in time, and the change record is complete, the result is conforming; otherwise it is non-conforming or partially conforming.

3.10.9 Password Management

[Standard Requirements]

This control point for Level Ⅲ includes that of L3-MMS1-29, L3-MMS1-30, and Level Ⅳ includes that of L4-MMS1-30, L4-MMS1-31, L4-MMS1-32.

[L3-MMS1-29/L4-MMS1-30 Interpretation and Description]

The main evaluation targets of the evaluation metric that "The relevant national and industry cryptography standards shall be followed" are the list of cryptography techniques and products related to the system. If the evaluated system does not involve cryptography techniques and products, then this evaluation metric is not applicable.

The key points of the evaluation implementation include: interviewing with security administrators to understand whether cryptography products are used and the types of cryptography products used (such as electronic document encryption systems, CA certificate certification systems); checking the category and model of cryptography products, and checking the sales license certificate of cryptography products or the test report issued by the relevant state authorities.

If the test results show that the cryptography techniques and products used by the evaluated organization follow the cryptography related national standards and industry standards, the result is conforming; otherwise it is non-conforming or partially conforming.

[L3-MMS1-30/L4-MMS1-31 Interpretation and Description]

The main evaluation targets of the evaluation metric that "Cryptography techniques and products certified and approved by the national cryptography administration authority shall be used" are cryptography technique and product lists, inspection reports, or model certificates of cryptography products. If cryptography technology and products are not used, then this evaluation metric is not applicable.

The key points of the evaluation implementation include: checking whether the cryptography product has obtained a valid inspection report or a cryptography product model certificate specified by the national cryptography administration authorities.

If the test results show that the cryptography products used by the evaluated organization have obtained the valid test reports or the cryptography product model certificates specified by the national cryptography administration authorities, the result is conforming; otherwise it is non-conforming or partially conforming.

[L4-MMS1-32 Interpretation and Description]

The main evaluation targets of the evaluation metric that "Hardware cryptographic modules shall be adopted to achieve cryptographic computation and key management" are relevant devices and products that use cryptographic techniques. If cryptographic techniques and products are not used, then this evaluation metric is not applicable.

The key points of the evaluation implementation include: interviewing with security administrators, reviewing the technical documentation of cryptographic products, and checking whether the functions of cryptographic computation and key management are implemented through hardware cryptographic modules.

If the test results show that the cryptographic products used by the evaluated organization use hardware cryptographic modules to implement cryptographic computation and key management, the result is conforming; otherwise it is non-conforming or partially conforming.

3.10.10 Change Management

[Standard Requirements]

This control point for Level III includes the evaluation units of L3-MMS1-31, L3-MMS1-32, L3-MMS1-33, and Level IV includes that of L4-MMS1-33, L4-MMS1-34, L4-MMS1-35.

[L3-MMS1-31/L4-MMS1-33 Interpretation and Description]

The main evaluation targets of the evaluation metric that "The change requirements shall be clearly specified, and the change plan shall be formulated according to the change requirements before the change, and the change plan can be implemented only after review and approval" include the change plan, review record, approval record, change process record, etc.

The key points of the evaluation implementation include: interviewing with the system administrator to understand the change management related system, reviewing the change implementation process, and checking whether the change requirements analysis, the formulation of the change plan, the review and approval of the change plan are included; checking the review record of the historical change plan to check whether the review time, participants, review results are included; checking the change process record to see whether it includes the change performer, execution time, operation content, change content, etc.

If the test results show that the evaluated organization has formulated the change plan based on the change requirements, and retained the change plan review records and change process records, the result is conforming; otherwise it is non-conforming or partially conforming.

[L3-MMS1-32/L4-MMS1-34 Interpretation and Description]

The main evaluation targets of the evaluation metric that "The change application and approval control procedures shall be established. All changes shall be controlled according to the procedures, and the change implementation process shall be recorded" are change application, control and approval procedure documents, and change record documents.

The key points of the evaluation implementation include: reviewing the change management system to check the change application and approval control procedures. The control procedure shall at least include the change application, change acceptance, change approval and other items; checking the change implementation process record to see whether the entire process of the change implementation is recorded.

If the tests result show that the evaluated organization has established the change application and approval control procedure, and retained the record of the change implementation process, the result is conforming; otherwise it is non-conforming or partially conforming.

[L3-MMS1-33/L4-MMS1-35 Interpretation and Description]

The main evaluation targets of the evaluation metric that "Procedures for suspending changes and recovering from failed changes shall be established, the process control methods and personnel responsibilities shall be specified, and the recovery process drill shall be conducted if necessary" are the recovery procedures and system documents after the failure of the change, and the recovery process drill record documents.

The key points of the evaluation implementation include: reviewing the change management system or change plan to check whether the procedures for terminating the change and recovering from the failed change are established, to see whether the process control method and the responsibilities of related personnel are specified, etc.; interviewing with the person in charge of operation and maintenance on the spot to understand the recovery process drill situation; checking the recovery process drill record to see whether it contains the drill content, the drill time, involved participants, the recovery process, the summary of the drill, etc.

If the test results show that the evaluated organization has established the recovery procedure after the failure of change, specified the process control methods and personnel responsibilities, and conducted the recovery process drill if necessary, the result is conforming; otherwise it is non-conforming or partially conforming.

3.10.11 Backup and Recovery

[Standard Requirements]

This control point for Level Ⅲ includes that of L3-MMS1-34, L3-MMS1-35, L3-MMS1-36, and Level Ⅳ includes that of L4-MMS1-36, L4-MMS1-37, L4-MMS1-38.

[L3-MMS1-34/L4-MMS1-36 Interpretation and Description]

The main evaluation targets of the evaluation metric that "Important business information, system data and software systems that need to be regularly backed up shall be identified" are network administrators, system administrators and other relevant personnel.

The key points of the evaluation implementation include: interviewing with network administrators, system administrators to understand whether important business information, system data and software systems that need to be backed up regularly are defined or determined; reviewing relevant documents and checking the specific definitions; reviewing relevant logs, records and other documents to check whether the relevant data is backed up as required.

If the test results show that the evaluated organization has defined important business information, system data and software systems that need to be backed up regularly, and has corresponding backup logs or records, the result is conforming; otherwise it is non-conforming or partially conforming.

[L3-MMS1-35/L4-MMS1-37 Interpretation and Description]

The main evaluation target of the evaluation metric that "The backup information such as the backup method, backup frequency, storage media, retention period shall be specified" is the backup and recovery management system.

The key points of the evaluation implementation include: reviewing the backup and recovery management system to check whether there are corresponding provisions on the backup method, backup frequency, storage media and retention period.

If the test results show that the evaluated organization has formulated the backup and recovery system with clear provisions on backup method, backup frequency, storage media and storage period, etc., the result is conforming; otherwise it is non-conforming or partially conforming.

[L3-MMS1-36/L4-MMS1-38 Interpretation and Description]

The main evaluation targets of the evaluation metric that "The data backup and recovery strategies, backup and recovery procedures shall be formulated according to the importance of the data and the impact of the data on the system operation" are the data backup and recovery management system and operating documents.

The key points of the evaluation implementation include: reviewing related backup management systems to understand the basis for determining backup strategies, recovery strategies, backup and recovery procedures, etc., checking whether the importance of data and the influence of data on system operations are fully considered; reviewing data backup strategies, backup procedures, whether it includes backup data type, backup data content, backup time, backup method, backup frequency, retention period, etc.; checking data recovery strategies, recovery procedures to see whether they include the application before data recovery, the backup before data recovery, the data recovery operation, the verification after data recovery, etc.

If the test results show that the evaluated organization has formulated data backup and recovery strategies, backup procedures and recovery procedures based on the importance of the data and the impact of the data on the operation of the system, the result is conforming; otherwise it is non-conforming or partially conforming.

3.10.12 Security Incident Handling

[Standard Requirements]

This control point for Level Ⅲ includes the evaluation units of L3-MMS1-37, L3-MMS1-38, L3-MMS1-39, L3-MMS1-40, and Level Ⅳ includes that of L4-MMS1-39, L4-MMS1-40, L4-MMS1-41, L4-MMS1-42, L4-MMS1-43.

[L3-MMS1-37/L4-MMS1-39 Interpretation and Description]

The main evaluation targets of the evaluation metric that "The security weaknesses and suspicious incidents discovered shall be reported to the security management department in a timely manner" are the list of security weaknesses and suspicious incidents, security weakness and suspicious incidents reporting process documents and report records, etc.

The key points of the evaluation implementation include: reviewing security incident specifications or procedures to check whether there are provisions for timely reporting of security weaknesses and suspicious incidents; reviewing the defined cope of security weaknesses and suspicious events and the reporting process; reviewing historical reports or records of security weaknesses and suspicious incidents, and checking whether the security incident report is timely and accurate.

If the test results show that the evaluated organization has established the security weakness and suspicious incident reporting system and process, the result is conforming; otherwise it is non-conforming or partially conforming.

[L3-MMS1-38/L4-MMS1-40 Interpretation and Description]

The main evaluation targets of the evaluation metric that "The security incidents reporting and handling management system shall be established to specify the reporting, handling and response processes for different security incidents, and specify the management responsibilities for on-site handling of security incidents, incident reporting and post-incident recovery, etc." include management system documents containing security incident reporting and handling content.

The key points of the evaluation implementation include: interviewing with the person in charge of system operation and maintenance to understand whether the security incident reporting and handling system is formulated; reviewing the management system to check whether the reporting, handling and response procedures of different security incidents are included, and whether the incident reporting and the management responsibilities of the later recovery are clearly stipulated; checking the historical records or documents related to the security incident reporting and handling.

If the test results show that the evaluated organization has formulated the security incident reporting and handling management system, the content of the system includes the reporting, handling and response processes of different security incidents, and has the content of management responsibilities for on-site handling of security incidents, incident reporting and later recovery, etc., the result is conforming; otherwise it is non-conforming or partially conforming.

[L3-MMS1-39/L4-MMS1-41 Interpretation and Description]

The main evaluation targets of the evaluation metric that "The analysis and identification of the cause of incident, collecting of the evidence, recording of the handling process, summarizing of the experiences and lessons learned shall be performed in the process of security incident reporting and handling" are record form documents containing the security incident report and response handling.

The key points of the evaluation implementation include: interviewing with the person in charge of operation and maintenance or system security to understand the entire process of security incident reporting and response handling, and checking whether the various links such as "identification of the cause of the incident, collection of evidence, comprehensive recording of the handling process, and summary of experience and lessons" are included; review the history records of security incident reporting and response handling, and checking whether each step is fully implemented.

If the test results show that the evaluated organization has the security incident reporting and response handling record, which are properly kept, the content of the record includes the cause of the security incident, the analysis of the situation, handling process and summary, etc., the result is conforming; otherwise it is non-conforming or partially conforming.

[L3-MMS1-40/L4-MMS1-42 Interpretation and Description]

The main evaluation targets of the evaluation metric that "Different handling procedures and reporting procedures shall be adopted for major security incidents that cause system interruption and information leakage" are documents and record form documents containing major security incident handling and reporting procedures.

The key points of the evaluation implementation include: reviewing the security incident handling management system to understand the classification regulations of security incidents, checking whether the "system interruption and information leakage" are listed as major security incidents, and whether different reporting and handling procedures from other security incidents are stipulated; reviewing major security incident reports and handling procedures and related history records, and checking whether each link is implemented in accordance with the procedures.

If the test results show that the evaluated organization has formulated different handling and reporting procedures for different types of major security incidents, and has kept relevant records and reports, the result is conforming; otherwise it is non-conforming or partially conforming.

[L4-MMS1-43 Interpretation and Description]

The main evaluation targets of the evaluation metric that "A joint protection and emergency response mechanism shall be established to be responsible for handling cross-unit security incidents" are joint protection and emergency response related systems, measures documents, and history records of cross-unit handling of security incidents.

The key points of the evaluation implementation include: interviewing with the person in charge of operation and maintenance to understand whether a joint protection and emergency response mechanism is established to deal with cross-unit security incidents; reviewing related documents to see whether the scope of cross-unit security incidents, the composition of joint handling organization, the joint protection and emergency plans are specified.

If the test results show that the evaluated organization has established the joint protection and emergency mechanism document for the cross-unit security incidents, and the content of the document specifies the incident's protection plan, emergency measures, handling methods, etc., the result is conforming; otherwise it is non-conforming or partially conforming.

3.10.13 Contingency Plan Management

[Standard Requirements]

This control point for Level III includes the evaluation units of L3-MMS1-41, L3-MMS1-42, L3-MMS1-43, L3-MMS1-44, and Level IV includes that of L4-MMS1-44, L4-MMS1-45, L4-MMS1-46, L4-MMS1-47, L4-MMS1-48.

[L3-MMS1-41/L4-MMS1-44 Interpretation and Description]

The main evaluation targets of the evaluation metric that "A unified emergency plan framework shall be specified, including the conditions for starting the plan, the emergency organization structure, the emergency resource assurance, post-incident education and training, etc." are the management system documents containing the emergency plan.

The key points of the evaluation implementation include: reviewing the emergency plan management system and understanding the main content of the emergency plan; checking whether the emergency plan includes the emergency organization structure, division of responsibilities, plan activation conditions, resource assurance, post-incident education and training.

If the test results show that the evaluated organization has formulated a unified emergency plan framework, and the content includes the conditions for activating the plan, emergency organization structure, emergency resource assurance, post-incident education

Chapter 3 Application and Interpretation of the General Security Evaluation Requirements at Level Ⅲ and Level Ⅳ

and training, etc., the result is conforming; otherwise it is non-conforming or partially conforming.

【L3-MMS1-42/L4-MMS1-45 Interpretation and Description】

The main evaluation targets of the evaluation metric that "Emergency plans for important incidents shall be developed, including emergency handling procedures, system recovery procedures" are the special emergency plan documents.

The key points of the evaluation implementation include: checking whether the special emergency plan are developed for important incidents, covering the content such as the equipment room (power supply, fire fighting, water leakage, etc.), network (network interruption, network attack, etc.), system (data breach, virus, business interruption, etc.), etc.; checking whether the special emergency plan includes emergency handling procedures, recovery procedures, etc.

If the test results show that the evaluated organization has formulated a special emergency plan for important incidents, and the content includes emergency handling procedures, system recovery procedures, etc., the result is conforming; otherwise it is non-conforming or partially conforming.

【L3-MMS1-43/L4-MMS1-46 Interpretation and Description】

The main evaluation targets of the evaluation metric that "The emergency plan training shall be conducted for the system related personnel on a regular basis, and the drill of emergency plan shall be run" are emergency plan training and drill related record documents.

The key points of the evaluation implementation include: interviewing with the person in charge of operation and maintenance to find out whether relevant personnel are regularly trained in emergency plans and drills are organized regularly; reviewing the emergency plan training records to check whether the training objects, training content, training results are included; reviewing the emergency plan records to see whether the drill time, main operation content, drill result are recorded.

If the test results show that the evaluated organization regularly conducts emergency plan training and drill activities, and relevant training and drill records are properly retained, the result is conforming; otherwise it is non-conforming or partially conforming.

【L3-MMS1-44/L4-MMS1-47 Interpretation and Description】

The main evaluation targets of the evaluation metric that "The original emergency plan shall be re-evaluated, revised and improved periodically" are the emergency plan revision record forms, etc.

The key points of the evaluation implementation include: reviewing the relevant management system to see whether there is the provision of "regular evaluation, revision and improvement of the emergency plan"; reviewing the revision record of the emergency plan, and checking whether the revision time, revision content and review information are included.

If the test results show that the evaluated organization regularly revises the emergency plan and keeps relevant revision records, the result is conforming; otherwise it is non-conforming or partially conforming.

[L4-MMS1-48 Interpretation and Description]

The main evaluation targets of the evaluation metric that "A cross-unit joint emergency plan for major security incidents shall be established and emergency plan drills shall be conducted" are cross-unit joint emergency plans, drill records, etc.

The key points of the evaluation implementation include: interviewing with the person in charge of operation and maintenance to understand whether a cross-unit joint emergency plan are formulated for major security incidents; reviewing the related emergency plan; understanding the implementation of the cross-unit joint emergency plan, especially the drill work of the emergency plan; checking whether there are relevant records of the joint emergency plan drill. The drill record includes the drill time, participating departments, main operation content, drill result, experience and summary, etc.

If the test results show that the evaluated organization has formulated a cross-unit joint emergency plan for major security incidents, related drill activities are run regularly, and drill records are kept properly, the result is conforming; otherwise it is non-conforming or partially conforming.

3.10.14 Outsourcing Operation and Maintenance Management

[Standard Requirements]

This control point for Level III includes the evaluation units of L3-MMS1-45, L3-MMS1-46, L3-MMS1-47, L3-MMS1-48, and Level IV includes that of L4-MMS1-49, L4-MMS1-50, L4-MMS1-51, L4-MMS1-52.

[L3-MMS1-45/L4-MMS1-49 Interpretation and Description]

The main evaluation targets of the evaluation metric that "It shall be ensured that the selection of outsourcing operation and maintenance service providers conforms to relevant national regulations" are the relevant record documents of the selection of outsourcing operation and maintenance services. If there is no outsourcing operation and maintenance service, then this evaluation metric is not applicable.

The key points of the evaluation implementation include: it is necessary to know whether there exists the outsourcing operation and maintenance in the evaluated organization, and understanding the outsourcing operation and maintenance methods, such as on-site outsourcing operation and maintenance, managed outsourcing operation and maintenance; checking whether the selection of outsourcing operation and maintenance service providers is in line with the relevant state regulations, such as the selection of outsourcing operation and maintenance service providers in accordance with the procurement law, checking the bidding documents, contracts and operation and maintenance qualification

Chapter 3 Application and Interpretation of the General Security Evaluation Requirements at Level Ⅲ and Level Ⅳ

and certification materials of outsourcing operation and maintenance services.

If the test results show that the selection of the outsourcing operation and maintenance service provider in the evaluated organization complies with the relevant national regulations, the result is conforming; otherwise it is non-conforming or partially conforming.

[L3-MMS1-46/L4-MMS1-50 Interpretation and Description]

The main evaluation targets of the evaluation metric that "The relevant agreement shall be signed with the selected outsourcing operation and maintenance service provider to clearly specify the work scope and work content of the outsourcing operation and maintenance" are related agreements or contracts signed by the outsourcing operation and maintenance service providers, and work records of the outsourcing operation and maintenance, etc. If there is no outsourcing operation and maintenance service, then this evaluation metric is not applicable.

The key points of the evaluation implementation include: checking whether the evaluated organization has signed a related agreement or contract with the selected outsourcing operation and maintenance service provider; checking whether the outsourcing operation and maintenance service agreement or contract includes the scope of outsourcing operation and maintenance, work content and other related information; the scope and content of the operation and maintenance agreed in the service agreement or contract shall be compared with the actual scope and content of the operation and maintenance.

If the test results show that the evaluated organization has signed a related agreement with the selected outsourcing operation and maintenance service provider, and the agreement includes the scope of outsourcing operation and maintenance, work content and other information, the result is conforming; otherwise it is non-conforming or partially conforming.

[L3-MMS1-47/L4-MMS1-51 Interpretation and Description]

The main evaluation targets of the evaluation metric that "shall ensure that the selected outsourcing operation and maintenance service provider shall have the ability to carry out safe operation and maintenance work in accordance with the level of protection requirements in terms of technology and management, and the ability requirements shall be specified in the signed agreement" include the agreement or contract signed by the outsourcing operation and maintenance service provider, the capability certificate documents of the outsourcing operation and maintenance service. If there is no outsourcing operation and maintenance service, then this evaluation metric is not applicable.

The main points of the evaluation implementation include: checking whether the agreement or contract signed with the outsourcing operation and maintenance service provider includes operation and maintenance work capability requirements; confirming whether the scope of operation and maintenance work covers technical and management aspects; checking whether the outsourcing operation and maintenance service provider has the ability to carry out security operation and maintenance work in accordance with the requirements of classified protection.

If the test results show that the outsourcing operation and maintenance service provider of the evaluated organization have the ability to perform security operation and maintenance work in accordance with classified protection requirements in terms of technology and management, and the capability requirements are specified in the signed agreement, the result is conforming; otherwise it is non-conforming or partially conforming.

[L3-MMS1-48/L4-MMS1-52 Interpretation and Description]

The main evaluation targets of the evaluation metric that "All relevant security requirements shall be specified in the agreement signed with the outsourcing operation and maintenance service provider, such as requirements for access, processing and storage of sensitive information, and emergency assurance requirements for IT infrastructure service interruption" are agreements or contracts signed by outsourcing service providers, records of outsourcing operation and maintenance assurance measures, etc. If there is no outsourcing operation and maintenance service, then this evaluation metric is not applicable.

The key points of the evaluation implementation include: checking the outsourcing operation and maintenance service agreement or contract, and confirming the requirements related to operation and maintenance security; checking whether the outsourcing service agreement or contract has the access, processing, and storage requirements for sensitive information, and emergency assurance requirements for IT infrastructure interruption; checking whether there are record documents of implementation in accordance with relevant security requirements in the actual operation and maintenance work, such as the signed confidentiality agreement, the formulated sensitive information operation and maintenance operating manual, and the emergency assurance plan for IT infrastructure service interruption.

If the test results show that all relevant security requirements are specified in the agreement signed between the evaluated organization and the outsourcing operation and maintenance service provider, the result is conforming; otherwise it is non-conforming or partially conforming.

Chapter 4 Application and Interpretation of the Extended Security Evaluation Requirements of Cloud Computing

4.1 Overview of Cloud Computing

4.1.1 Basic Concepts

1. Cloud Computing

The model of accessing a scalable and flexible pool of physically or virtually shared resources through the network with on-demand and self-service acquisition and administration of resources.

2. Cloud Service Provider

Supplier of cloud computing services.

Cloud service provider manages, operates, supports the infrastructure and software of cloud computing, and delivers cloud computing resources through the network.

3. Cloud Service Customer

Participants who establish business relationships with cloud service providers for the use of cloud computing services.

4. Cloud Computing Platform/System

A collection of cloud computing infrastructure provided by cloud service providers and service software on it.

5. Virtual Machine Monitor

The middle software layer that runs between the basic physical server and the operating system, which allows multiple operating systems and applications to share hardware.

6. Cloud Computing Platform

A collection of cloud computing infrastructure provided by cloud service providers and service software on it.

7. Cloud Computing Environment

A collection of cloud computing platforms provided by cloud service providers, and software and related components deployed by customers on cloud computing platform.

8. Cloud Infrastructure

The infrastructure that supports cloud computing consisting of hardware resources and resource abstraction controlling components. Hardware resources refer to all physical computing resources, including servers (CPU, memory, etc.), storage components (hard disks, etc.), network components (routers, firewalls, switches, network links and interfaces, etc.) and other basic elements of physical computing. The resource abstraction controlling component perform software abstraction on physical computing resources, and cloud service providers manages the access to the physical computing resources through these components.

9. Cloud Computing Service

Cloud computing service includes the ability to provide one or more resources with the help of cloud computing through using defined interfaces.

10. Business Application System of Cloud Service Customer

Business application system of cloud seriice customer includes the business applications deployed by cloud service customers on the cloud computing platform and the application services provided by cloud service providers to cloud service customers through the network.

11. Business Application System Built via Cloud Computing Technology

Business application system built via cloud computing technology includes business applications and a collection of underlying cloud computing services and hardware resources that are independently provided for business applications. There is no cloud service customer in this type of system.

12. Cloud Products (Services)

Cloud products(services) include the use of software, hardware or services provided by cloud computing service.

4.1.2 Characteristics of Cloud Computing System

Based on the application background of cloud computing, the characteristics of cloud computing can be summarized into the following five categories.

1. Self-Service on Demand

Without or requiring only a small number of cloud service provider personnel to participate, customers can obtain the required computing resources, such as determining the time and quantity of resources. For example, for IaaS services, customers can self-select the number of virtual machines to buy through the cloud service provider's website, the configuration of each virtual machine (including the number of CPU, memory capacity, disk space, external network bandwidth, etc.), service usage time and so on.

2. Ubiquitous Network Access

Through the standard access mechanism, customers use computers, mobile phones, tablets and other terminal devices to access services anytime and anywhere through the network. For customers, the ubiquitous access characteristic of cloud computing enable customers to access services in different environments (such as working or non-working environments), increasing the availability of services.

3. Resource Pooling

Cloud service providers provide resources (such as computing resources, storage resources, network resources, etc.) to multiple customers, and these physical, virtual resources are dynamically allocated or redistributed according to the needs of customers.

4. Rapid Resilience

Customers can quickly, flexibly and easily access and release computing resources as needed. For customers, this kind of resource is "unlimited" and the required amount of resources can be obtained at any time.

5. Measurable Service

Cloud computing can automatically control or quantify resources according to a variety of metering methods (such as pay-per-use or pre-paid), and metering objects range from storage space, computing capability, network bandwidth to the number of active accounts.

4.1.3 Deployment Model of Cloud Computing

Cloud computing can be divided into private cloud, public cloud and hybrid cloud by deployment type.

1. Public Cloud

The public cloud computing service is fully operated and managed by a third party provider, offering users with affordable access service to computing resources. Users do not need to buy hardware, software or supporting infrastructure, but only pay per usage for the resources they consume.

2. Private Cloud

The private cloud is defined as a cloud service in which businesses purchase their own infrastructure, build cloud platform, and develop applications on it.

3. Hybrid Cloud

The hybrid cloud is generally created by users with management and operation responsibilities shared by both users and cloud computing providers, which uses private clouds as the basis and combines the public cloud service strategies. Users can easily switch between public cloud and private cloud according to the degree of their business privacy.

4.1.4 Service Model of Cloud Computing

The service model of cloud computing is still evolving, but the three categories of cloud computing according to how services are provided are generally accepted by the industry:

SaaS (Software as a Service), PaaS (Platform as a Service), IaaS (Infrastructure as a Service).

1. Infrastructure as a Service

Infrastructure as a Service (IaaS) consists of automated, reliable and highly scalable dynamic computing resources, and mainly provides some basic resources, including services such as server, network, storage. Without having to manage or control any cloud computing infrastructure, users can deploy and run any software while having the choice of controlling operating systems, storage space, deployed applications, and may also gain control over network components. Cloud services such as virtual machines, object-based storage, virtual networks, etc.

2. Platform as a Service

Platfrom as a Service (PaaS) mainly provides a development and operation platform as a service to users, including customized development middleware platform, database and big data applications, etc. Developers only need to focus on the business logic of their own system, can quickly and easily create web applications without worrying about CPU, memory, disk, network and other infrastructure resources.

3. Software as a Service

Software as a Service (SaaS) provides application services to end users through the network. Most SaaS applications run directly in browsers and do not require users to download and install any programs. For users, software development, management and deployment are handed over to third parties, and users do not need to care about technical issues, and service is immediately ready to use.

4.1.5 Cloud Computing Evaluation

Cloud computing test and evaluation refers to the classified protection evaluation agency (hereinafter referred to as the "evaluation agency") performs inspection and evaluation activities on the status of the classified protection for the system built with cloud computing technology in accordance with the regulations of the national cybersecurity classified protection system, which is commissioned by the relevant organization. Cloud computing test and evaluation is an important component of the classified protection of cybersecurity.

Control points mainly involved in the extended requirements for cloud computing security include infrastructure location, network architecture, access control of network boundary, intrusion prevention of network boundary, security audit of network boundary, centralized management and control, identity authentication of computing environment, access control of computing environment, intrusion prevention of computing environment, mirroring and snapshot protection, data security, data backup and recovery, residual information protection, selection of cloud service provider, supply chain management and cloud computing environment management.

1. Use Scenario of Extended Requirements for Cloud Computing Security

General security requirements are proposed for the generalized protection requirements. Regardless of the form of the classified protection objects, the corresponding general security requirements must be realized according to the level of security protection. The extended requirements for cloud computing security are special protection requirements based on the characteristics of cloud computing. When conducting test and evaluation for cloud computing platforms/systems, the relevant requirements of the general security requirements part and the extended requirements part of the cloud computing security shall be used at the same time, and the use of extended requirements for cloud computing security alone shall not suffice.

In addition, the evaluation content of extended requirements for the cloud computing security itself is an overall capability requirement, which is not considered as the requirement for a certain evaluation target or device, and which shall be used as the overall metric in the cloud computing evaluation. However, considering that the specific evaluation index are still implemented by specific objects, the evaluation targets will still be given for each evaluation metric in the subsequent chapters.

2. Scope of Applicability of Extended Requirements for Cloud Computing Security

The extended requirements for Cloud computing security apply to information systems built by adopting cloud computing technology, including cloud computing platforms, business application systems of cloud service customer, and business application systems built with cloud computing technology.

The extended requirements for cloud computing security apply to three mainstream cloud computing service models, including Infrastructure as a Service (IaaS), Platform as a Service (PaaS), and Application as a Service (SaaS).

The extended requirements for cloud computing security apply to multiple deployment models, such as public cloud, private cloud, community cloud, hybrid cloud, etc.

3. Principles of Test and Evaluation Implementation

When performing test and evaluation for cloud computing platforms/systems, the following two basic principles shall be followed:

(1) Principle of Responsibility Sharing

Different from traditional information systems, the cloud computing environment involves one or more security responsibility entities. For example, for public clouds, there are at least two responsibility entities: cloud service providers and cloud service customers. Each security responsibility entity divides the security responsibility boundary according to the scope of management permissions.

The sum of the security protection capabilities of multiple security responsibility entities in the cloud computing environment constitutes the security protection capability of the entire cloud computing environment.

The main security responsibility of cloud service providers is to develop and operate the cloud platform, to ensure the security of cloud platform infrastructure, and to provide various infrastructure services and built-in security functions for each service. Cloud service providers have different security responsibilities in different service models. In the infrastructure-as-a-service (IaaS) model, cloud service providers need to ensure that the infrastructure is free of vulnerabilities. The cloud service provider's infrastructure includes the physical environment that supports cloud services. The service provider's self-developed software and hardware, operation and maintenance operations, including computing, storage, database and virtual machine mirroring and other cloud service system facilities. At the same time, the cloud service provider is also responsible for the security protection of the underlying infrastructure and virtualization technology from external attacks and internal abuse, and share the protection of network access control strategies together with cloud service customers. In the platform-as-a-service (PaaS) model, in addition to protecting the security of the underlying infrastructure, cloud service providers also need to protect virtual machines they provided, cloud application development platform and network access control, etc., and provide basic security reinforcement for the database and middleware. In the software-as-a-service (SaaS) model, cloud service providers need to assume security protection responsibilities for the entire cloud computing environment protection responsibility.

The main responsibility of cloud service customers is to customize the configuration on the cloud platform infrastructure and services and to operate and maintain the virtual networks, platforms, applications, data, management and other services they needed. In the IaaS model, cloud service customers need to securityly configure various controllable resources deployed on the cloud, configure security strategies for cloud platform related accounts, and implement permissions management and separation of duties for operation and maintenance personnel, and configure reasonable security strategies for virtual machines, security groups, advanced security services provided by cloud service providers, and security protection software deployed by cloud service customers themselves. In addition, the business applications, databases, and middleware that cloud service customers deploy on the cloud by themselves need to be managed by cloud service customers. Cloud service customers need to perform security management of data confidentiality, availability, integrity and data access verification and authorization. In the PaaS model, cloud service customers need to ensure the security of their business applications and data deployed on the cloud platform, and perform security configuration for various services provided by cloud service providers and perform security management for various accounts to prevent unauthorized damage to their business applications, resulting in data leakage or loss. In the SaaS model, cloud service customers only need to perform security configuration for their selected applications and do a good job of security protection for their own business data.

Regardless of the service model, cloud service providers shall provide tenants with data protection means and implement data protection related functions, but cloud service providers must never allow operation and maintenance personnel to access cloud service customer data without authorization. Cloud Service customers have ownership and control over their business data, and are responsible for various specific data security configurations.

(2) Applicability Principle of Cloud Service Model

The cloud computing environment may carry one or more cloud service models. Each cloud service model provides different cloud computing services and corresponding security protection measures. When conducting test and evaluation for cloud computing platforms/systems, focus shall only be put on the effectiveness of security protection measures corresponding to the cloud services provided in each specific cloud service model. As shown in Figure 4.1, in different service models, cloud service providers and cloud service customers have different control ranges over computing resources, and the control range determines the boundaries of security responsibilities. In the infrastructure-as-a-service model, the cloud computing platform/system consists of facilities, hardware, and resource abstraction control layer. In the PaaS model, the cloud computing platform/system includes facilities, hardware, resource abstraction control layer, and virtualized computing resources and software platform. In the SaaS model, cloud computing platform/system includes facilities, hardware, resource abstraction control layer, virtualized computing resources, software platform and application software. The security management responsibilities of cloud service providers and cloud service customers are different in different service models.

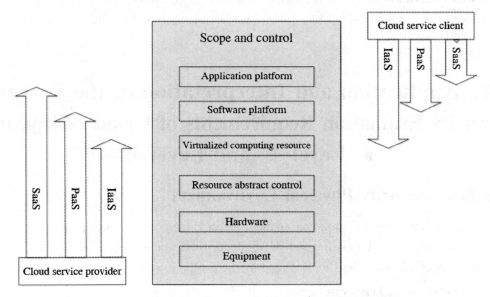

Figure 4.1 The Relationship Between Cloud Computing Service Model and Control Range

Regarding the selection of evaluation index, especially in platform as a service (PaaS) and software as a service (SaaS), it needs to be determined by scenarios. For example, the selection of evaluation targets in the PaaS model requires different scenarios for cloud service providers. Such as:

① If the cloud service provider that provides PaaS services is directly responsible for the construction of the underlying infrastructure, in addition to the general requirements during the evaluation, the evaluation needs to be performed in accordance with the corresponding extended requirements for cloud computing.

② If the platform of the cloud service provider that provides PaaS services is deployed on the IaaS platform, it is recommended to split IaaS and PaaS services on objects, that is, different service models need to be graded separately. At this time, when conducting PaaS evaluation, there is no need to consider the protection objects of the IaaS layer, only the protection objects of the PaaS layer and the use of corresponding extended requirements for cloud computing need to be considered. However, in the actual evaluation, some cloud platforms provided by cloud service providers provide both IaaS services and PaaS services. It is recommended that the cloud platforms be graded separately according to the service model, but often the evaluation organization does not split the grading. Then, it is necessary to consider the protection objects of the IaaS layer and the protection objects of the PaaS layer at the same time during the evaluation.

In addition, because cloud computing is only a computing model, it is not technically unified, especially PaaS and SaaS models, there are too many possibilities, so the selection of evaluation targets in the subsequent chapters of this book is mainly based on examples, offering some methods based on principles. And the applicability of related index to cloud service providers and cloud service customers is also mainly based on the IaaS model in public cloud.

4.2 Application and Interpretation of the Extended Security Evaluation Requirements of Cloud Computing at Level Ⅲ and Level Ⅳ

4.2.1 Security Physical Environment

There are more physical data center facilities in the cloud computing environment than traditional systems, which may include self-built facilities and managed facilities. Therefore, different types of facilities shall be included in the evaluation of facilities.

Location of Infrastructure

[Standard Requirements]

This control point for Level Ⅲ includes the evaluation unit of L3-PES2-01, and Level Ⅳ includes that of L4-PES2-01.

[L3-PES2-01/L4-PES2-01 Interpretation and Description]

The main evaluation targets of the evaluation metric that "It shall be ensured that the cloud computing infrastructure is located within the territory of China" are physical server rooms, server room administrators, and platform construction plans. If the classified protection objects under test and evaluation have multiple data center facilities, then the evaluation targets shall cover all facilities.

The key points of the evaluation implementation include: interviewing with the facility administrators to understand the location of software and hardware that run services and host data, such as cloud computing servers, storage devices, network devices, cloud management platform, information systems, etc.; checking the facility list and cloud computing platform construction plan and on-site checking the location of the facility, selecting and confirming the specific locations of some physical devices in the facility.

If the test results show that the cloud computing platform infrastructure such as software and hardware that run services and host data, including cloud computing servers, storage devices, network devices, cloud management platform, information systems are all located within China, the result is conforming; otherwise it is non-conforming or partially conforming.

4.2.2 Security Communication Network

Compared with the communication network of the traditional system, the cloud platform has more virtual networks, and thus more extended evaluation requirements for at the virtual network layer; and for cloud service customers, the network is essentially part of the virtual network of the cloud platform, and when conducting the evaluation, the applicability of the evaluation index needs to be determined according to the specific conditions of the cloud platform and cloud service customers.

Network Architecture

[Standard Requirements]

This control point for Level III includes the evaluation units of L3-CNS2-01, L3-CNS2-02, L3-CNS2-03, L3-CNS2-04, L3-CNS2-05, and Level IV includes that of L4-CNS2-01, L4-CNS2-02, L4-CNS2-03, L4-CNS2-04, L4-CNS2-05, L4-CNS2-06, L4-CNS2-07, L4-CNS2-08.

[L3-CNS2-01/L4-CNS2-01 Interpretation and Description]

The main evaluation targets of the evaluation metric that "It shall be ensured that the cloud computing platform does not host business application systems higher than its corresponding level of security protection" are cloud computing platform grading and filing materials and business application system grading and filing materials, as well as cloud service providers' management and control measures for cloud service customers' business systems before moving to the cloud.

The key points of the evaluation implementation include: checking the grading and filing materials of the cloud computing platform to confirm the security protection level; and checking whether the cloud service provider has relevant control measures or system regulations for the security level of the cloud service customer's business system; checking the grading and filing materials of the cloud service customer's business systems hosted by the cloud computing platform, and confirming the security protection level of the cloud service customer's business systems.

If the test results show that the cloud service provider has relevant management and control measures for the security level of the cloud service customer's business system moving the cloud to ensure that the security protection level of the cloud computing platform is not lower than that of the business application system it hosts, the result is conforming; otherwise it is non-conforming.

【L3-CNS2-02/L4-CNS2-02 Interpretation and Description】

The main evaluation targets of the evaluation metric that "The isolation between virtual networks of different cloud service customers shall be realized" are cloud computing platform network resource isolation measures, integrated network management systems and cloud management platforms.

The key points of the evaluation implementation include: interviewing with cloud platform administrators and reviewing the network topology diagram to determine network isolation boundaries; checking whether isolation measures are taken at the boundaries of virtual networks of different cloud service customers, such as cloud firewall, VPC, VxLAN and other isolation measures; checking the isolation technical documents, network design schemes and isolation test reports of boundary network isolation measures; opening a test account to simulate a cloud service customer, and verify whether the isolation measures are effective through ping, telnet different ports, or remote connections between different cloud service customer servers, or verifying the effectiveness of the network isolation measures through testing tools.

If the test results show that the virtual network architecture of the cloud service provider has divided different network areas for different cloud service customers, and the network resource isolation strategy is effective, and the isolation between the virtual networks of different cloud service customers can be realized, the result is conforming; otherwise it is non-conforming or partially conforming.

【L3-CNS2-03/L4-CNS2-03 Interpretation and Description】

The main evaluation targets of the evaluation metric that "The ability to provide security mechanisms such as communication transmission, boundary protection, intrusion prevention according to business needs of cloud service customers shall be available" are security devices such as firewalls (including virtual firewalls), intrusion detection systems, intrusion protection systems, and anti-APT systems, etc.

The key points of the evaluation implementation include: checking whether the cloud computing platform provides security devices such as firewalls (including virtual firewalls), intrusion detection systems, intrusion protection systems, and anti-APT systems; checking whether the above security devices can enable different security protection strategies according to different security requirements; vulnerability scanning and security penetration shall be performed to verify the effectiveness of various security protection measures.

If the test results show that security devices such as firewalls (including virtual firewalls), intrusion detection systems, intrusion protection systems, and anti-APT systems of the cloud computing platform under test and evaluation can provide communication transmission, boundary protection, and intrusion prevention security protection mechanisms to meet the needs of customers, the result is conforming; otherwise it is non-conforming or partially conforming.

L3-CNS2-04/L4-CNS2-04 Interpretation and Description

The main evaluation targets of the evaluation metric that "It shall have the ability to independently set up security strategies according to business needs of cloud service customers, including defining access paths, selecting security components, and configuring security strategies" are cloud management platforms, network management platforms, cloud firewalls, security components and security access paths, etc.

The key points of the evaluation implementation include: checking the deployment of cloud computing platform firewalls, cloud management platforms and security components based on the network topology diagram; checking whether the scope of permissions includes self-defining access paths, selecting security components, configuring security policies, etc. ; checking various security services and security components provided by the cloud management platform, logging into the operation interface to review the relevant security strategy configuration, whether the cloud service customer is provided with the corresponding admin permissions, and to confirm whether the cloud service customer can use the management platform to independently select the needed security services or components; checking cloud computing platform security services and security components related design documents, and reviewing related security strategy testing reports; verifying whether the security strategies implemented by each security service and security component are authentic and effective.

If the test results show that the relevant platform or components of the cloud computing platform provide various security services to meet the security needs of cloud service customers, and the security policies of each security service are authentic and effective, the result is conforming; otherwise it is non-conforming or partially conforming.

L3-CNS2-05/L4-CNS2-05 Interpretation and Description

The main evaluation targets of the evaluation metric that "Open interfaces or open security services shall be provided to allow cloud service customers to access third-party security products or select third-party security services on the cloud computing platform" are related open interfaces and security services, and security design documents.

The key points of the evaluation implementation include: interviewing with the security administrators of cloud service providers to understand the open interfaces or open security services provided by the cloud computing platform; checking the cloud computing platform interface design documents or open service technical documents to confirm whether the provided interfaces and security services meet the openness and security requirements; checking whether the cloud computing platform supports cloud service customers to access third-party security products, checking the content of access configuration, access list, access specifications, etc.

If the test results show that the cloud computing platform can provide open interfaces and open security services that meet security requirements, when cloud service customers access third-party security products, the cloud service provider provides access configuration, access list, and access specifications for customers' reference, the result is conforming; otherwise it is non-conforming or partially conforming.

[L4-CNS2-06 Interpretation and Description]

The main evaluation targets of the evaluation metric that "It shall provide the ability to set up security tags to the subject and object of virtual resources, and ensure that cloud service customer can determine the subject's access to the object based on the security tags and mandatory access control rules" are virtual routers, virtual switches, virtual Firewall, virtual WAF, virtual VPN, etc.

The main points of the evaluation include: checking whether the cloud computing platform provides the ability to set security tags on the subject and object of virtual resources; checking the situation of setting security tags on the subject and object of virtual resources; testing to verify whether the subject's access to the object can take effect based on security tags and mandatory access control rules.

If the test results show that the cloud computing platform has the ability to set security tags on the subject and object of virtual resources, and which can ensure that the cloud service customer determines the subject's access to the object according to the security tags and mandatory access control rules, the result is conforming; otherwise it is non-conforming or partially conforming.

[L4-CNS2-07 Interpretation and Description]

The main evaluation targets of the evaluation metric that "Data exchange models such as communication protocol conversion or communication protocol isolation shall be provided to ensure that cloud service customers can choose boundary data exchange model independently according to their business needs" are devices or relevant components that provide communication protocol conversions or communication protocol isolation functions such as

Chapter 4　Application and Interpretation of the Extended Security Evaluation Requirements of Cloud Computing

network gatekeepers.

The key points of the evaluation implementation include: It shall be noted that the gatekeeper and the firewall are two different products in terms of function and principle. In a system classified as Level IV, attention shall be paid to whether devices or related components that provide communication protocol conversion or communication protocol isolation functions such as gatekeepers are adopted at the network boundaries of the IaaS cloud platform to implement physical security isolation of the network.

Checking the cloud platform construction plan to understand the cloud platform network architecture design; checking whether devices or related components such as gatekeepers that provide communication protocol conversion or communication protocol isolation for data exchange are deployed at the system network boundary; checking whether the cloud computing platform supports cloud services customers to independently choose the boundary data exchange method; by means of penetration testing or other methods, checking what measures the cloud computing platform adopts to isolate the communication protocol, and testing to verify whether the isolation methods are effective by sending data with common protocols (such as Telnet, FTP, HTTP, etc.).

If the test results show that gatekeepers are deployed at the network boundaries of the cloud computing platform, and the isolation measures of the gatekeepers are confirmed through the verification test to be effective or equivalent measures are taken, the result is conforming; otherwise it is non-conforming or partially conforming.

[L4-CNS2-08 Interpretation and Description]

The main evaluation targets of the evaluation metric that "Separate resource pools shall be allocated for business application systems graded as Level IV" include the network topology and cloud computing platform construction plan.

The key points of the evaluation implementation include: Independent resource pools mainly include independent computing resources, independent storage resources, and independent network resources. For example, the use of independent host machines to host virtual machines of the business application system classified as Level IV is considered to be the independence of computing resources; the use of independent storage devices are distributed storage systems is considered to be the independence of data storage resources; the use of independent access control measures and boundary protection measures belongs to the network resource independence.

Checking the design and construction plan of the cloud computing platform to see if the resource pools for Level IV business applications are separately divided; checking the network topology of the cloud computing platform to see if the resource pools are divided according to the design and construction plan, and the networks are isolated; testing to verify whether the resource pool isolation of the cloud computing platform is effective.

If the test results show that the cloud computing platform divides the independent resource pools for business application systems classified as Level IV, the result is

conforming; otherwise it is non-conforming or partially conforming.

4.2.3 Security Area Boundary

1. Access Control

[Standard Requirements]

This control point for Level Ⅲ includes the evaluation units of L3-ABS2-01, L3-ABS2-02, and Level Ⅳ includes that of L4-ABS2-01, L4-ABS2-02.

[L3-ABS2-01/L4-ABS2-01 Interpretation and Description]

The main evaluation targets of the evaluation metric that "Access control mechanisms shall be deployed at virtualized network boundaries and access control rules shall be set up" are network boundary devices and virtualized network boundary devices and their access control rules; for cloud service providers, the evaluation targets are devices that implement access control to the virtualized network boundaries, such as physical switches, virtual switches, physical firewalls, virtual firewalls, etc.; for cloud service customers, the evaluation targets are devices that implement access control to the virtualized network boundaries, such as virtual switches, virtual firewalls, cloud security protection components, bastion hosts, etc.

The key points of the evaluation implementation include: The concept of "virtual network boundary" needs to be clarified during the evaluation, which mainly includes the boundary between the cloud computing platform and the Internet, the boundary between the virtual network of cloud service customers and the external network, and the virtual network boundary between different cloud service customers and the boundary between different virtual subnets in the same cloud service customer virtual network.

Interviewing with the security administrator of the cloud computing platform and reviewing the network topology diagram to check whether virtualized network boundaries are clearly divided between the cloud computing platform and the Internet, between the virtual network of cloud service customers and the external network, between cloud service customers and cloud service customers, and between different virtual subnets in the same cloud service customer virtual network; checking whether virtualized network boundary devices are deployed at the virtual network boundary, and whether access control rules are set; checking the virtual network boundary devices to check the reasonableness of their access control rules, and testing to verify whether their access control rules are valid.

If the test results show that there is a clear division of virtual network boundary between the cloud computing platform and the Internet, between the cloud service customer and the cloud platform, between the cloud service customer and the cloud service customer, and between different virtual subnets in the same cloud service customer virtual network, between the virtual network of cloud service customers and external networks, between other cloud service customers, and between different virtual subnets, and border access control measures are deployed, and access control rules are set for these measures, the result is conforming; otherwise it is non-conforming or partially conforming.

【L3-ABS2-02/L4-ABS2-02 Interpretation and Description】

The main evaluation targets of the evaluation metric that "Access control mechanisms shall be deployed at different levels of network area boundaries and access control rules shall be set up" are devices that provide access control functions such as network boundary devices and virtualized network boundary devices and their access control rules.

The key points of the evaluation implementation include: when conducting test and evaluation for cloud service providers, checking whether areas with different security levels are established for cloud service customers, and access control mechanisms are deployed at the boundaries of different levels of network areas, and whether devices that provide access control functions such as gatekeepers, firewalls, routers, and switches are deployed for access control strategy settings. When conducting test and evaluation for cloud service customers, if there are different levels of network areas within the same VPC, it is also necessary to check whether access control rules are deployed. In addition, the systems with different levels of security protection in the cloud computing environment shall have access control measures and clear virtual boundaries.

Taking the cloud computing platform (the service model is IaaS) as an example, the evaluation implementation steps mainly include: interviewing with the security administrators and reviewing the network topology diagrams to check whether the cloud computing platform has different security domains for different security levels; checking whether different security domains are isolated by boundary access control devices, such as by deploying VPC boundary firewalls and other boundary access control devices to isolate them; checking whether the VPC boundary firewalls and other boundary access control devices are configured with access control strategies and rules between different security domains; testing to verify the effectiveness of access control strategies and access rules. If the test results show that the cloud computing platform has deployed boundary access control measures between different security domains, and these measures have set access control strategies and rules, the result is conforming; otherwise it is non-conforming or partially conforming.

Taking the business system of the cloud service customer as an example, the evaluation implementation steps mainly include: interviewing with the security administrator of the cloud service customer and reviewing the network topology diagram to check whether different levels of systems are deployed in the same VPC; if systems of different levels are deployed in the same VPC, checking whether the security group of the virtual machine of the high-level system has set access control strategies and rules to prevent the access of the virtual machine of the low-level system; if systems of different levels are deployed in different VPCs, checking whether access control strategies and rules are set on the boundary firewall of the VPC where the high-level system is located to prevent access to virtual machines in the VPC where the low-level system is located; testing to verify the effectiveness of access control strategies and access rules, such as logging in to any virtual machine in the VPC where the low-level system is located, and verifying whether Ping can pass the high-level system.

If the test results show that the cloud service customer has deployed boundary access control measures at different levels of network area boundaries, and these measures have set access control strategies and rules, the result is conforming; otherwise it is non-conforming or partially conforming.

2. Intrusion Prevention

[Standard Requirements]

This control point for Level III includes the evaluation units of L3-ABS2-03, L3-ABS2-04, L3-ABS2-05, L3-ABS2-06, and Level IV includes that of L4-ABS2-03, L4-ABS2-04, L4-ABS2-05, L4-ABS2-06.

[L3-ABS2-03/L4-ABS2-03 Interpretation and Description]

The main evaluation targets of the evaluation metric that "It shall be able to detect network attacking behavior initiated by cloud service customers and be able to record attack type, attack time, attack traffic, etc." are security protection devices (including virtual equipment) related to intrusion prevention, including anti-APT attack systems, network backtracking systems, threat intelligence detection systems, anti-DDoS attack systems, intrusion protection systems or related components, VPC boundary firewalls, Internet boundary firewalls, and other related devices used to detect network attacks initiated by cloud service customers.

Taking the cloud computing platform (the service model is IaaS) as an example, the key points of the evaluation implementation include: interviewing with security administrators of cloud service providers (take IaaS as an example) to check cloud computing platform design documents, construction plans and the network topology diagram to specify the deployed intrusion prevention devices or relevant components (hereinafter referred to as "intrusion prevention measures"); checking the update method of the intrusion prevention device or related component rule base and whether it has been updated to the latest version; reviewing the product manual of the intrusion prevention measure or in combination with the evaluator's cognitive experience of the intrusion prevention measure to check whether it has the detection function of abnormal traffic, large-scale attack traffic, advanced persistent attack, and whether it has alerting and cleaning disposal functions; logging into any virtual machine and deploying penetration testing tools on it, launching simulated attack actions on the cloud computing platform such as DDos attacks, APT attacks, SQL injection, cross-site scripting; logging into the relevant intrusion prevention platform or product to check whether the alert information about above-mentioned simulation attacks can be generated in the first place; logging into the cloud management platform to check whether there are detailed records of the attack type, attack time, attack traffic and other details of the above-mentioned simulated attacks. It is best to have a record of blocking the part of above-mentioned simulated attacks; logging into the virtual machine that initiated the simulation attacks to check its logs or cloud tenant network attack log records in the cloud management platform.

If the test results show that the cloud computing platform has deployed protection measures in both north-south and east-west directions, firewalls deployed at the boundaries of each security domain and virtual network can detect and record cross-regional attack behaviors, cross-VPC and attack behaviors within the same VPC can be detected by security devices such as virtual firewalls and virtual IPS. The relevant platform or product can record the type of attack initiated by the cloud service customer, attack time, attack traffic, etc., the result is conforming; otherwise it is non-conforming or partially conforming.

[L3-ABS2-04/L4-ABS2-04 Interpretation and Description]

The evaluation targets of the evaluation metric that "It shall be able to detect the network attacking behavior on virtual network nodes, and can record the attack type, attack time, attack traffic, etc." are mainly intrusion prevention related security protection devices (including virtual devices), including anti-APT attack systems, network backtracking systems, threat intelligence detection systems, anti-DDoS attack systems, intrusion protection systems or related components, VPC boundary firewalls, Internet boundary firewalls, etc., filtering out related devices used to detect virtual network nodes under network attacks.

Taking the cloud computing platform (the service model is IaaS) as an example, the evaluation implementation steps mainly include: interviewing with the security administrator of the cloud service provider to check the cloud computing platform design documents and construction plan, in combination of the interview and on-site check results and the network topology diagrams to clarify the relevant devices deployed to detect network attacks at virtual network nodes; checking whether the monitoring strategy of intrusion prevention measures can prevent attacking behavior against virtual network nodes, and record the attack types, attack time, attack traffic, etc.; checking whether the rule base of intrusion prevention measures has been updated to the latest version; launching a short-term Dos simulation attack on a virtual machine IN the cloud computing platform through the Internet; logging into the relevant network attack detection device to check whether its log records contain details such as the attack type, attack time, and attack traffic of the above-mentioned simulation attack.

If the test results show that the cloud computing platform has deployed protective measures in both north-south and east-west directions, firewalls deployed at the boundaries of each security domain and virtual network can detect and record cross-regional attacking behavior, and attacking behaviors across VPCs and within the same VPC can be detected by security devices such as virtual firewalls and virtual IPS. The relevant platform or product can record the attack type, attack time, attack traffic of the virtual network nodes, the result is conforming; otherwise it is non-conforming or partially conforming.

[L3-ABS2-05/L4-ABS2-05 Interpretation and Description]

The main evaluation targets of the evaluation metric that "It shall be able to detect abnormal traffic between virtual machines and host machines, and between virtual machines" are mainly security protection devices, traffic security monitoring devices, etc. The security protection devices here refers to anti-APT attack system, network backtracking system, threat intelligence detection system, anti-DDoS attack system and intrusion protection system or related components.

The key points of the evaluation implementation include: when conducting test and evaluation for cloud service providers, the virtual machines of the evaluation target include at least the management virtual machine on the cloud computing platform side and the business virtual machine on the cloud service customer side. Therefore, the abnormal traffic here shall include the traffic between the host machine and the management virtual machine, between the host machine and the business virtual machine, between business virtual machines, between management virtual machines, between the business virtual machine and the management virtual machine, therefore, the network architecture of the entire platform needs to be analyzed during the evaluation, clarifying the adopted protection methods, such as isolation through the physical network, through VLAN, etc.; when conducting test and evaluation for cloud service customers, it is also necessary to check the abnormal traffic detection method between virtual machines. In addition, in the implementation of the evaluation, it is necessary to clarify what traffic flows between the virtual machine and the host machine, and between virtual machines. Considering that different platforms have different implementation methods, such as establishing a management network platform and a business network platform, these two platforms are isolated from each other, and the host protection system can be installed on the virtual machine, which can detect the abnormal traffic between virtual machines. At present, there are generally two mainstream methods for detecting abnormal traffic between virtual machines and hosts, and between virtual machines: one is to divert east-west traffic to traffic security monitoring device in real time, then perform detection; the other is to deploy a proxy Client on the virtual machine to monitor the access traffic of the virtual machine where it is located. In the implementation of the evaluation, the traffic security monitoring device or proxy Client can be checked respectively according to the actual detection method adopted by the cloud computing platform.

Taking the cloud computing platform (the service model is IaaS) as an example, the evaluation implementation steps mainly include: interviewing with the security administrator to confirm the type of virtual machines and host machines of the cloud computing platform, and the means of communication between the two, checking the cloud computing platform design documents and construction plans, in combination of interview results, on-site checking and the network topology diagrams, clarifying the relevant device that has been deployed to detect abnormal traffic between the virtual machine and the host machine, and between virtual machines; checking whether the monitoring strategies of abnormal traffic detection measures are reasonable and effective; with the cooperation of the security administrator of the cloud service provider, logging into any virtual machine and deploying penetration testing tools on it, such as launching short-term Dos simulation attacks on its host and another virtual machine respectively; logging into the relevant abnormal traffic detection measures to check whether there is a monitoring log record of the above-mentioned simulation attack.

If the test results show that the host, the management virtual machine and the business virtual machine of the cloud computing platform belong to different network segments, they are not connected by default. For abnormal traffic, the external traffic probe will be used to detect; the host protection system on the host machine can detect the traffic of virtual machines and host machines; the access traffic between virtual machines across VPCs needs to be detected through virtual firewalls and virtual IPS; virtual machine traffic on different network segments in the same VPC is detected through virtual firewalls and virtual IPS; the access to virtual machines on the same network segment in the same VPC needs to redirect the traffic through devices such as Switch, leading the traffic to the traffic probe, and performing abnormal traffic detection. Or if there are similar measures that can detect abnormal traffic between the virtual machine and the host, and between virtual machines, the result is conforming; otherwise it is non-conforming or partially conforming.

[L3-ABS2-06/L4-ABS2-06 Interpretation and Description]

The main evaluation targets of the evaluation metric that "Security alerts shall be generated when network attack behaviors or abnormal traffics are detected" are security protection devices (including virtual devices) related to intrusion prevention, including anti-APT attack systems, network backtracking systems, threat intelligence detection systems, Anti-DDoS attack system, intrusion protection system or related components, VPC boundary firewalls, Internet boundary firewalls, etc.

Taking the grading object of the cloud computing platform (the service model is IaaS) as an example, the evaluation implementation steps mainly include: interviewing with the security administrator of the cloud computing platform, logging in and opening the monitoring configuration interface of the relevant device, and checking the sending method of its alert information, such as local log alerts, mobile phone and SMS alerts for the operation and maintenance personnel, email alerts, etc.; initiating a short-term Dos simulation attack on a virtual machine of the cloud computing platform through the Internet to check the alerting method.

If the test results show that the intrusion prevention devices, abnormal traffic detection devices or related components deployed on the cloud computing platform can generate alerts in time when they detect network attack behaviors or abnormal traffic, the result is

conforming; otherwise it is non-conforming or partially conforming.

3. Security Audit

[Standard Requirements]

This control point for Level Ⅲ includes the evaluation units of L3-ABS2-07, L3-ABS2-08, and Level Ⅳ includes that of L4-ABS2-07, L4-ABS2-08.

[L3-ABS2-07/ L3-ABS2-07 Interpretation and Description]

The main evaluation targets of the evaluation metric that "All privileged commands executed by cloud service providers and cloud service customers in remote management shall be audited, including at least virtual machine deletion and virtual machine reboot" are bastion hosts or related components.

The key points of the evaluation implementation include: for operations such as deleting and rebooting virtual machines without using the cloud computing platform, checking whether there is a record of the privileged commands executed by the bastion hosts or related components, and whether the audit record is kept properly; for cloud service customers, the remote management of virtual machines can be done through the cloud computing platform, directly through RDP, SSH and other protocols, or through the bastion hosts. Therefore, during the evaluation, it is necessary to verify the management method of the virtual machine. Checking the relevant audit records.

Taking the cloud computing platform (the service model is IaaS) as an example, the evaluation implementation steps mainly include: checking the remote privileged commands executed by the cloud service provider (including the third-party operation and maintenance service provider) during the remote management, and what are the access paths; checking whether there are related audit record of each access path; testing to verify whether the deletion or reboot of the test virtual machines can be audited.

If the test results show that there are related devices or platforms to provide complete audit playback and permission control services for privileged operations such as the virtual machine deletion, the virtual machine reboot and the virtual machine shutdown, and important operations can be recorded, the result is conforming; otherwise it is non-conforming or partially conforming.

[L3-ABS2-08/L4-ABS2-08 Interpretation and Description]

The main evaluation targets of the evaluation metric that "It shall be ensured that the cloud service provider's operations on the cloud service customer system and data can be audited by the cloud service customer" are the integrated audit system or related components.

Take the cloud computing platform (the service model is IaaS) as an example, the implementation steps of the evaluation mainly include: checking the specific measures of the cloud computing platform to protect the cloud service customer's systems and data, testing to verify whether the cloud service provider's operations on the cloud service customer systems and data can be audited by the cloud service customer.

If the test results show that all administrators of the cloud computing platform cannot obtain the virtual machine login account of the cloud service customer, and the administrator cannot directly obtain the data on the storage device, the cloud service provider can access the systems and data of the cloud service customer only under the authorization of the cloud service customer, and these operations can be audited by the cloud service customer, and the audit log is properly retained, the result is conforming; otherwise it is non-conforming or partially conforming.

4.2.4 Security Computing Environment

For cloud computing platforms/systems, in addition to networks, hosts, and applications, the Security computing environment also includes virtual machines, virtual firewalls and other virtual devices, and this part is also the part where cloud service customers rather than the cloud platforms play a major role. The situation is rather complicated, and there are many evaluation items, which need to be carefully distinguished according to the actual situation of the system. For example, for cloud service customer business systems, virtual machines are generally the most important product delivered by the IaaS platform to cloud service customers, which are controlled by cloud service customers. Therefore, the selection of cloud service customers' host servers as protection objects is no different from ordinary systems. However, for network devices and security devices, in many cases they are controlled and protected by the cloud platform. In this case, the cloud management console at the cloud service customer side (bastion hosts, VPNs, and jump hosts provided by other cloud platforms also have some similar functions) can be evaluated as a collection of network and security device services.

1. Identity Authentication

[Standard Requirements]

This control point for level III includes the evaluation unit of L3-CES2-01, and Level IV includes that of L4-CES2-01.

[L3-CES2-01/L4-CES2-01 Interpretation and Description]

The main evaluation targets of the evaluation metric that "when remotely managing devices in the cloud computing platform, a two-way identity authentication mechanism shall be established between the management terminal and the cloud computing platform" include the management terminal and the cloud computing platform.

The key points of the evaluation implementation include: For cloud service providers, the devices in the cloud computing platform can be various systems, devices and servers in the platform, or it can be the unified operation and maintenance portal (such as the bastion host) and an internal operation and maintenance terminal; for cloud service customers, this requirement is largely dependent on the cloud service provider/the cloud platform. If the cloud platform does provide, cloud service customers shall also try their best to meet this requirement through other means, so cloud service customers shall also be evaluated. In the implementation of the evaluation, it is necessary to pay attention to the concept of "two-way authentication", which means that the client needs to verify the identity of the server, and the server also needs to verify the identity of the client. Due to the different cloud computing service models and deployment models, the "cloud computing platform" in the requirements involves different objects, which may provide a unified interface, or may be very scattered. At the same time, the scope of applicability of the evaluation metric to different levels of protection objects is different, and the evaluation targets can be determined according to the actual situation during the implementation of the evaluation.

Taking the cloud computing platform (the service model is IaaS) as an example, the evaluation implementation steps mainly include: interviewing with security administrators to understand whether a two-way authentication mechanism is established during remote management; testing whether the two-way authentication mechanism is effective, for example, when remotely managing cloud computing platform devices, using 3A authentication server (unified identity authentication center) for device management, using SSL technology to ensure the encryption security of the transmission channel, and establishing a two-way identity verification mechanism in the form of two-way SSL certificates.

If the test results show that the cloud service provider's management terminals use a two-way authentication mechanism when accessing the cloud computing platform, the result is conforming; otherwise it is non-conforming or partially conforming.

2. Access Control

[Standard Requirements]

This control point for Level Ⅲ includes the evaluation units of L3-CES2-02, L3-CES2-03, and Level Ⅳ includes that of L4-CES2-02, L4-CES2-03.

[L3-CES2-02/L4-CES2-02 Interpretation and Description]

The main evaluation targets of the evaluation metric that "When the virtual machine is migrated, the access control strategy shall be migrated with it" are the virtual machine or container and related migration records and related configurations.

The key points of the evaluation implementation include: the emergence and rise of container technology is much later than virtual machines, therefore, in the PaaS model, containers must also be evaluated as the evaluation targets. In addition, for the mixed application of IaaS, PaaS, SaaS technologies, the evaluation target can be determined according to the actual situation during the evaluation implementation process.

Taking the cloud computing platform (the service model is IaaS) as an example, the evaluation implementation steps mainly include: interviewing with the security administrator of the cloud platform to check whether the security group strategies are migrated when the virtual machines are migrated, and whether the strategies are normal and effective; verifying the security group strategies after the virtual machine is migrated by creating a new test virtual machine to check whether the access control measures are migrated with it after the migration. If the test results show that the access control strategies of the cloud computing platform virtual machine is migrated along with it and is effective, the result is conforming; otherwise it is non-conforming or partially conforming.

Taking the cloud computing platform (the service model is PaaS) as an example, the PaaS platform is built on the IaaS platform (on top of another public cloud platform). The evaluation implementation steps are the same as above, and attention shall be paid when conducting evaluation: the PaaS cloud platform is viewed as the IaaS tenant of the public cloud platform, and the virtual machine migration strategy synchronization is guaranteed by the IaaS public cloud platform. It is not applicable to the cloud service provider PaaS cloud platform; when evaluating, only need to pay attention to the container, container migration record and related configuration to determine whether the access control strategies are migrated with them.

[L3-CES2-03/L4-CES2-03 Interpretation and Description]

The main evaluation targets of the evaluation metric that "Cloud service customers shall be allowed to set access control strategies between different virtual machines" are virtual machines, containers, security groups or related components.

The key points of the evaluation implementation include: containers in the PaaS model also need to be considered as the evaluation targets. In addition, for the mixed application of IaaS, PaaS, and SaaS technologies, the evaluation targets can be determined according to the actual situation during the evaluation implementation process.

Take the cloud computing platform (the service model is IaaS) as an example: reviewing the cloud computing platform design documents and construction plans to check whether cloud service customers are allowed to set access control strategies between different virtual machines; a new test virtual machine can be created to test and verify whether the access control strategies between different virtual machines can be set.

If the test results show that the cloud computing platform allows the cloud service customer to set access control policies between different virtual machines, the result is conforming; otherwise it is non-conforming or partially conforming.

3. Intrusion Prevention

[Standard Requirements]

This control point for Level III includes the evaluation units of L3-CES2-04, L3-CES2-05, L3-CES2-06, and Level IV includes that of L4-CES2-04, L4-CES2-05, L4-CES2-06.

[L3-CES2-04/L4-CES2-04 Interpretation and Description]

The main evaluation targets of the evaluation metric that "It shall be able to detect the failure of resource isolation between virtual machines and generate the security alert" include the cloud management platform or related components that can provide the detection and alert mechanism for resource isolation failures between virtual machines.

The key points of the evaluation implementation include: the failure of resource isolation between virtual machines generally refers to the failure of the isolation of the virtual machine's CPU, memory, internal network, disk I/O, and user data; alert methods include but not limited to SMS, email, voice, etc. In addition, the isolation failure of containers also needs to be considered.

Taking the cloud computing platform (the service mode is IaaS) as an example, the evaluation implementation steps mainly include: checking whether the cloud management platform can detect the failure of resource isolation between virtual machines, checking alert measures and corresponding alert records.

If the test results show that the cloud management platform of the cloud computing platform has measures to detect the failure of resource isolation between virtual machines, and can generate alerts by email, SMS, etc., the result is conforming; otherwise it is non-conforming or partially conforming.

[L3-CES2-05/L4-CES2-05 Interpretation and Description]

The main evaluation targets of the evaluation metric that "It shall be able to detect unauthorized new virtual machines or re-enabled virtual machines and generate the security alarm" are the cloud management platform or related components that can provide virtual machine abnormality detection and alert mechanisms.

The key points of the evaluation implementation include: paying attention to whether the cloud platform can detect unauthorized creation, deletion, shutdown, and reboot of virtual machines, and generate alerts when such illegal actions are detected. In addition, due to technological development, in the PaaS model, unauthorized creation and reboot of containers also need to be considered.

Taking the cloud computing platform (the service model is IaaS) as an example, the evaluation implementation steps mainly include: checking whether the cloud management platform or related components have measures to detect unauthorized creation or reboot of virtual machines, checking alert measures, and establish a test account for verification. If the test results show that the cloud management platform or related components of the cloud computing platform have the ability to detect unauthorized creation or reboot of virtual machines, and generate alerts via email, SMS, etc., then the result is conforming; otherwise it is non-conforming.

Taking the cloud computing platform (the service model is PaaS) as an example, the PaaS platform is built on the IaaS platform (on top of another public cloud platform). The evaluation implementation steps are the same as above. Note when evaluating: PaaS cloud platform is the IaaS of the public cloud platform Tenants, virtual machine restart and new creation detection alarms are guaranteed by the IaaS public cloud platform. It is not applicable to the PaaS cloud platform. During the evaluation, we only need to check the detection and alarm measures of their PaaS platform for unauthorized container creation and restart.

[L3-CES2-06/L4-CES2-06 Interpretation and Description]

The main evaluation targets of the evaluation metric that "It shall be able to detect malicious code infections and the condition of spreading among virtual machines, and generate the security alert" are the cloud management platform or related components that can detect malicious code infections and spread between virtual machines.

The key points of the evaluation implementation include: For cloud service providers, malicious code detection can only be implemented at the operating system (virtual machine/host) level, or through network anti-malicious code products, Therefore, the evaluation mainly focuses on the malicious code detection measures at the operating system (virtual machine/host) level and the network level.

Taking the cloud computing platform (the service model is IaaS) as an example, the evaluation and implementation steps mainly include: interviewing and checking whether the cloud management platform or related components detect malicious code infection and malicious code spread between all virtual machines; checking the alerting measures; checking the malicious code scanning records and the generated warning records.

If the test results show that the cloud computing platform has a dynamic malicious code detection and alert mechanism between virtual machines, the result is conforming; otherwise it is non-conforming.

4. Mirroring and Snapshot Protection

[Standard Requirements]

This control point for Level III includes the evaluation units of L3-CES2-07, L3-CES2-08, L3-CES2-09, and Level IV includes that of L4-CES2-07, L4-CES2-08, L4-CES2-09.

[L3-CES2-07/L4-CES2-07 Interpretation and Description]

The main evaluation targets of the evaluation metric that "Reinforced OS mirroring or OS security reinforcement services shall be provided for critical business systems" are virtual machine image files, including image files provided by cloud computing platforms to cloud service customers and image files of the cloud service customer business systems.

The key points of the evaluation implementation include: Generally speaking, cloud service customers use the mirroring and reinforced system of the cloud platform, but if the cloud service customers use their own mirroring and system, they shall also provide mirroring and system reinforcement.

Taking the cloud computing platform (service model as IaaS) as an example, the evaluation and implementation steps mainly include: interviewing with the operation and maintenance administrator to understand whether the provided operating system image file is a reinforced image file, and whether regular security reinforcement service is provided for the operating system on a regular basis; reviewing security reinforcement records, such as regular vulnerability scanning of operating systems and software upgrade records.

If the test results show that the cloud computing platform provides the reinforced operating system images, the result is conforming; otherwise it is non-conforming.

[L3-CES2-08/L4-CES2-08 Interpretation and Description]

The main evaluation targets of the evaluation metric that "The integrity verification function for virtual machine image, snapshot shall be provided to prevent virtual machine mirroring from being maliciously tampered with" include the virtual machine image, snapshot or related components.

The key points of the evaluation implementation include the following: Generally, cloud service customers use the mirroring and reinforcement system of the cloud platform, but if the cloud service customer uses their own mirroring and system, they shall also provide integrity verification functions.

Taking the cloud computing platform (service model is IaaS) as an example, the evaluation implementation steps mainly include: interviewing with security administrators to check whether the image and snapshot files are verified, and whether there exists the verification recording mechanism; conducing on-site test to verify the verification effects of image and snapshot files. If the test results show that the cloud computing platform has the integrity check mechanism for image and snapshot files, adopts MD5 to check the integrity (or equivalent technology), and has a strict integrity verification record, then the result is conforming; otherwise it is non-conforming.

[L3-CES2-09/L4-CES2-09 Interpretation and Description]

The main evaluation targets of the evaluation metric that "Cryptographic or other technical measures shall be adopted to prevent sensitive resources that may exist in virtual machine images and snapshots from being illegally accessed" are virtual machine images, snapshots or related components.

The key points of the evaluation implementation include the following: In general, cloud service customers use the mirroring and reinforced system of the cloud platform, but if cloud service customers use their own mirror images and system, encryption technology shall also be adopted. In addition, when evaluating this requirement, it is required to encrypt virtual machine images and snapshots with cryptographic techniques. If access control is used to restrict users' illegal access to virtual machine images and snapshots, their security will also be protected to a certain extent. Therefore, it cannot be directly judged as non-conformity if no cryptographic techniques are adopted in the evaluation.

Taking the cloud computing platform (the service model is IaaS) as an example, the evaluation and implementation steps mainly include: interviewing with the cloud computing platform administrator and verifying whether encryption or access control measures are taken for image and snapshot files; verifying whether the encryption measures or access control measures are effective.

Chapter 4　Application and Interpretation of the Extended Security Evaluation Requirements of Cloud Computing

If the test results show that the cloud computing platform adopts cryptographic techniques or access control measures to prevent the sensitive resources that may exist in the virtual machine image or snapshot from being illegally accessed, the result is conforming; otherwise it is non-conforming.

5. Data Integrity and Confidentiality

[Standard Requirements]

This control point for Level Ⅲ includes the evaluation units of L3-CES2-10, L3-CES2-11, L3-CES2-12, L3-CES2-13, and Level Ⅳ includes that of L4-CES2-10, L4-CES2-11, L4-CES2-12, L4-CES2-13.

[L3-CES2-10/L4-CES2-10 Interpretation and Description]

The main evaluation targets of the evaluation metric that "It shall be ensured that cloud service customer data, users' personal information are stored within the territory of China, and cross-border data transmission shall comply with relevant national regulations" are database servers, data storage devices, and management documents.

The key points of the evaluation implementation include: whether for cloud service providers or cloud service customers, no matter what service model, as long as cloud service customer data, user personal information are involved, the data shall be stored in China, and the relevant national regulations shall be followed if the cross-border data transfer is needed.

Taking the cloud computing platform (service model is IaaS) as an example, the evaluation implementation steps mainly include: interviewing with security administrators to understand the storage situation of cloud service customers and users' personal information, verifying whether the data storage location is within the country; whether there exists the management system of cross-border data transfer and whether it complies with relevant national laws and regulations.

If the test results show that cloud service customer data, user's personal information are all stored in China, and there exists the management system that complies with relevant national regulations for the cross-border data transfer, the result is conforming; otherwise it is non-conforming or partially conforming.

[L3-CES2-11/L4-CES2-11 Interpretation and Description]

The main evaluation targets of the evaluation metric that "It shall be ensured that only under the authorization of cloud service customers, cloud service providers or third parties have the administrative permission over the cloud service customers' data" are the cloud management platform, database, related authorization documents and management documents.

Taking the cloud computing platform (the service model is IaaS) as an example, the implementation steps of the evaluation mainly include: interviewing with security administrators and reviewing related management systems and other documents to see whether it is clearly stipulated that only under the authorization of the cloud service customer, the cloud service provider or third party has the administrative permission over the cloud service customer's data; checking the authorization process, authorization method and authorization content of cloud service customer data management permission.

If the test results show that the cloud service provider or a third party has the permission to manage cloud service customer's data only under the authorization of the cloud service customer, the result is conforming; otherwise it is non-conforming or partially conforming.

[L3-CES2-12/L4-CES2-12 Interpretation and Description]

The main evaluation targets of the evaluation metric that "Verification techniques or cryptographic techniques shall be adopted to ensure the integrity of important data during the process of virtual machine migration, and necessary recovery measures shall be taken when any damage to the data integrity is detected" are virtual machines and containers.

The main points of the evaluation implementation include: the virtual machine or container migration process is rather complicated, and it is also difficult to carry out the evaluation. It is necessary to conduct on-site interviews with developers and designers based on the relevant security design documents to understand the migration process, and then comprehensively make the judgment based on its specific technical means and on-site verification.

Taking the cloud computing platform (the service model is IaaS) as an example, the evaluation and implementation steps mainly include: checking the related development and design documents of virtual machine migration, and interviewing with developers to understand the specific method of virtual machine migration and the integrity protection measures taken.

If the test results show that verification techniques or cryptographic techniques are adopted to ensure the integrity of important data during the migration of the virtual machine of the computing platform, then the result is conforming; otherwise it is non-conforming.

[L3-CES2-13/L3-CES2-13 Interpretation and Description]

The main evaluation target of the evaluation metric that "Cloud service customers shall be supported to deploy key management solutions to ensure that cloud service customers can implement the process of data encryption and decryption by themselves" is the key management solution.

The main points of the evaluation implementation include: the key management solution can be implemented by the cloud platform or through third-party solution. Cloud service customers shall be able to choose and configure by themselves, and which can take effect in a timely manner.

Taking the cloud computing platform (the service model is IaaS) as an example, interviewing with security administrators to understand whether the cloud computing platform supports cloud service customers to deploy key management solutions, checking related technical documents; test accounts can be used to test to verify whether it is effective.

If the test results show that the cloud computing platform supports cloud service customers to deploy key management solutions to ensure that the cloud service customers implement the data encryption and decryption process by themselves, the result is conforming; otherwise it is non-conforming or partially conforming.

6. Data Backup and Recovery

[Standard Requirements]

This control point for Level Ⅲ includes the evaluation units of L3-CES2-14, L3-CES2-15, L3-CES2-16, L3-CES2-17, and Level Ⅳ includes that of L4-CES2-14, L4-CES2-15, L4-CES2-16, L4-CES2-17.

[L3-CES2-14/L4-CES2-14 Interpretation and Description]

The main evaluation targets of the evaluation metric that "Cloud service customers shall save backups of their business data locally" include the cloud computing platform or related components.

The key points of the evaluation implementation include: cloud service providers provide local data download services, and cloud service customers perform local backup configuration.

Take the cloud service customer business system as an example, interviewing with cloud service customers to find out whether the business data of the business system are stored locally; checking the local backup record of the cloud service customers' data.

If the test results show that the cloud computing platform provides the backup function for cloud service customers to store their business data locally, then the result is conforming; otherwise it is non-conforming.

[L3-CES2-15/L4-CES2-15 Interpretation and Description]

The main evaluation targets of the evaluation metric that "The ability to query the cloud service customer data and the location of backup storage shall be provided" include the cloud management platform or related components.

Take the cloud computing platform (service model is IaaS) as an example, interviewing with the cloud service provider to find out whether it provides cloud tenants with data and backup storage location query interfaces or other technologies and management methods; using test accounts to query and verify the query results.

If the test results show that the cloud platform provides the function for cloud service customers to query their backup storage location, then the result is conforming; otherwise it is non-conforming.

[L3-CES2-16/L4-CES2-16 Interpretation and Description]

The main evaluation targets of the evaluation metric that "The cloud storage service of cloud service providers shall ensure that there are several available copies of cloud service customers' data, and the contents of each copy shall be consistent" include the cloud management platform or cloud storage system or related components.

Taking the cloud computing platform (service model is IaaS) as an example, the evaluation and implementation steps mainly include: interviewing with the cloud computing platform administrator and checking the cloud management platform, checking the storage strategy and storage model of the cloud storage service, and checking whether the multi-copy storage model is set; interviewing with cloud computing platform administrators and reviewing relevant technical documents to check whether the integrity detection measures of multiple copies of data can ensure the integrity and consistency of each copy; on-site testing of the number of copies of data stored by cloud service customers and verifying whether the copies are consistent.

If the test results show that the cloud storage service provided by the cloud computing platform can achieve multiple copies of the stored data of the cloud service customer, and the contents of the multiple copies are guaranteed to be consistent, the result is conforming; otherwise it is non-conforming or partially conforming.

[L3-CES2-17/L4-CES2-17 Interpretation and Description]

The main evaluation targets of the evaluation metric that "It shall provide technical measures for cloud service customers to migrate their business systems and data to other cloud computing platforms and local systems, and assist them in completing the migration process" are technical measures and means to provide relevant capabilities.

The main points of the evaluation implementation include: due to the different scales of the cloud platform and different service models, the technical measures during the migration and the service models of each service provider will inevitably be diverse. Generally speaking, in the implementation of the evaluation, it is necessary to consider both technical and management aspects, and at the same time make judgments based on the specific conditions on the spot.

Taking the cloud computing platform (the service model is IaaS) as an example, the evaluation and implementation steps mainly include: checking whether there are relevant technical means to ensure that cloud service customers can migrate business systems and data to other cloud computing platforms and local systems, and understand their technical implementation methods, types and management regulations in the migration data; interviewing and checking whether the cloud service provider server administrator provides measures, means or personnel to assist the cloud service customers to complete the migration process; using the platform tenants' test accounts for data migration to check whether it is consistent with related techniques and management documents.

If the test results show that the cloud computing platform assists cloud service customers' data migration to the local cloud platform through server migration tools or other equivalent means, and at the same time provides technical support for cloud service customers to migrate to other cloud computing platforms, then the result is conforming; otherwise it is non-conforming.

7. Residual Information Protection

[Standard Requirements]

This control point for Level Ⅲ includes the evaluation units of L3-CES2-18, L3-CES2-19, and Level Ⅳ includes that of L4-CES2-18, L4-CES2-19.

[L3-CES2-18/L4 -CES2-18 Interpretation and Description]

The main evaluation targets of the evaluation metric that "It shall be ensured that the memory and storage space used by virtual machines are completely cleared when recycled" are technical measures and means that provide related capabilities.

Taking the cloud computing platform (the service model is IaaS) as an example, the implementation steps of the evaluation mainly include: checking whether the memory and storage space of the cloud service provider's virtual machine is completely cleared when the virtual machine is recycled; checking whether the data and backup data (such as image files, snapshot files, etc.) are cleaned up after the virtual machine is migrated or deleted; interviewing with developers, reviewing design documents, and conducting verification tests: such as using memory and disk writing tools to fill up with specific data, and then conducting recycle and redistribution to read the disk or memory to see if there is any previously written specific data. If the test results show that the cloud computing platform can ensure that the memory and storage space used by the virtual machine are completely cleared when the virtual machine is recycled, then the result is conforming; otherwise it is non-conforming.

Taking the cloud computing platform (the service model is PaaS) as an example, the PaaS platform is built on the IaaS platform (another public cloud platform). The evaluation and implementation steps are the same as above. During the evaluation implementation, attention shall be paid to: whether the memory and storage space of the container is completely cleared when the container is recycled; whether the data and backup data are cleansed after the container is migrated or deleted.

[L3-CES2-19/L4-CES2-19 Interpretation and Description]

The main evaluation targets of the evaluation metric that "When cloud service customers delete their business application data, cloud computing platforms shall delete all copies in the cloud storage" include the cloud computing platform, cloud storage, or technical measures and methods that provide related capabilities.

The key points of the evaluation implementation include: the business application data in the above evaluation indicators does not only refer to the deletion of the cloud service customer data of the SaaS platform, but also the deletion of data such as Docker and virtual machines. Therefore, it is necessary to consider comprehensively when implementing the evaluation.

Taking the cloud computing platform (the service model is IaaS) as an example, the evaluation and implementation steps mainly include: interviewing with the security administrators of the cloud computing platform to understand the cloud computing platform's handling measures for the backup and copies of cloud service customer business application data; checking whether all the copies in the cloud storage are deleted when the cloud service customer deletes their business application data.

If the test results show that when the cloud service customer deletes the business application data, the cloud computing platform can ensure that the corresponding copies in the cloud storage are also deleted, then the result is conforming; otherwise it is non-conforming.

4.2.5 Security Management Center

There are many types of devices involved in the cloud computing environment, including centralized management of virtual devices. For cloud service customers, the cloud management console at the cloud service customer side provides certain centralized security management capabilities. It is necessary to understand the actual usage of the cloud service customer, and also consider whether the cloud service customer uses a third party or self-built security mechanism, and whether this mechanism and the security mechanism provided by the cloud platform can be managed in a centralized and unified manner.

1. Centralized Management and Control

[Standard Requirements]

This control point for Level III includes the evaluation units of L3-SMC2-01, L3-SMC2-02, L3-SMC2-03, L3-SMC2-04, and Level IV includes that of L4-SMC2-01, L4-SMC2-02, L4-SMC2-03, L4-SMC2-04.

[L3-SMC2-01/L4-SMC2-01 Interpretation and Description]

The main evaluation targets of the evaluation metric that "It shall be able to perform the unified management, scheduling and allocation of physical resources and virtual resources in accordance with strategies" include the cloud management platform or related components and resource scheduling platforms.

The key points of the evaluation implementation include: due to different cloud computing service models and deployment models, the scope of application of the evaluation index for different levels of protection objects is different. For example, in the PaaS service model, Docker may also be involved in the unified deployment of resources, but whether SaaS is involved shall be judged according to the actual situation, or whether it based on the IaaS or PaaS environment.

Taking the cloud computing platform (the service model is IaaS) as an example, the implementation steps of the evaluation mainly include: checking whether there is the resource scheduling platform that provides unified resource management, scheduling and allocation strategies; checking whether the physical resources and virtual resources can be managed, scheduled and allocated in a unified manner according to the above strategies.

If the test results show that the cloud computing platform performs the unified management, scheduling and allocation of physical resources and virtual resources in accordance with the strategies, then the result is conforming; otherwise it is non-conforming.

[L3-SMC2-02/L4-SMC2-02 Interpretation and Description]

The main evaluation targets of the evaluation metric that "It shall be ensured that the management traffic of the cloud computing platform is separated from the business traffic of the cloud service customers" include the cloud management platform and its network architecture.

The key points of the evaluation implementation include: the cloud computing platform management traffic refers to the traffic generated by cloud computing platform management, platform deployment, system loading, etc., including traffic generated by administrators for resource management, policy issuance, platform deployment, log collection, event alerts, etc, the cloud service customer business traffic refers to the traffic generated by providing users with business applications, including the traffic generated by cloud service customers logging in to the management console and logging in to virtual machines for application deployment. The cloud computing platform management traffic and the cloud service customer business traffic need to be separated. "Separation" here refers to the use of different network cards for business and management, and different physical switching network devices to host them.

Taking the cloud computing platform (the service model is IaaS) as an example, the implementation steps of the evaluation mainly include: checking whether the network architecture and configuration strategy can adopt out-of-band management or strategy configuration to realize the separation of management traffic and business traffic; by capturing packets on different network platforms, verifying whether the cloud computing platform management traffic and business traffic are truly separated.

If the test results show that the management traffic and business traffic of the cloud computing platform use separate network platforms and separate network cards are used on the server, the cloud computing platform meets the requirements for separation of management traffic and cloud service customer business traffic, the result is conforming; otherwise it is non-conforming or partially conforming.

[L3-SMC2-03/L4-SMC2-03 Interpretation and Description]

The main evaluation targets of the evaluation metric that "According to the division of responsibilities between cloud service providers and cloud service customers, the audit data of their respective control parts shall be collected and their centralized audit shall be realized respectively" are cloud management platforms, integrated audit systems or related components.

The main points of the evaluation implementation include: when the evaluation is implemented, first of all, it is necessary to clarify what the audit data of the cloud service provider and the cloud service customer are in their respective control parts, and whether the centralized management of all logs are realized. Regarding this evaluation metric, it is recommended to conduct evaluation from the cloud service customer side, but the focus can be put on whether the division of responsibilities is clear and whether it is fully implemented in place. In addition, different service models have different audit data types. For example, in the IaaS service model, for cloud service providers, the audit data includes host logs, platform related virtual machine logs, network device logs, and security device Logs, logs of various IaaS management platforms, and cloud service customers' IaaS service login logs, etc. However, it shall be noted that without authorization, cloud service providers cannot view the operation logs of cloud service customers for their own virtual machines, the Client management platform (if any) provided by IaaS, and their own business systems. For cloud service customers of IaaS services, the audit data includes virtual machine logs, logs of the Client management platform provided by IaaS (if any), and logs of business systems; in the PaaS service model, for cloud service providers, the audit data includes virtual machine logs, logs of the Client management platform provided by IaaS (if any), logs of various PaaS management platforms, and PaaS service login logs of cloud service customers. However, it shall be noted that without authorization, cloud service providers cannot view the logs (if any) of the client management platform provided by cloud service customers for PaaS and the operation logs of their own business systems. For cloud service customers of PaaS services, the audit data includes the logs of the Client management platform provided by PaaS (if any) and business system logs; in the SaaS service model, for cloud service providers, the audit data includes the logs of the Client management platform provided by PaaS (if any), the logs of various SaaS management platforms, and the SaaS service login logs of cloud service customers. It shall be noted that cloud service providers cannot view the operation logs of cloud service customers without authorization. For cloud service customers of SaaS services, the audit data includes operation (audit) logs of their own business systems.

Taking the cloud computing platform (the service model is IaaS) as an example, the implementation steps of the evaluation mainly include: checking whether the cloud service provider and the cloud service customers are divided according to their responsibilities to achieve the collection of audit data under their respective control; checking whether the cloud service provider can achieve centralized audit, whether it establishes the comprehensive log audit system for centralized log auditing.

If the test results show that the cloud computing platform can collect the audit data of the respective control parts according to the division of responsibilities of the cloud service provider and the cloud service customer, and realize the centralized audit respectively, the result is conforming; otherwise it is non-conforming or partially conforming.

Chapter 4 Application and Interpretation of the Extended Security Evaluation Requirements of Cloud Computing

〚L3-SMC2-04/L4-SMC2-04 Interpretation and Description〛

The main evaluation targets of the evaluation metric that "According to the division of responsibilities between cloud service providers and cloud service customers, the centralized monitoring of the operation of their respective control parts, including virtualized networks, virtual machines, virtualized security devices, shall be realized" include the cloud management platform or related components.

The key points of the evaluation implementation include: for cloud service providers, the evaluation target are the platform or other components that can centrally monitor virtual machines, virtual network devices, virtual security devices, virtualized networks, etc. The virtualized network within the cloud service customer shall not be the monitoring content of the cloud service provider, unless the cloud service provider is authorized. For this evaluation metric, it is recommended to conduct evaluation from the cloud service customers' side. The focus of evaluation can be put on whether the division of responsibilities is clear and whether it is fully implemented in place. The evaluation target is the platform that can centrally monitor its own virtual machines, virtual network devices, virtual security devices, virtualized network conditions, etc. This monitoring platform can be provided by the cloud service provider, or a centralized monitoring system can be built by themselves.

Taking the cloud computing platform (the service model is IaaS) as an example, the evaluation and implementation steps mainly include: interviewing with the cloud platform security administrator to see whether the the respective control parts are realized according to the division of responsibilities between cloud service providers and cloud service customers, including the centralized monitoring of the operating status of virtual private networks, virtual machines, virtualized security devices, etc. ; checking the related design scheme and the network topology diagram, checking technical measures to achieve centralized monitoring of the operating conditions of their respective parts. If the test results show that the cloud computing platform implements centralized monitoring of the operating conditions of their respective control parts, including virtual private networks, virtual machines, virtualized security equipment, etc. , according to the division of responsibilities between cloud service providers and cloud service customers, then the result is conforming; otherwise it is non-conforming.

Taking the business application system of a cloud service customer as an example, the evaluation and implementation steps mainly include: checking the service agreement signed by the cloud service customer and the cloud service provider, the cloud service customer's business application system design document and the construction plan; checking whether the management console from the end the cloud service customer side can realize the security monitoring of its own virtualized network, virtual machine, cloud firewall, cloud host protection, cloud WAF service, etc. , or other centralized monitoring methods. If the test results show that the management console of the cloud service client can realize the security monitoring of its own virtualized network, virtual machine, cloud firewall, cloud host protection, cloud WAF service, etc., then the result is conforming; otherwise it is non-conforming.

4.2.6 Security Management System

Please refer to the relevant chapters of this book for the interpretation of general requirements for security evaluation.

4.2.7 Security Management Organization

Please refer to the relevant chapters of this book for the interpretation of general requirements for security evaluation.

4.2.8 Security Management Personnel

Please refer to the relevant chapters of this book for the interpretation of general requirements for security evaluation.

4.2.9 Security Development Management

1. The Selection of Cloud Service Provider

[Standard Requirements]

This control point for Level III includes the evaluation units of L3-CMS2-01, L3-CMS2-02, L3-CMS2-03, L3-CMS2-04, L3-CMS2-05, and Level IV that of L4-CMS2-01, L4-CMS2-02, L4-CMS2-03, L4-CMS2-04, L4-CMS2-05.

[L3-CMS2-01/L4-CMS2-01 Interpretation and Description]

The main evaluation targets of the evaluation metric "select secure and compliant cloud service provider, and the cloud computing platform provided by it shall provide the corresponding level of security protection capability for the business application system carried by it" are the person in charge of system development and the service contract signed with the cloud service provider and so on.

The key points of the evaluation implementation include: when cloud service customers choose a cloud service provider, they shall regard the security compliance of the cloud service provider as one of the evaluation metric. The security-compliant cloud service providers here mainly focus on the qualifications of the cloud service providers and the security protection capabilities of the cloud computing platform. When choosing a cloud service provider, whose evaluation conclusion is determined as "bad" shall not be selected. It is recommended to choose a cloud service provider whose evaluation conclusion is "excellent" or "good".

Taking the cloud service customer business system as an example, the evaluation and implementation steps mainly include: interviewing with the person in charge of the cloud service customer business system construction, understanding the procedures and requirements for screening cloud service providers, and whether the qualifications of cloud service providers and the security protection level of the cloud platform are rigorously reviewed; checking the grading status of the cloud computing platform of the cloud service provider and the evaluation results of the classified security protection, and determining that the cybersecurity protection level of the cloud computing platform is equal to or higher than the cloud service customer application system.

If the test results show that the cloud service customer has selected a security-compliant cloud service provider, and the cybersecurity protection level of the cloud computing platform is equal to or higher than that of the business application system, the result is conforming; otherwise it is non-conforming or partially conforming.

【L3-CMS2-02/L4-CMS2-02 Interpretation and Description】

The main evaluation targets of the evaluation metric that "The service content and specific technical indicators of cloud services shall be specified in the service level agreement" are service contracts or service level agreements signed with cloud service providers.

The key points of the evaluation implementation include: the legal documents (such as contracts, agreements, etc.) signed by the cloud service customer and the cloud service provider, which stipulate the service level and the performance level that the service must meet (including service level measurement, service level report, reputation and fees, etc.), making the cloud service provider responsible for completing these predetermined service levels; mature cloud service providers generally provide a fixed template service level agreement (SLA), and the corresponding SLA needs to be confirmed according to the specific services purchased by the cloud service customer. When the evaluation is implemented, the focus shall be put on whether a personalized SLA is signed. SLA shall include at least two key technical indicators. One is the uptime promised by the cloud service provider. It is recommended to achieve 99.9% or higher availability. The second is the response time for the most serious service problems promised by cloud service providers. It is recommended to respond within half an hour.

Take the cloud service customer business system as an example: checking whether the service contract signed by the cloud service customer and the cloud service provider specifies the service content and specific indicators of the cloud service; checking whether the contract includes the two key technical indicators of the cloud service provider's commitment of the uptime and the response time for the most serious service problems.

If the test results show that the service contract signed by the cloud service customer and the cloud service provider includes the service content and specific technical indicators of the cloud service, and the promised uptime and response time for the most serious service problems are not lower than the recommended values in the evaluation implementation points, the result is conforming; otherwise it is non-conforming or partially conforming.

【L3-CMS2-03/L4-CMS2-03 Interpretation and Description】

The main evaluation targets of he evaluation metric that "The rights and responsibilities of cloud service providers shall be stipulated in the service level agreement, including management scope, division of responsibilities, access authorization, privacy protection, code of conducts, liability for breach of contract, etc." are the service contract or the service level agreement signed with the cloud service provider.

The key points of the evaluation implementation include: focus shall be put on whether the cloud service customer has signed a service level agreement (or contract) with the cloud service provider, and checking whether the service level agreement (or contract) contains the cloud service provider's rights and responsibilities, including the scope of management, division of responsibilities, access authorization, privacy protection, code of conducts, liability for breach of contract, etc.

If the test results show that the service contract between the cloud service customer and the cloud service provider specifies the rights and responsibilities of the cloud service provider, including the content such as the management scope, division of responsibilities, access authorization, privacy protection, code of conducts, liability for breach of contract, etc., the result is conforming; otherwise it is non-conforming or partially conforming.

[L3-CMS2-04/L4-CMS2-04 Interpretation and Description]

The main evaluation targets of the evaluation metric that "It shall be stipulated in the service level agreement that when the service contract expires, the cloud service customer data shall be fully provided, and the relevant data shall be cleared on the cloud computing platform" are the service contract or service level agreement signed by cloud service customers and cloud service providers, etc.

The key points of the evaluation implementation include: focus shall be put on the service level agreement (or contract) signed by the cloud service customer and the cloud service provider, and checking whether the content of the service level agreement (or contract) stipulates the obligations when the service contract expires. At least the following two commitments shall be included: being able to provide complete cloud service customer data; clearing relevant data in a timely manner on the cloud computing platform.

If the test results show that the service contract signed by the cloud service customer and the cloud service provider stipulates that when the service contract expires, the cloud service customer data shall be provided in full, and the commitment that the relevant data shall be cleared on the cloud computing platform is made, and the data migration methods provided are proved too be effective, then the result is conforming; otherwise it is non-conforming.

[L3-CMS2-05/L4-CMS2-05 Interpretation and Description]

The main evaluation targets of the evaluation metric that "A confidentiality agreement shall be signed with the selected cloud service provider, requiring them not to cause any leakage of cloud service customer's data" are the service contract or service level agreement signed with the cloud service provider.

The key points of the evaluation implementation include: focus shall be put on checking the confidentiality agreement signed by the cloud service customer and the cloud service provider, and at least the following commitments shall be included: It is forbidden to leak cloud service customer data; and necessary measures shall be taken to prevent cloud service customer data from being stolen.

If the test results show that the cloud service customer has signed a non-disclosure agreement with the cloud service provider, requiring the cloud service provider not to disclose cloud service customers' data, then the result is conforming; otherwise it is non-conforming.

2. Supply Chain Management

[Standard Requirements]

This control point for Level Ⅲ includes the evaluation units of L3-CMS2-07, L3-CMS2-08, L3-CMS2-09, and Level Ⅳ includes that of L4-CMS2-07, L4-CMS2-08, L4-CMS2-09.

[L3-CMS2-07/L4-CMS2-07 Interpretation and Description]

The main evaluation targets of the evaluation metric that "It shall be ensured that the selection of suppliers complies with relevant national regulations" are qualification documents of the corresponding suppliers.

The key points of the evaluation implementation include: suppliers mainly include software suppliers, hardware suppliers, service suppliers, etc. Among them, software includes the operating system, middleware, database, cloud platform special software, etc.; hardware includes the server, network device, security device, etc.; services include security services, emergency services, telecommunications services, and talent services.

Interviewing with the supplier management personnel and reviewing system documents related to supplier management, checking whether there are clauses "It shall be ensured that the selection of suppliers meets relevant national regulations", and checking the process record documents of supplier selection and relevant qualification documents of suppliers.

If the test results show that the selection of supplier complies with relevant national regulations, then the result is conforming; otherwise it is non-conforming.

[L3-CMS2-08/L4-CMS2-08 Interpretation and Description]

The main evaluation targets of the evaluation metric that "Supply chain security incident information or security threat information shall be communicated to cloud service customers in a timely manner" are supply chain security incident reports or threat reports, etc.

The key points of the evaluation implementation include: reviewing the service level agreement, service contract and other documents to check whether it contains the clause of "cloud service providers shall promptly communicate supply chain security incident information or threat information to cloud service customers" or similar clauses; checking the definition and scope of "supply chain security incident information or threat information", as well as the notification method, timeliness, etc.; checking the supply chain security incidents or threat reports and other documents, and checking the history records of the customer notifications.

If the test results show that the cloud service provider can convey the supply chain security incident information or threat information to cloud service customers in a timely manner, the result is conforming; otherwise it is non-conforming or partially conforming.

[L3-CMS2-09/L4-CMS2-09 Interpretation and Description]

The main evaluation targets of the evaluation metric that "It shall be ensured that important changes of the supplier are communicated to cloud service customers in a timely manner, and security risks caused by changes shall be evaluated, and measures shall be taken to control these risks" are the supplier's important change records, security risk assessment reports and risks emergency plans and so on.

The key points of the evaluation implementation include: important changes of suppliers generally include the migration of physical data center facilities, network bandwidth reduction or expansion, hardware reduction or upgrade, software version upgrade, vulnerability patch repair, etc. Reviewing relevant documents such as the service level agreement, service contract, important change record, risk assessment report, risk emergency plan, etc., and checking whether the contract contains relevant provisions such as timely notification of important changes, assessment of security risks caused by changes, and measures to control risks; whether the content of notification form and invalidation is stipulated; checking the risk assessment history record and emergency response plan, and checking whether the risk control measures are reasonable and effective.

If the test results show that the cloud service provider has communicated important changes of the supplier to the cloud service customer in a timely manner, and carried out a risk assessment for each important change of the supplier, and took measures to control the risks, the result is conforming; otherwise it is non-conforming or partially conforming.

4.2.10 Security Operation and Maintenance Management

Cloud Computing Environment Management

[Standard Requirements]

This control point for Level III includes the evaluation unit of L3-MMS2-01, and Level IV includes that of L4-MMS2-01

[L3-MMS2-01/L4-MMS2-01 Interpretation and Description]

The main evaluation targets of the evaluation metric that "The operation and maintenance location of the cloud computing platform shall be located within the territory of China, and the operation and maintenance operation of cloud computing platform from abroad shall follow the relevant national regulations" are operation and maintenance devices, operation and maintenance locations, operation and maintenance records and related management documents, etc.

Chapter 4 Application and Interpretation of the Extended Security Evaluation Requirements of Cloud Computing

The key points for the implementation of the evaluation include: "within the territory of China" refers specifically to Mainland China, excluding Hong Kong, Macao, and Taiwan. The operation and maintenance of the domestic cloud computing platform from overseas shall comply with the requirements of relevant regulations of China, such as *the Cybersecurity Law of the People's Republic of China*, *"Measures for the Security Evaluation of Cross-Border Transfer of Personal Information and Important Data"*, *"Information Security Technology-Guidelines for Cross-Border Data Transfer Security Assessment"*, etc.

Take the cloud computing platform (the service model is PaaS) as an example: interviewing with the operation and maintenance administrator; checking whether the operation and maintenance locations of all cloud computing platforms are located within the territory of China; checking whether the operation and maintenance operations from overseas are carried out and implemented in accordance with relevant national regulations.

If the test results show that all the operation and maintenance locations of the evaluated cloud computing platforms are located with in the territory of China, or that the remote operation and maintenance of the domestic cloud computing platforms from abroad are implemented in accordance with relevant national regulations, then the result is conforming; otherwise it is non-conforming.

Chapter 5 Application and Interpretation of the Extended Security Evaluation Requirements of Mobile Internet

5.1 Basic Concepts

5.1.1 Mobile Interconnection

The process of using wireless communication techniques to connect mobile terminals to the wired network. The mobile internet part of the information system that adopts mobile internet technology is composed of three parts: mobile terminals, mobile applications and wireless networks. The mobile terminal connects to the wireless access device through the wireless channel, and the wireless access gateway restricts the mobile terminal's access behavior through the access control strategy. The mobile terminal management system in the backend is responsible for the management of mobile terminals, including sending mobile device management, mobile application management, and mobile content management strategies to the Client software.

5.1.2 Mobile Terminals

Mobile terminals refer to terminal devices used in mobile services, including general-purpose terminals and special-purpose terminal devices, such as smart phones, tablets, and personal computers.

5.1.3 Wireless Access Gateway

The device that is deployed between the wireless network and the wired network to protect the wired network.

5.1.4 Mobile Application Software

Mobile application software refers to applications running on smartphones, tablets, and other mobile devices. It can be divided into Native App running locally (operating system) on the mobile device and Web App running in the browser of high-end terminals. Mobile application software in the broad sense includes personal and enterprise-level applications; in the narrow sense, it refers to enterprise-level business application software, which can be divided into messaging applications, field applications, management applications, and autonomous applications.

5.1.5 Mobile Terminal Management System

Products that provide the unified management and security access control to mobile terminal devices and mobile application software through customized security strategies in order to enhance the security and controllability of mobile terminals.

5.2 Application Interpretation of Extended Security Evaluation Requirements of Mobile Internet at Level Ⅲ and Level Ⅳ

5.2.1 Security Physical Environment

Physical Location of the Wireless Access Point

[Standard Requirements]

Level Ⅲ of the control point includes the evaluation unit L1-PES3-01, and Level Ⅳ includes the evaluation unit L2-PES3-01.

[L3-PES3-01/ L4-PES3-01 Interpretation and Description]

The targets of the evaluation index "the installation location of wireless access equipment should be selected reasonably to avoid over-coverage and electromagnetic interference" are wireless router and wireless AP.

The key points of evaluation implementation include: interview the security administrator or network administrator to ask the coverage requirements of wireless access equipment; check the deployment location of wireless access equipment such as wireless router and wireless AP, use wireless signal detection tool to detect the coverage of wireless signal, and check whether there is excessive coverage of wireless signal. If the wireless signal cannot be searched outside the range that the wireless service needs to cover, or the wireless strength is very weak (for example, less than -80 dBm), it is difficult to be used by malicious personnel; check whether there is electromagnetic energy interference caused by electromechanical or other artificial devices around the wireless access equipment, and test and verify whether the wireless signal can avoid electromagnetic interference. If there is no other wireless signal on the channel used by the service network, or the detection result of the wireless signal detection tool indicates whether the signal strength of the SSID is less than -80 dBm, the packet loss rate is detected through the ping command.

If the test results show that the installation location of the wireless access equipment is selected reasonably, and there is no wireless signal over-coverage and electromagnetic interference problems during the test, the result of unit evaluation is conforming; otherwise it is non-conforming or partially conforming.

5.2.2 Security Communication Network

Taking into account the characteristics of the mobile interconnection system, this chapter lists the characterized interpretation of some general security requirement evaluation in the mobile interconnection environment.

Network Architecture

[Standard Requirements]

Level Ⅲ of the control point includes evaluation units L3-CNS1-01, L3-CNS1-02, L3-CNS1-03, and Level Ⅳ includes evaluation units L4-CNS1-01, L4-CNS1-02 and L4-CNS1-03.

[L3-CNS1-01/ L4-CNS1-01 Interpretation and Description]

The targets of the evaluation index "should ensure that the service processing capacity of the network equipment meets the requirements of the service peak" are the wireless access gateway, wireless router, wireless AP, etc.

The key points of evaluation implementation include: know the peak service period and check whether the service processing capability of the wireless access equipment meets the requirements during this period; if cannot know the peak service period and the equipment never suffered shutdown, use tool test to verify whether the equipment meets the requirements of the business peak period, such as performing stress test on the system.

If the results show that the peak utilization rate of the CPU and the memory of the radio access gateway is not more than 70%, and the network management platform does not have an alarm log of equipment shutdown or abnormal restart, the result of unit evaluation is conforming; otherwise it is non-conforming or partially conforming.

[L3-CNS1-02/ L4-CNS1-02 Interpretation and Description]

The targets of the evaluation index "ensure that the bandwidth of each part of the network meets the needs of the business peak" are the wireless access gateway, wireless router, wireless AP, etc.

The key points of evaluation implementation include: understand the time period of business peak, and specify the actual bandwidth demand of each part of the network during this period; ensure that the performance of equipment of each important network switching node and boundary meets the requirements of business peak period; the bandwidth occupied by each part of the network should not exceed 70% of the designed bandwidth of switching, access and security devices of each layer of the network during peak operation of the system, ensure that the wireless access network has certain bandwidth redundancy to guarantee the normal operation of the service system in case of emergency; if conditions permit, the wireless access equipment can pass the verification test (bandwidth pressure test) mode, and the upper limit of the test flow pressure should be less than or equal to 70% of the designed bandwidth.

If the results show that the bandwidth of each part of the network meet the requirements of the service peak, the result of unit evaluation is conforming; otherwise it is non-conforming or partially conforming.

[L3-CNS1-03/ L4-CNS1-03 Interpretation and Description]

The targets of the evaluation index "different network areas should be divided and addresses should be allocated for each network area according to the principle of convenient management and control" are wireless access gateway, wireless AP, wireless router, etc.

The key points of evaluation implementation include: interview with network administrators to confirm whether different network areas are divided on the main network equipment according to the organization's IP address planning principles and security area protection requirements, including physical network division and logical division, for example, all wireless networks should be independently networked; check whether the division of wireless network area (or subnet) is reasonable to meet the security requirements such as convenient control and broadcast area reduction, and focus on whether "visitors" and "internal employees" are divided into different network areas for management; check the configuration information of relevant network equipment to verify whether the divided network areas are consistent with the division principles.

If the results show that the unit undergoing test divides different network areas according to the importance, department and other factors, and verify that the divided network areas are consistent with the division principle by check the configuration information of the relevant network equipment, the result of unit evaluation is conforming; otherwise it is non-conforming or partially conforming.

5.2.3 Security Area Boundary

Taking into account the characteristics of the mobile interconnection system, this chapter lists the characterized interpretation of some general security requirement evaluation in the mobile interconnection environment.

1. Boundary Protection

[Standard Requirements]

Level Ⅲ of the control point includes evaluation units L3-ABS1-01, L3-ABS1-02, L3-ABS1-04, L3-ABS3-01, and Level Ⅳ includes evaluation units L4-ABS1-01, L4-ABS1-02, L4-ABS1-04, L4-ABS1-05, L4-ABS1-06, L4-ABS3-01.

[L3-ABS1-01/ L4-ABS1-01 Interpretation and Description]

The targets of the evaluation index "should ensure that the cross-boundary access and data flow communicate through the controlled interface provided by the boundary equipment" are the wireless access gateway, wireless AP, wireless router, etc.

The key points of evaluation implementation include: interviewing the security administrator or network administrator to inquire about the wireless network networking mode and whether the boundary access control device is deployed between the wired network and the wireless network; checking the configuration strategy of the boundary access control device between the wired network and the wireless network to ensure that the network communication across the wireless network boundary is controlled, and the granularity reaches the service port or communication interface level; checking whether the wireless network SSID issued by the configuration of the wireless access gateway, wireless AP and wireless router is the designated SSID, and check the security policy configuration of the network communication across the wireless network boundary; using other technical means (such as illegal wireless network device location, check the device configuration information, etc.) to check or test to verify whether there is no other unspecified SSID and uncontrolled network communication across the wireless network boundary.

If the results show that there is no unassigned SSID signal and the device or relevant component of the boundary access control function is configured with perfect policy, the result of unit evaluation is conforming; otherwise it is non-conforming or partially conforming.

[L3-ABS1-02/ L4-ABS1-02 Interpretation and Description]

The targets of the evaluation index "should be able to check or restrict the behavior of unauthorized devices connecting to the internal network" are the wireless access gateway, wireless AP and wireless router.

The key points of the evaluation include: simulating the unauthorized mobile device accessing the wireless network for test verification, and verifying the effectiveness of the wireless network identity authentication method; verifying whether the access authentication method of the guest mode is valid If the wireless network starts the guest mode; replaying and bypassing by modifying the URL parameter or the login session parameter (such as cookie value) to check whether the guest account can be granted as the service account. It is necessary to analyze and determine the boundary of the wireless network, and determine whether the wireless access gateway, wireless AP and wireless router restrict and authorize the device access behavior, for example, enable the access authentication or adopt the white list setting mode (such as terminal MAC address binding).

If the results show that the wireless network undergoing test can perform corresponding access check or restriction on the mobile terminal, wireless AP and other equipment connected to the internal network, the result of unit evaluation is conforming; otherwise it is non-conforming or partially conforming.

[L3-ABS1-04/ L4-ABS1-04 Interpretation and Description]

The targets of the evaluation index "the use of wireless network should be restricted to ensure that the wireless network accesses the internal network through the controlled boundary device" are the wireless access gateway, wireless AP, wireless router, etc.

The key points of evaluation implementation include: check whether the existing security management system has management regulations for wireless network access, including but not limited to equipment type and general requirements for wireless policy configuration; determine the wireless network networking mode, such as wireless access gateway + wireless AP, independent deployment of wireless AP, wireless router, etc. ; check whether the wireless network SSID issued by the wireless access gateway, wireless AP and wireless router are all designated SSID, and check the security policy configuration of the network communication across the wireless network boundary; check or test to verify whether there is no other unspecified SSID and uncontrolled network communication across the wireless network boundary by other technical means (such as illegal wireless network equipment location, equipment configuration information verification, etc.).

If the results show that there is no unassigned SSID signal and the device or relevant component of the boundary access control function is configured with perfect policy, the result of unit evaluation is conforming; otherwise it is non-conforming or partially conforming.

[L4-ABS1-05 Interpretation and Description]

The evaluation targets of evaluation index "should be able to effectively block unauthorized equipment connecting to the internal network or internal users connecting to the external network without authorization" are the wireless access gateway, wireless AP and wireless router.

The key points of evaluation implementation include: interview the security administrator or network administrator, understand the identity authentication mode when the terminal device is accessed, and know whether the wireless network guest access function is enabled; it is necessary to analyze and determine the boundary of the wireless network, and determine whether the wireless access gateway, wireless AP and wireless router restrict and authorize the access behavior of the device, for example, enable the access authentication (portal authentication, WeChat, etc.) or adopt the white list setting mode (such as terminal MAC address binding), and check whether the wireless network access device has the access alarm function for illegal mobile terminal equipment, such as alarm log, SNMP Trap, SMS, email, etc. ; check whether the wireless network access device has the forced offline and blacklist functions for the mobile terminal equipment, whether the guest account access validity period is set for the guest access mode, and whether the interview network administrator or security administrator timely clears (deletes or disables) the redundant and expired internal accounts regularly; test and verify the effectiveness of wireless network identity authentication mode, forced offline, blacklist and other functions by simulating unauthorized mobile devices accessing the wireless network. If the guest mode is enabled by the wireless network, verify whether the access authentication method and the guest account access validity period of the guest mode are valid.

If the results show that the wireless network undergoing test adopts technical means to check or restrict the access of the mobile terminal, wireless AP and other equipment connected to the internal network, and at the same time, the illegal mobile terminal can be effectively blocked from accessing the wireless network through forced offline or blacklist, the result of unit evaluation is conforming; otherwise it is non-conforming or partially conforming.

[L4-ABS1-06 Interpretation and Description]

The targets of the evaluation index "trusted verification mechanism should be adopted to authenticate the equipment connected to the network to ensure the authenticity and credibility of the equipment connected to the network" are the wireless access gateway, wireless AP, wireless router, mobile terminal equipment, etc.

The key points of evaluation implementation include: check the instructions of equipment such as the wireless access gateway, wireless AP, wireless router and mobile terminal or the approval of the national cryptography management department of the trusted chip to check whether the trusted chip is preset or using other trusted verification methods; issue the ID certificate of the test platform to the trusted chip to check whether the chip responds correctly; and check whether the security management center has reliable verification audit records.

If the results show that the trust root exists and the response is normal, the result of unit evaluation is conforming; otherwise it is non-conforming or partially conforming.

[L3-ABS3-01/ L4-ABS3-01 Interpretation and Description]

The targets of the evaluation index "the access between the wired network and the wireless network boundary and the data flow through the wireless access gateway equipment should be guaranteed" are the wireless access gateway equipment, etc.

The key points of evaluation implementation include: check whether the wireless network is divided into independent network areas (Vlan or subnets) andis allocated independent network segments; check the actual network topology to confirm whether access control is carried out between the wireless network and the wired network through the boundary gateway device, and whether there are multiple wireless network access points in the whole service network, and whether all wireless access points access the wired network through the wireless access gateway equipment.

If the results show that the wireless network is independently organized or divided into independent network areas, and allocated with independent network segments, and the wireless network accesses the wired network through the wireless access gateway device, the result of unit evaluation is conforming; otherwise it is non-conforming or partially conforming.

2. Access Control

[Standard Requirements]

Level Ⅲ of the control point includes the evaluation unit L3-ABS3-02, and Level Ⅳ includes the evaluation unit L4-ABS3-02.

[L3-ABS3-02/ L4-ABS3-02 Interpretation and Description]

The evaluation targets of the evaluation index "the wireless access equipment should enable the access authentication function and support authentication by using authentication server authentication or the cryptography module approved by the national cryptography management organization" are the wireless router and wireless AP, etc.

The key points of the evaluation implementation include: interview the security administrator or network administrator to understand the authentication mode of the wireless access device; check whether the wireless access device is enabled with the authentication function; the form can be common wireless cryptography authentication (WPA/ WPA2), which is authenticated by the authentication server (portal, radius, ldap, etc.), other third-party authentication (Wechat authentication, etc.) or the cryptography module approved by the national cryptography administration; check whether the authentication protocol with potential security hazards is used for authentication, for example, whether the wireless network authentication adopts the WEP encryption mode, which is very easy to be cracked; and check whether the authentication information of the relevant authentication method has weak cryptography (consisting of a single combination of pure letters/ pure numbers/ pure characters, which is less than 8 digits).

If the results show that the access wireless AP need to carry out identity authentication, and the identity authentication means itself does not have a security hazard, and the identity authentication method does not have a weak cryptography, the unit determines that the result is conforming; otherwise it is non-conforming or partially conforming.

3. Intrusion Prevention

[Standard Requirements]

Level Ⅲ of the control point includes evaluation units L3-ABS3-03, L3-ABS3-04, L3-ABS3-05, L3-ABS3-06, L3-ABS3-07, L3-ABS3-08, and Level Ⅳ includes evaluation units L4-ABS3-03, L4-ABS3-04, L4-ABS3-05, L4-ABS3-06, L4-ABS3-07, L4-ABS3-08.

[L3-ABS3-03/ L4-ABS3-03 Interpretation and Description]

The targets of the evaluation index "should be able to detect the access behavior of the unauthorized wireless access device and the unauthorized mobile terminal" are the wireless access gateway, wireless AP, wireless router, etc.

The key points of evaluation implementation include: interview the security administrator or network administrator to understand the technical measures for preventing the unauthorized wireless access device and unauthorized mobile terminal from accessing;

check whether the wireless access device has the terminal access control function (wireless EAD); and check the detection log of the access of the unauthorized wireless access device and the mobile terminal.

If the results show that the wireless access equipment and the mobile terminal perform access management through the corresponding equipment, and can prevent the access of the unauthorized wireless access equipment and the unauthorized mobile terminal, the unit determines that the result is conforming; otherwise it is non-conforming or partially conforming.

[L3-ABS3-04/ L4-ABS3-04 Interpretation and Description]

The targets of the evaluation index of "should be able to detect such acts as network scanning, DDoS attack, key cracking, man-in-the-middle attack and deception attack against the wireless access device" are the wireless access gateway, wireless intrusion detection system, etc.

The key points of the evaluation implementation include: check whether the wireless access device has the wireless intrusion detection and protection function, if so, check whether the configuration strategy is reasonable, check the detection log to determine whether it can detect or block network attack behaviors such as network scanning, DDoS attack, key cracking, man-in-the-middle attack and deception attack; check whether other devices or components with intrusion detection or blocking function are deployed to detect or block the wireless network attack behavior; and check whether the version of the rule base is updated in a timely manner.

If the results show that the wireless network attack behavior can be detected or blocked through the wireless access gateway or other equipment, and the rule base version is updated in time, the evaluation determine that the result is conforming; otherwise it is non-conforming or partially conforming.

[L3-ABS3-05/ L4-ABS3-05 Interpretation and Description]

The targets of the evaluation index "should be able to detect the startup status of high-risk functions such as SSID broadcast and WPS of the wireless access equipment" are mainly the wireless AP and wireless router.

The key points of the evaluation implementation include: check whether the radio access gateway has the SSID broadcast and WPS function status management functions of the wireless access equipment, and whether the openness of corresponding function can be detected.

If the results show that the SSID broadcast and WPS functions of the radio access equipment can be uniformly managed through the radio access gateway, and their status can be detected, the evaluation determine that the result is conforming; otherwise it is non-conforming or partially conforming.

[L3-ABS3-06/ L4-ABS3-06 Interpretation and Description]

The targets of the evaluation index "the functions of wireless access equipment and wireless access gateway with risks should be disabled, such as SSID broadcast, WEP authentication, etc. ", are the wireless access gateway and wireless router.

The key points of evaluation implementation include: check whether the SSID broadcast of the wireless access equipment and the wireless access gateway is disabled, and whether the WEP authentication function is disabled.

Check whether the radio access gateway has the SSID broadcast and WEP authentication function status management functions of the wireless access device; check the configurationof wireless access device SSID broadcast and WEP authentication functions closing.

If the results show that the wireless network device doe not enable the SSID broadcast and the WEP authentication function, the result is conforming; otherwise it is non-conforming or partially conforming.

[L3-ABS3-07/ L4-ABS3-07 Interpretation and Description]

The targets of evaluation index "multiple AP should be prohibited from using the same authentication key" are mainly the wireless access gateway and wireless router.

The key points of evaluation implementation include: check whether wireless access devices such as wireless router and wireless AP are allocated with different authentication keys. Log in the wireless access gateway to check the authentication mode selected by the wireless access gateway when accessing the wireless AP. The authentication methods include but are not limited to MAC authentication and SN authentication. Check whether different wireless APs use the same authentication key.

If the results show that the authentication keys of all the wireless AP accessing the radio access gateway are not the same, the result is conforming; otherwise it is non-conforming or partially conforming.

[L3-ABS3-08/ L4-ABS3-08 Interpretation and Description]

The targets of the evaluation index "should be able to block unauthorized wireless access devices or non-authorized mobile terminals" are the wireless access gateway, wireless AP, wireless router, terminal access control system, mobile terminal management system or relevant components.

The key points of the evaluation implementation include: check the configuration file of the wireless access gateway to confirm whether it has configured the access authentication policy for the wireless access; check the configuration policy of the terminal management system to confirm whether it is configured for the wireless access equipment or the mobile terminal; check the configuration file of the wireless access gateway to confirm whether it enables the guest mode and whether the controllable identity authentication mode is enabled in the guest mode; check the configuration file of the wireless access gateway to confirm whether the guest mode has an access control policy corresponding to the authority level.

If the results show that the system or equipment undergoing test can block and control the unauthorized radio access behavior, the evaluation result is conforming; otherwise it is non-conforming or partially conforming.

5.2.4 Security Computing Environment

According to the characteristics of the mobile interconnection system, this chapter lists the characterized interpretation of several General security requirement evaluation in the context of mobile interconnection.

1. Identification

[Standard Requirements]

Level Ⅲ of the control point includes evaluation units L3-CES1-01, L3-CES1-02, L3-CES1-03, L3-CES1-04, and Level Ⅳ includes evaluation units L4-CES1-01, L4-CES1-02, L4-CES1-03 and L4-CES1-04.

[L3-CES1-01/ L4-CES1-01 Interpretation and Description]

The targets of the evaluation index "should identify and authenticate the logged-in users, the identity identification should be unique, and the identity authentication information should be subject to the complexity requirements and be replaced regularly" are wireless access equipment (wireless AP, wireless access gateway), mobile terminals (general terminals and special terminals), mobile terminal management system, mobile application software, etc.

The key points of evaluation implementation include: check whether the user has identity authentication measures when logging in the mobile interconnection access device, mobile terminal and mobile terminal management system, and especially check whether the user has the access and operation authority of the mobile interconnection access device, mobile terminal and mobile terminal management system in the non-login state; check whether the user ID is unique, especially whether the account with the same name or the same ID can be established, whether the background database of the wireless AP has the same SSID number, and whether the background database of the mobile terminal management system has the same terminal number or terminal name; check and verify whether there are weak cryptography and null cryptography users in the mobile terminal management system; check whether the user identity authentication information of the mobile terminal management system, the mobile terminal, the wireless AP, the wireless gateway and the like has set the policy of complexity and replacement cycle, and verify the validity of the cryptography complexity by check the function modules such as the cryptography modification; check whether the wireless AP can be cracked by software such as master key.

If the results show that the system undergoing test has the identity authentication function module, the login authentication module cannot be bypassed, the null cryptography can not be used to log in after the factory setting is restored, the user's identification in the

Chapter 5 Application and Interpretation of the Extended Security Evaluation Requirements of Mobile Internet

system is unique, the cryptography complexity meets the requirement and is changed regularly, and there is no weak cryptography, null cryptography and default cryptography, the result is conforming; otherwise it is non-conforming or partially conforming.

【L3-CES1-02/ L4-CES1-02 Interpretation and Description】

The targets of the evaluation index "should have the login failure processing function, should be configured with and enabled such measures as session ending, restriction on the numbers of illegal login, and automatically exit when the login connection is overtime" are wireless access equipment (wireless AP, wireless access gateway), mobile terminal (general terminal, special terminal), mobile terminal management system, mobile application software, etc.

The key points of evaluation implementation include: check whether the mobile access device (wireless AP and wireless gateway), mobile terminal and mobile terminal management system have the login failure processing function and the login failure handling strategy. Focus on trying to log in with the same account through different mobile terminals with wrong cryptographys to check whether the number of login failures is superimposed on different mobile terminals; continuously log in to the access device (wireless AP, wireless gateway), mobile terminal and mobile terminal management system with correct user name and wrong cryptography, and check the response of the system, i. e., take specific actions after the illegal login reaches a certain number of times, such as locking the account; check the waiting time for automatic disconnection when connection timeout occurs when the system performs identity authentication during user login. After connection timeout, check the system status, such as whether it is in the automatic exit state after session disconnection, and test whether the user can log in successfully when the waiting time is not over, and terminate the locking time in advance.

If the results show that the system or equipment undergoing test has locking or log-out function, the automatic connection exits after the connection time-out, and the timeout time is reasonable, the result is conforming; otherwise it is non-conforming or partially conforming.

【L3-CES1-03/ L4-CES1-03 Interpretation and Description】

The targets of the evaluation index "when conducting remote management, necessary measures should be taken to prevent authentication information from being eavesdropped during network transmission" are wireless access equipment (wireless AP, wireless access gateway), mobile terminal management system, mobile application software, etc.

The key points of evaluation implementation include: verify whether the system is remotely managed by security measures such as encryption, so as to prevent the authentication information from being eavesdropped in the process of network transmission. For the accessing device (wireless AP, wireless gateway, wireless router) and the mobile terminal management system, check whether the authentication information is prevented from being eavesdropped during transmission through encrypted transmission such as HTTPS or SSH, and send the authentication information to the mobile terminal management system through the mobile terminal, and analyze the encryption measures.

If the results show that the system or equipment undergoing test adopts the encryption mechanism such as HTTPS and SSH for data transmission, and the packet capture is effective encryption, the evaluation result is conforming; otherwise it is non-conforming or partially conforming.

[L3-CES1-04/ L4-CES1-04 Interpretation and Description]

The targets of the evaluation index "two or more identification technologies such as cryptography, cryptography technology and biotechnology should be adopted to authenticate the user's identity, and at least one of the authentication technologies should be implemented by cryptography technology" are the wireless access equipment (wireless AP, wireless access gateway), mobile terminals (general terminals and special terminals), mobile terminal management system, mobile application software, etc.

The key points of evaluation implementation include: check whether the wireless AP, mobile terminal and mobile terminal management system adopt two or more combination authentication technologies such as dynamic cryptography, digital certificate and biotechnology to authenticate the identity of the management user, and the two authentication mechanisms must be heterogeneous, such as "cryptography + fingerprint"; check whether at least one authentication technology of the wireless AP, mobile terminal and mobile terminal management system is realized by cryptography technology.

If the test results show that the administrator of the system or equipment undergoing test and the newly-built test account have two or more combination of identity authentication modes, and the other authentication mechanism is realized by using cryptography technology, the evaluation result is conforming; otherwise it is non-conforming or partially conforming.

2. Access Control

[Standard Requirements]

Level Ⅲ of the control point includes evaluation units L3-CES1-05, L3-CES1-06, L3-CES1-07, L3-CES1-08, L3-CES1-09, L3-CES1-10, L3-CES1-11, and Level Ⅳ includes evaluation units L4-CES1-05, L4-CES1-06, L4-CES1-07, L4-CES1-08, L4-CES1-09, L4-CES1-10, L4-CES1-11.

[L3-CES1-05/ L4-CES1-05 Interpretation and Description]

The targets of the evaluation index "should allocate account and authority to logged-in users" are wireless access equipment (wireless AP, wireless access gateway), mobile terminals (general terminals and special terminals), mobile terminal management system, mobile application software, etc.

The key points of evaluation implementation include: check whether the access device type, mobile terminal and mobile terminal management system have the user account and authority configuration function, and whether the new user can be allocated with account and authority; log in the system with users of different roles and check whether the user's authority is consistent with the allocated one; try to use the login user to access unauthorized functions and check whether the access control policy is effective; check whether the access authority of anonymous and default accounts has been disabled or restricted.

If the results show that the system or equipment has allocated corresponding account and authority for each logged-in user, the authority is valid, and the operation between different users cannot exceed the authority, and the access authority of anonymous and default accounts has been disabled or restricted, the evaluation result is conforming; otherwise it is non-conforming or partially conforming.

[L3-CES1-06/ L4-CES1-06 Interpretation and Description]

The targets of the evaluation index "rename or delete the default account and modify the default cryptography of the default account" are the wireless access device class (wireless AP, wireless access gateway), mobile terminal (general terminal, special terminal), mobile terminal management system, mobile application software, etc.

The key points of the evaluation include: check whether the default account has been renamed or deleted; if there is a default account, check whether the default cryptography of the default account has been modified; if the device has the RESET function, try to reset the device and check whether there is a default account.

If the results show that the system undergoing test does not have a default account name or a default cryptography, or the default account is renamed, and the default account name cannot be restored after reset, the result is conforming; otherwise it is non-conforming or partially conforming.

[L3-CES1-07/ L4-CES1-07 Interpretation and Description]

The targets of the evaluation index "redundant and expired accounts should be deleted or disabled in time to avoid the existence of shared accounts" are wireless access equipment (wireless AP, wireless access gateway), mobile terminals (general terminals, special terminals), mobile terminal management system, mobile application software, etc.

The key points of evaluation implementation include: check whether there are redundant and expired accounts in the access equipment, mobile terminal and mobile terminal management system; check the user status identifiers in the user table of the access equipment, mobile terminal and mobile terminal management system database, and try to log in if there are redundant and expired accounts in the system; check whether there is one-to-one correspondence between the administrator user and the account, check whether there is a shared account, especially check whether the management account of the previous manager is used by the successor due to factors such as resignation or post adjustment; check whether the access equipment (such as wireless AP) has temporary guest account access, and check whether the expired or redundant guest account exists and whether it is valid.

If the results show that there are no redundant and expired management account and temporary guest account in the system under test, and there is one-to-one correspondence between the administrator user and the account, and there is no shared account, the result is conforming; otherwise it is non-conforming or partially conforming.

[L3-CES1-08/ L4-CES1-08 Interpretation and Description]

The targets of the evaluation index "the minimum authority required by the management user should be granted to realize the authority separation of the management user" are the wireless access equipment (wireless AP, wireless access gateway), mobile terminal (general terminal, special terminal), mobile terminal management system, mobile application software, etc.

The key points of evaluation implementation include: login with administrator account, check whether the background of the system has role setting function, or which roles are set by default, and whether there are multiple roles corresponding to one administrator account; check whether the authority of the management user has been separated and whether there is authority restriction among management accounts; log in the management account with different authorities and verify the authority restriction; check whether the authority of the access device, mobile terminal management system and mobile terminal business user is the minimum authority required for their work tasks. Select users of different roles, check the corresponding duties of the users, and log in the system to check whether the actual authority assignment of the users is consistent with the work responsibilities.

If the results show that the system undergoing test has the function of role division and assignment, and multiple roles cannot be assigned to one user, various management users (such as system administrator, security administrator and audit administrator) are provided with authority separation mechanism, and the authority of all kinds of business users is the minimum required for their work tasks, the result is conforming; otherwise it is non-conforming or partially conforming.

[L3-CES1-09/ L4-CES1-09 Interpretation and Description]

The targets of the evaluation index "the access control policy should be configured by the authorized subject, and the access control policy should specify the access rules of the subject to the object" are wireless access equipment (wireless AP, wireless access gateway), mobile terminal (general terminal, special terminal), mobile terminal management system, mobile application software, etc.

The key points of evaluation implementation include: check whether the access control policy is configured by the authorized subject to manage the user login viewing authority management function and access control policy, and log in the non-authorized entity to check whether it has the assignment function of the access control policy; check whether the authorization subject has configured the access rules of the subject to the object according to the security policy, whether the access rules are reasonable, and check whether the new

Chapter 5 Application and Interpretation of the Extended Security Evaluation Requirements of Mobile Internet

account has the default access permission; check whether the user has unauthorized access, and attempt unauthorized access by modifying wireless SSID parameters and URL parameters.

If the test results show that the access control policy is configured by the authorized subject, the authorized entity grants the corresponding access permission to the user according to the reasonable security policy, the new account does not have the default access permission, and the unauthorized access cannot be realized by modifying the SSID or URL parameters, the result is conforming; otherwise it is non-conforming or partially conforming.

[L3-CES1-10/ L4-CES1-10 Interpretation and Description]

The targets of the evaluation index "the granularity of access control should reach user at level or process level, object at file or database table level" are wireless access equipment (wireless AP and wireless access gateway), mobile terminal (general terminal and special terminal), mobile terminal management system, mobile application software, etc.

The key points of evaluation implementation include: check accessing device, mobile terminal and mobile terminal management system, and test whether the control granularity of access control policy reaches that user at level or process level, object at file or database table level.

Take the mobile terminal management system as an example: log in accounts with different permissions and try to access unauthorized files and databases, and the access objects of accounts with different permissions are different, for example, the login system administrator should be able to view the user's authority assignment (files stored in the background of authority assignment), and the login audit administrator should be able to view log information (stored in log files in the background); check whether the system administrator can view the log information, and whether the audit administrator can view the permission assignment.

If the results show that the account subjects with different permissions in the system undergoing test can only access the files, database tables and other objects corresponding to the permissions, the result is conforming; otherwise it is non-conforming or partially conforming.

[L3-CES1-11 Interpretation and Explanation]

The targets of the evaluation index of "set security marks for important subjects and objects, and control the access of subjects to information resources with security marks" include wireless access devices (wireless AP, wireless access gateway), mobile terminals (general terminals, special terminals), mobile terminal management systems, mobile application software, etc.

The key points of evaluation implementation include: check whether the access device, mobile terminal and mobile terminal management system set security marks for important subjects and objects according to the security policy, for example, check whether important files and fields are set according to guest attributes in the database; establish new users with different security marks, and verify whether the subject-to-object mandatory access control function is realized according to the security marks after login, that is, whether the subjects (users) with different security marks can only access the objects (files, database tables, fields, etc.) with corresponding security marks.

If the results show that the system undergoing test has the security marks corresponding to the security policy, and the mandatory access control is realized according to the security marks, the result is conforming; otherwise it is non-conforming or partially conforming.

[L4-CES1-11 Interpretation and Description]

The targets of the evaluation index of "set security marks for subjects and objects, and determine the subject-to-object access according to security marks and mandatory access control rules" are wireless access devices (wireless AP, wireless access gateway), mobile terminals (general terminals, special terminals), mobile terminal management systems, mobile application software, etc.

The key points of evaluation implementation include: check whether all subjects and objects in the access device, mobile terminal and mobile terminal management system have set security marks for important subjects and objects according to the security policy, for example, check whether the file and field have set the corresponding object security mark according to the visitor attribute in the database, and whether the corresponding user has set the subject security mark; check whether the mobile terminal management system has set the access control rules between the subject and the object, and whether the mapping relationship between the subject and the object security mark is valid; create a user with different security marks (covering all types of users with security marks), verify whether the subject-to-object mandatory access control function is realized according to the security marks after login, that is, whether the subject (user) with different security marks can only access the objects (files, database tables, fields, etc.) with corresponding security marks.

If the results show that the system undergoing test has the security mark corresponding to the security policy, and the mandatory access control is realized according to the security mark, the result is conforming; otherwise it is non-conforming or partially conforming.

3. Security Audit

[Standard Requirements]

Level Ⅲ of the control point includes evaluation units L3-CES1-12, L3-CES1-13, L3-CES1-14, L3-CES1-15, and Level Ⅳ includes evaluation units L4-CES1-12, L4-CES1-13, L4-CES1-14 and L4-CES1-15.

[L3-CES1-12/ L4-CES1-12 Interpretation and Description]

The targets of the evaluation index "security audit function should be enabled, the audit should cover each user, and audit important user behaviors and important security events" are wireless access equipment (wireless AP and wireless access gateway), mobile terminal (general terminal and special terminal), mobile terminal management system, mobile application software, etc.

The key points of evaluation implementation include: check whether the mobile terminal management system and wireless AP management device have the security audit function and whether the security audit function is enabled; check the content of the audit log and check whether the user body in the log covers each user recorded in the system; check whether the security audit configuration or audit log contains important user behaviors and important security events, including but not limited to mobile terminal online/ offline, mobile terminal installation/operation application software, administrator addition/deletion/modification of mobile terminal information, and administrator addition/deletion/modification of software whitelist configuration.

If the test results show that the system or equipment undergoing test is enabled with security audit function, the audit scope covers all users, and the audit event includes important user behaviors and important security events, including but not limited to mobile terminal online/offline, administrator adding/deleting/modifying mobile terminal information, etc., the result is conforming; otherwise it is non-conforming or partially conforming.

[L3-CES1-13 Interpretation and Description]

The targets of the evaluation index "the audit record should include the date and time of the event, user, event type, success of the event and other audit-related information" are wireless access equipment (wireless AP, wireless access gateway), mobile terminal (general terminal, special terminal), mobile terminal management system, mobile application software, etc.

The key points of evaluation implementation: check whether the audit records of the mobile terminal management system include the date and time of the event, mobile terminal identifier (ID, IMEI or MAC, etc.), IP address, user name, event type (software installation, whitelist configuration modification, etc.), event content, execution result (success or failure), etc.; check whether the audit record of the wireless AP management equipment includes the online time, offline time, MAC address, IP address and terminal name of the terminal.

If the results show that the log records of the system or equipment undergoing test include the online time, offline time, MAC address, IP address and terminal name of the terminal, the result is conforming; otherwise it is non-conforming or partially conforming.

[L4-CES1-13 Interpretation and Description]

The targets of the evaluation index "the audit record should include the date and time of the event, event type, subject identification, object identification and result" are wireless access equipment (wireless AP, wireless access gateway), mobile terminal (general terminal, special terminal), mobile terminal management system, mobile application software, etc.

The key points of evaluation implementation: check whether the audit records of the mobile terminal management system include the date and time of the event, mobile terminal identifier (ID, IMEI or MAC, etc.), IP address, user name, event type (software installation, whitelist configuration modification, etc.), subject identification access control audit content, object identification access control audit content, execution result (success or failure), etc.; check whether the audit record of wireless AP management equipment includes the terminal online time, offline time, terminal MAC address, IP address, terminal name, etc.

If the results show that the log records of the system or equipment undergoing test include the online time, offline time, MAC address, IP address and terminal name of the terminal, the result is conforming; otherwise it is non-conforming or partially conforming.

【L3-CES1-14/ L4-CES1-14 Interpretation and Description】

The targets of evaluation index "audit records should be protected, backed up regularly to avoid unexpected deletion, modification or coverage" mainly include wireless access equipment (wireless AP, wireless access gateway), mobile terminal (general terminal, special terminal), mobile terminal management system, mobile application software, etc.

The key points of evaluation implementation include: check whether the log storage space of the mobile terminal management system and wireless AP management device meets the requirement of saving for more than 6 months; check whether the log forwarding function is available. If a special log server is configured, check whether the log forwarding configuration is valid (IP, port, running status, etc.); check whether the locally stored logs have a backup strategy, and adopting the mode of regular manual full-volume backup, regular automatic full-volume backup or multi-machine hot backup; the special log server requires that the stored files are backed up automatically or by multi-machine hot backup at regular intervals; the periodic backup cycle is determined according to the business volume, but should not exceed 7 days; and whether the audit function of the mobile terminal management system and the wireless AP management device or the special log audit platform allows the user to delete and modify the audit records.

If the test results show that the system or equipment undergoing test has the log export function, and the remote backup is saved manually or automatically, the periodic backup cycle is not more than 7 days, the storage time is more than 6 months, and the log is not allowed to be deleted and modified by the user, the result is conforming; otherwise it is non-conforming or partially conforming.

【L3-CES1-15/ L4-CES1-15 Interpretation and Description】

The targets of the evaluation index "the audit process should be protected against unauthorized interruption" are wireless access devices (wireless AP and wireless access

Chapter 5   Application and Interpretation of the Extended Security Evaluation Requirements of Mobile Internet

gateway), mobile terminals (general terminals and special terminals), mobile terminal management system, mobile application software, etc.

The key points of evaluation implementation include: check the instructions of the mobile terminal management system to know whether the audit process is independent or integrated with the management system process; check whether the security measures are taken to prevent the user from interrupting the audit process, such as binding the audit process with the system process and prohibiting the mobile terminal from using the super administrator to log in; generally, the wireless AP management device is a hardware device, and the audit process cannot be interrupted separately; if the wireless AP management device is installed on the server in the form of software, it is necessary to check the software instruction to determine whether the audit process is independent or integrated with the AP management software process, and check whether the security measures are adopted to prevent the user from interrupting the audit process; and check whether the server and client of the mobile terminal management system can terminate the audit process by terminating the service process, exiting the software, and unloading the software.

If the results show that the audit process of the system or equipment undergoing test is protected by security measures and can not be interrupted by the unauthorized user, the result is conforming; otherwise it is non-conforming or partially conforming.

4. Intrusion Prevention

[Standard Requirements]

Level III of the control point includes evaluation units L3-CES1-17, L3-CES1-18, L3-CES1-19, L3-CES1-20, L3-CES1-21, L3-CES1-22, and Level IV includes evaluation units L4-CES1-16, L4-CES1-17, L4-CES1-18, L4-CES1-19, L4-CES1-20, L4-CES1-21.

[L3-CES1-17/ L4-CES1-16 Interpretation and Description]

The targets of the evaluation index "minimum installation principle should be followed, and only the required components and applications should be installed" are wireless mobile terminals (general terminals and special terminals), mobile terminal management system and mobile application software, etc.

The key points of evaluation implementation include: interview with relevant administrators to provide list of applications and components and corresponding functions required for business and terminal management and control, and judge their necessity; check whether mobile terminal devices are installed with other non-essential applications and components against the list.

If the test results show that the system or equipment undergoing test has no non-essential application programs and components, the result is conforming; otherwise it is non-conforming or partially conforming.

[L3-CES1-18/ L4-CES1-17 Interpretation and Description]

The targets of the evaluation index "Unneeded system services, default sharing and high-risk ports should be closed" are wireless mobile terminals (general terminals and special terminals), mobile terminal management system, mobile application software, etc.

The key points of evaluation implementation include: check the system service list of the mobile intelligent terminal and the information of shared resources, check whether the non-essential system services and default sharing are closed by the mobile terminal and other devices; check the port occupied by the process on the mobile intelligent terminal, and check whether there are no unnecessary high-risk ports in the mobile terminal and other devices, such as the tcp 5555 port of the remote adb.

If the results show that the system or equipment undergoing test does not have open default sharing, unnecessary system service or high-risk port, the result is conforming; otherwise it is non-conforming or partially conforming.

[L3-CES1-19/ L4-CES1-18 Interpretation and Description]

The targets of the evaluation index "the management terminals managed through the network should be limited by setting the terminal access mode or network address range" are wireless mobile terminals (general terminals and special terminals), mobile terminal management system, mobile application software, etc.

The key points of evaluation implementation include: check whether the security policy, configuration file and parameters on the mobile terminal management system limit the access range of the mobile terminal; try to access the service system without authorization and check whether it is successful.

If the results show that the system or equipment undergoing test is set with the terminal access control strategy and the policy is valid, the result is conforming; otherwise it is non-conforming or partially conforming.

[L3-CES1-20/ L4-CES1-19 Interpretation and Description]

The targets of the evaluation index "should provide the data validity test function to ensure that the contents input through the man-machine interface or through the communication interface conform to the system setting requirements" are the wireless mobile terminal management system and mobile application software.

The key points of evaluation implementation include: ask the application administrator whether the system mobile application software has the software fault tolerance capability, check whether the system design document includes the content or module of the data validity check function; input different data (data format or data length) into the mobile application software, and check whether the application software can check and verify the length and format of the input data.

If the test results show that the function of check and verify the length and format of input data provided by the system or software undergoing test is effective, the result is conforming; otherwise it is non-conforming or partially conforming.

Chapter 5　Application and Interpretation of the Extended Security Evaluation Requirements of Mobile Internet

[L3-CES1-21/ L4-CES1-20 Interpretation and Description]

The targets of the evaluation index "should be able to find possible known vulnerabilities and repair them in time after full test and evaluation" are wireless mobile terminals (general terminals and special terminals), mobile terminal management system, mobile application software, etc.

The key points of evaluation implementation include: check the current patch package and current version in the "System Update" of the mobile terminal, and whether there is any new update version; inquire the system administrator whether to regularly scan the vulnerability of the mobile intelligent terminal connected to the business system, and whether to evaluate and update the vulnerabilities found in the scanning; check whether there are no high-risk vulnerabilities in the mobile terminal and other devices through vulnerability scanning and penetration test; check whether the mobile intelligent terminal is repaired in time after full test and evaluation, and refer to the vulnerability scanning report and patch update record for the mobile intelligent terminal.

If the test results indicate that the system or equipment undergoing test has no known serious bug problem, the result is conforming; otherwise it is non-conforming or partially conforming.

[L3-CES1-22/ L4-CES1-21 Interpretation and Description]

The targets of the evaluation index "should be able to detect the intrusion of important nodes and provide alarm in case of serious intrusion" are wireless mobile terminals (general terminals and special terminals), mobile terminal management system, mobile application software, etc.

The key points of evaluation implementation include: inquire the system administrator whether the system provides intrusion detection measures for mobile terminal and other devices and the specific measures; check the alarm records when a serious intrusion event occurs to the mobile terminal and other devices.

If the results show that the system or equipment undergoing test provides the intrusion alarm measure and alarm record of the mobile intelligent terminal, the result is conforming; otherwise it is non-conforming or partially conforming.

5. Malicious Code Prevention

[Standard Requirements]

Level Ⅲ of the control point includes the evaluation unit L3-CES1-23, and Level Ⅳ includes the evaluation unit L4-CES1-22.

[L4-CES1-22 Interpretation and Description]

The targets of the evaluation index that "the active immune trusted verification mechanism should be adopted to timely identify the intrusion and virus behaviors and effectively block them" are the application program or relevant component that provides anti-malicious code function for the mobile terminal.

The key points of the evaluation implementation include: if the malicious code technology protection measures are adopted to prevent the virus intrusion of the mobile terminal, check whether the anti-malicious code product installed on the mobile terminal operates normally, and whether the malicious code database has been updated to the latest; if the active immune trusted verification mechanism is adopted to prevent the virus intrusion against the mobile terminal, check the instruction or approval document of the mobile intelligent terminal, check whether the trusted chip is preset or other trusted verification modes are adopted, and the intrusion and virus behavior can be actively immune.

If the results show that the mobile terminal adopt a trusted verification mechanism to effectively block the intrusion and virus behavior when the intrusion and virus behavior are found, the result is conforming; otherwise it is non-conforming or partially conforming.

[L3-CES1-23 Interpretation and Description]

The targets of the evaluation index that "technical measures against malicious code attack or active immune trusted verification mechanism should be adopted to timely identify intrusion and virus behaviors and effectively block them" is the application program or relevant component that provides anti-malicious code function for the mobile terminal.

The key points of the evaluation implementation include: if the malicious code technology protection measures are adopted to prevent the virus intrusion of the mobile terminal, check whether the anti-malicious code product installed on the mobile terminal operates normally, and whether the malicious code database has been updated to the latest; if the active immune trusted verification mechanism is adopted to prevent the virus intrusion against the mobile terminal, check the instruction or approval document of the mobile intelligent terminal, check whether the trusted chip is preset or other trusted verification modes are adopted, and the intrusion and virus behavior can be actively immune.

If the results show that the mobile terminal adopts trusted verification mechanism to effectively block the intrusion and virus behavior when the intrusion and virus behavior are found, the result is conforming; otherwise it is non-conforming or partially conforming.

6. Trusted Verification

[Standard Requirements]

Level Ⅲ of the control point includes the evaluation unit L3-CES1-24, and Level Ⅳ includes the evaluation unit L4-CES1-23.

[L4-CES1-23 Interpretation and Description]

The targets of the evaluation index "can conduct trusted verification on the system boot program, system program, important configuration parameter and application program of the computing device based on the trusted root, conduct dynamic trusted verification in all the execution links of the application program, give an alarm when the credibility is detected to be damaged, form an audit record of the verification result and send it to the security management center, and conduct dynamic association perception".

The key points of evaluation implementation include: check the technical specification of mobile intelligent terminal or the national cryptography management approval document of the trusted chip to check whether the trusted chip is preset or adopt other trusted verification methods; issue the ID certificate of the test platform to the trusted chip to check whether the chip responds correctly; check whether the security management center has trusted verification audit records; and check whether dynamic association perception can be performed.

If the results show that the trust root exists and the response is normal and the dynamic association perception can be carried out, the result is conforming; otherwise it is non-conforming or partially conforming.

[L3-CES1-24 Interpretation and Description]

The targets of the evaluation index "should be able to verify the system boot program, system program, important configuration parameters and application programs of the computing device based on the trusted root, and perform dynamic trusted verification in the key execution links of the application program, give an alarm after detecting that its credibility is damaged, and send the verification results into audit records and send them to the security management center" are the trusted verification modules or components on the mobile terminals.

The key points of evaluation implementation include: check the technical specification of mobile intelligent terminal or the national cryptography administration approval document of the trusted chip to check whether the trusted chip is preset or adopt other trusted verification methods; issue the ID certificate of the test platform to the trusted chip to check whether the chip responds correctly; check whether the security management center has reliable verification audit records.

If the results show that the trust root exists and the response is normal, the result is conforming; otherwise it is non-conforming or partially conforming.

7. Data Integrity

[Standard Requirements]

Level III of the control point includes evaluation units L3-CES1-25, L3-CES1-26, and Level IV includes evaluation units L4-CES1-24, L4-CES1-25 and L4-CES1-26.

[L4-CES1-24 Interpretation and Description]

The evaluation targets of the "cryptography technology should be adopted to ensure the integrity of important data during transmission, including but not limited to authentication data, important business data, important audit data, important configuration data, important video data and important personal information" are wireless access equipment (wireless AP, wireless access gateway), mobile terminal (general terminal, special terminal), mobile terminal management system, mobile application software, etc.

The key points of evaluation implementation include: check the management mode of mobile terminal management system, confirm whether B/S mode or C/S mode is adopted for management, especially check whether the mobile terminal management system uses cryptography technology, such as HTTPS, SSL and other technologies to ensure the integrity of data transmission; check whether the important data (including important business data, important authentication data, important audit data, important management data and important personal information) between mobile terminal and mobile terminal management system adopts cryptography technology (such as SSL protocol) in the transmission process to ensure the integrity.

If the results show that the mobile terminal management system adopt the cryptographic techniques to protect the integrity of the self-managed and service-accessed communication data, the result is conforming; otherwise it is non-conforming or partially conforming.

[L3-CES1-25 Interpretation and Description]

The evaluation targets of "verification technology orcryptography technology should be adopted to ensure the integrity of important data during transmission, including but not limited to authentication data, important business data, important audit data, important configuration data, important video data and important personal information" are wireless access equipment (wireless AP, wireless access gateway), mobile terminal (general terminal, special terminal), mobile terminal management system, mobile application software, etc.

The key points of evaluation implementation include: check the management mode of the mobile terminal management system, confirming whether B/S mode or C/S mode is adopted for management, especially check whether the mobile terminal management system uses verification technology or cryptography technology, such as HTTPS, SSL and other technologies to ensure the integrity of data transmission; verify whether the important data (including important business data, important authentication data, important audit data, important management data and important personal information) between the mobile terminal and the mobile terminal management system adopts verification technology or cryptographic techniques (such as SSL protocol) in the transmission process to ensure the integrity.

If the results show that the mobile terminal management system adopt the verification technology or the cryptographic techniques to protect the integrity of the communication data of its own management and service access, the result is conforming; otherwise it is non-conforming or partially conforming.

[L4-CES1-25 Interpretation and Description]

The evaluation targets of the "cryptography technology should be adopted to ensure the integrity of important data in the storage process, including but not limited to authentication

Chapter 5 Application and Interpretation of the Extended Security Evaluation Requirements of Mobile Internet

data, important business data, important audit data, important configuration data, important video data and important personal information" are wireless access equipment (wireless AP, wireless access gateway), mobile terminal (general terminal, special terminal), mobile terminal management system, mobile application software, etc.

The key points of evaluation implementation include: confirm the storage location of important data (including authentication data, important business data, important audit data, important configuration data, important video data and important personal information), check whether cryptography technology is used to protect the integrity of relevant data during storage of important data; check whether the system can detect and timely recover the modified data after the important data stored in the mobile terminal, mobile terminal management system and mobile terminal management client are modified.

If the test results show that the important data of the mobile terminal management system adopts the cryptography technology for integrity protection in the storage process, the result is conforming; otherwise it is non-conforming or partially conforming.

【L3-CES1-26 Interpretation and Description】

The evaluation targets of "verification technology or cryptography technology should be adopted to ensure the integrity of important data in the storage process, including but not limited to authentication data, important business data, important audit data, important configuration data, important video data and important personal information" are wireless access equipment (wireless AP, wireless access gateway), mobile terminal (general terminal, special terminal), mobile terminal management system, mobile application software, etc.

The key points of evaluation implementation include: confirm the storage location of important data (including authentication data, important business data, important audit data, important configuration data, important video data and important personal information), verify whether the integrity of relevant data is protected by verification technology or cryptography technology during the storage of important data; verify whether the system can detect and timely recover the modified data after the important data stored in the mobile terminal, mobile terminal management system and mobile terminal management client are modified.

If the test results show that the important data of the mobile terminal management system adopts the verification technology or the cryptography technology for integrity protection in the storage process, the result is conforming; otherwise it is non-conforming or partially conforming.

【L4-CES1-26 Interpretation and Description】

The evaluation targets or the evaluation index "in the application that may involve the legal responsibility identification, cryptography technology should be adopted to provide data original evidence and data receiving evidence, and realize non-repudiation of data original behavior and non-repudiation of data receiving behavior" are mobile application software, etc.

The key points of evaluation include: interview and check whether cryptography technology is adopted to ensure non-repudiation of data sending and receiving operations; interview and check whether technical measures are adopted to ensure non-repudiation of data sending and data receiving operations; and check whether mobile application system can tamper with data during transmission.

If the test results show that the important data of the mobile application system is protected by the cryptographic techniques during transmission, the result is conforming; otherwise it is non-conforming or partially conforming.

8. Data Confidentiality

[Standard Requirements]

Level III of the control point includes evaluation units L3-CES1-27 and L3-CES1-28, and Level IV includes evaluation units L4-CES1-27 and L4-CES1-28.

[L3-CES1-27/ L4-CES1-27 Interpretation and Description]

The targets of the evaluation index "cryptography technology should be adopted to ensure the confidentiality of important data during transmission, including but not limited to authentication data, important business data and important personal information" are wireless access equipment (wireless AP, wireless access gateway), mobile terminal (general terminal, special terminal), mobile terminal management system, mobile application software, etc.

The key points of evaluation implementation include: confirm the management mode of wireless access equipment such as wireless router and wireless AP, especially check whether the management is carried out by the unsafe transmission protocol such as telnet or http, and whether the self-management data of the wireless access equipment is plaintext transmission; confirm the wireless network encryption technology used by wireless access equipment such as wireless router and wireless AP, and check whether WPA2 or more security encryption standard is adopted.

If the test results show that the tested equipment or system adopts an effective encryption method, the result is conforming; otherwise it is non-conforming or partially conforming.

[L3-CES1-28/ L4-CES1-28 Interpretation and Description]

The targets of the evaluation index "cryptography technology should be adopted to ensure the confidentiality of important data during storage, including but not limited to authentication data, important business data and important personal information, etc." mainly include wireless access equipment (wireless AP, wireless access gateway), mobile terminal (general terminal and special terminal), mobile terminal management system, mobile application software, etc.

The key points for implementation of the assessment include: confirm the storage location of important data (including identification data, important business data, important audit data, important configuration data, important video data and important personal information), and verify whether relevant data is encrypted and stored by cryptography technology during the storage process of important data.

If the results show that the important data (including identification data, important business data, important audit data, important configuration data, important video data and important personal information) of the system or equipment undergoing test has adopted effective cryptography technology for storing, the result is conforming; otherwise it is non-conforming or partially conforming.

9. Data Backup Recovery

[Standard Requirements]

Level Ⅲ of the control point includes evaluation units L3-CES1-29, L3-CES1-30, L3-CES1-31 and Level Ⅳ includes evaluation units L4-CES1-29, L4-CES1-30, L4-CES1-31.

[L3-CES1-29/ L4-CES1-29 Interpretation and Description]

The targets of the evaluation index "local data backup and recovery function of important data should be provided" are wireless access equipment (wireless AP, wireless access gateway), mobile terminal (general terminal, special terminal), mobile terminal management system, mobile application software, etc.

The implementation points of the evaluation include: check whether the mobile application server and the important authentication information, important business data and important personal information of the mobile application installed on the smart phone are backed up locally according to the backup strategy; especially check the rationality and effectiveness of the backup strategy and the consistency between the backup results and the backup strategy; check whether the recent recovery test records have been subject to normal data recovery operations, focus on the consistency between the data recovery operation and the original data.

If the results indicate that the important data is backed up locally on a regular basis and the administrator periodically perform data recovery tests, and the file permission of the backup script is limit, the result is conforming; otherwise it is non-conforming or partially conforming.

[L3-CES1-30/ L4-CES1-30 Interpretation and Description]

The targets of the evaluation index "remote real-time backup function should be provided, and important data should be backed up to the backup site in real time through the communication network" are wireless access equipment (wireless AP and wireless access gateway), mobile terminal (general terminal and special terminal), mobile terminal management system, mobile application software, etc.

The key points of evaluation implementation include: check whether the system provides remote real-time backup function, especially the adopted remote backup mechanism and measures; check whether the system backs up the important configuration data and important business data of the mobile terminal management system and mobile intelligent client to the backup site through the network, and focus on the real-time performance of remote backup.

If the results show that if the system undergoing test provides the remote real-time backup function of the important configuration data and the important service data of the mobile terminal equipment and the backup is valid according to the backup strategy, the result is conforming; otherwise it is non-conforming or partially conforming.

[L3-CES1-31/ L4-CES1-31 Interpretation and Description]

The targets of the evaluation index "hot redundancy of important data processing system should be provided to ensure high availability of the system" are wireless access equipment (wireless AP and wireless access gateway), mobile terminal (general terminal and special terminal), mobile terminal management system, mobile application software, etc.

The key points of evaluation implementation include: interview the network administrator with important components of the mobile terminal management system, check whether the important devices in the mobile terminal management system, such as application server, database server and database, are deployed redundantly, and especially check whether the backup equipment keeps the hot redundancy status after startup.

If the test results show that all important data processing equipment in the system undergoing test are deployed in hot redundancy mode, the result is conforming; otherwise it is non-conforming or partially conforming.

10. Residual Information Protection

[Standard Requirements]

Level Ⅲ of the control point includes evaluation units L3-CES1-32 and L3-CES1-33, and Level Ⅳ includes evaluation units L4-CES1-33 and L4-CES1-34.

[L3-CES1-32/ L4-CES1-33 Interpretation and Description]

The targets of the evaluation index "should ensure that the storage space in which the authentication information is stored is completely cleared before being released or reallocated" are mobile terminal management system, mobile application software, etc.

The key points of evaluation implementation include: check whether the operation system, database and mobile application software of the mobile terminal have the function of clearing user authentication information according to the system design document; especially check whether the user can continue to log in the mobile terminal operation system, database or mobile application software by using the residual temporary cache after the user exits; check whether there is residual user authentication information after the user exits.

If the test results show that the system or software undergoing test has the function of clearing user authentication information, the previous page cannot be accessed after exiting, and there is no residual user authentication information, the result is conforming; otherwise it is non-conforming or partially conforming.

[L3-CES1-33/ L4-CES1-34 Interpretation and Description]

The targets of the evaluation index "should ensure that the storage space with sensitive data is completely cleared before being released or reallocated" are mobile terminal management system, mobile application software, etc.

The key points of evaluation implementation include: check whether the operation system, database and mobile application software of the mobile terminal have the function of clearing user-sensitive data according to the system design document; check whether there is residual user-sensitive data after the user exits.

If the results show that the application softwareundergoing test has the function of clearing the sensitive data of the user, and there is no residual sensitive data information, the result is conforming; otherwise it is non-conforming or partially conforming.

11. Protection of Personal Information

[Standard Requirements]

Level Ⅲ of the control point includes evaluation units L3-CES1-34 and L3-CES1-35, and Level Ⅳ includes evaluation units L4-CES1-35 and L4-CES1-36.

[L3-CES1-34/ L4-CES1-35 Interpretation and Description]

The targets of the evaluation index "only collect and save the user's personal information necessary for the business" are the mobile terminal management system, mobile application software, etc.

The key points of evaluation implementation include: check whether the unit undergoing test has formulated the management system and process for the protection of user's personal information; especially check whether the management system and flow list in detail the types, methods and storage locations of all personal information that must be collected by the business, and whether the system and process cover all the business of the system, including but not limited to the server, mobile device, APP, data management, etc. ; check whether the user's personal information actually collected by each module of the system is consistent with the regulations and procedures; combine the system business process, especially check whether the user's personal information collected by each module is necessary for business application; check the storage location of the collected personal information, such as the corresponding directory of the server, the terminal storage path and database where the APP is located, and check whether the stored personal information is consistent with those listed in the system and process.

If the results show that the unit undergoing test has formulated the management system and process for the protection of user's personal information, enumerates in detail the types, methods and storage locations of all personal information that must be collected by the business, and the actually collected and stored personal information of the user is consistent with the regulations and does not exceed the business scope, the result is conforming; otherwise it is non-conforming or partially conforming.

[L3-CES1-35/ L4-CES1-36 Interpretation and Description]

The targets of the evaluation index "unauthorized access to and illegal use of user's personal information should be prohibited" are the mobile terminal management system, mobile application software, etc.

The key points of evaluation implementation include: check whether the unit undergoing test has formulated the management system and process for the protection of user's personal information; especially check whether the management system and flow describe the technical measures restricting the access and use of the user's personal information, which should cover all the business of the system, including but not limited to the server, mobile device, APP, data management, etc.; check the technical measures described in the management system and flow, and try to access or use the personal information by bypassing the measures, such as directly accessing the terminal storage path of the APP, check whether the relevant data can be accessed without authorization, and check whether the measures actually taken by the system are consistent with and effective with the provisions of the management system and procedures.

If the test results show that the tested unit has formulated the management system and process for the protection of user's personal information, defined the technical measures for restricting the access and use of the user's personal information, and the actually adopted measures are consistent with the management system and procedures and come into effect, the result is conforming; otherwise it is non-conforming or partially conforming.

12. Mobile Terminal Control

[Standard Requirements]

Level III of the control point includes evaluation units L3-CES3-01, L3-CES3-02, and Level IV includes evaluation units L4-CES3-01, L4-CES3-02, L4-CES3-03.

[L3-CES3-01/ L4-CES3-01 Interpretation and Description]

The main evaluation targets of the evaluation index "the mobile terminal should be ensured to install, register and operate terminal management client software" are mobile terminal and mobile terminal management system, etc.

The implementation points of the evaluation include: check whether the mobile terminal has installed the terminal management client software, especially check whether the software runs normally, log in the terminal client management software to check whether the communication with the mobile terminal management system is normal; check whether the equipment information of the mobile terminal is registered in the mobile terminal management system, and whether the mobile terminal is online. And check whether the device information is consistent with the mobile terminal device and whether the control policy is effective.

If the results show that the terminal management client software has been installed in the system under test or the equipment, and the communication between the terminal management client software and the mobile terminal management system is normal, and the mobile terminal to be sampled has been register in the mobile terminal management system at the same time, the registration information is consistent with the mobile terminal, and the management and control policy is effective, the result is conforming; otherwise it is non-conforming or partially conforming.

【L3-CES3-02/ L4-CES3-02 Interpretation and Description】

The targets of the evaluation index "the mobile terminal should be subject to the equipment life cycle management and remote control of the mobile terminal management server, such as remote locking, remote erasing, etc." are mobile terminal and mobile terminal management system, etc.

The key points of evaluation implementation include: log in the mobile terminal management system as an administrator, check whether the security policies such as device life cycle management (such as controlled, log-out, elimination, etc.), device remote control (such as remote locking, remote erasing, peripheral interface control, application management and control, etc.) are set for the mobile terminal, and focus on the successful issuance of the policy; open the mobile terminal receiving the security policy, check whether the life cycle of the mobile terminal is consistent with the life cycle management policy issued by the mobile terminal management system, check whether the device remote control strategy for the terminal is effective, especially check whether the remotely locked terminal is locked, whether the data in the remotely erased terminal is erased, and whether the peripheral interface and mobile application control policy are effective.

If the results show that the management system of the mobile terminal undergoing test is set with the device life cycle management and the remote control strategy of the device for the mobile terminal, and the policy is effective, the result is conforming; otherwise it is non-conforming or partially conforming.

【L4-CES3-03 Interpretation and Description】

The targets of the evaluation index "should ensure that the mobile terminal is only used for processing the specified business" are the mobile terminal.

The key points of evaluation implementation include: check whether the management system stipulates that the mobile terminal can only be used to process the specified business; check the service type allowed by the mobile terminal, check whether the mobile terminal is installed with other applications, and check whether the mobile terminal has handled other services.

If the results show that the management system specify that the mobile terminal can only be used to process the specified service, and specifies the service type that the mobile terminal is allowed to handle in detail, check that the mobile terminal is not installed with other unnecessary application program, check the processing log of the terminal, and no data for processing other services is found, the result is conforming; otherwise it is non-conforming or partially conforming.

13. Mobile Application Management and Control

[Standard requirements]

Level III of the control point includes evaluation units L3-CES3-03, L3-CES3-04, L3-CES3-05, and Level IV includes evaluation units L4-CES3-04, L4-CES3-05, L4-CES3-06 and L4-CES3-07.

[L3-CES3-03/ L4-CES3-04 Interpretation and Description]

The targets of the evaluation index "should have the function of selecting the installation and operation of application software" are the mobile terminal management system, etc.

The key points of evaluation implementation include: check whether the mobile terminal has installed the mobile terminal management client, verify whether the client operates normally (service process, resident taskbar, etc.), especially check whether the function of preventing the user from closing the security function or uninstalling the client (for example, inputting the management cryptography when modifying the configuration); check whether the security policy set by the mobile terminal management server has redundant or expired rules, especially check whether the software installation/ operation policy can only be configured by the administrator to allow or prevent the installation and operation of the software, and the user cannot modify these security policies by himself; and check whether the mobile terminal management client accepts the policy of the server and restricts or controls the application software that can be installed/ operated by the mobile terminal, especially check whether only the application software in the whitelist is allowed to be installed and run, while the application software other than the whitelist cannot be installed and run, and whether relevant error information is displayed. Check whether the server has the audit function, whether it can record the log records of the above behaviors, and whether it supports viewing details.

If the results show that the mobile terminal undergoing test has been installed with the mobile terminal management client, the client end user can not close the client terminal by itself, the software installation/operation strategy is reasonable and effective and can be configured by the system administrator, the result is conforming; otherwise it is non-conforming or partially conforming.

[L3-CES3-04/ L4-CES3-05 Interpretation and Description]

The targets of the evaluation index "only application software with specified certificate signature should be allowed to be installed and run" are the mobile terminal management system, etc.

The key points of evaluation implementation include: check whether the mobile terminal management system has the certificate management function, especially check whether the administrator can only import and delete certificates; check whether the administrator can only configure the certificate list of software that can be installed and run, and the user cannot modify the certificate list by himself; check whether the mobile terminal management system can verify the validity of the certificate signature of the application software, especially when installing the application software signed by the certificate in the certificate list, whether the certificate is verified to pass the verification; when installing the application software not in the certificate list, whether the certificate verification fails and whether the client prompts relevant error information; check whether the server records the log records of the above behaviors, and can view the details.

If the results show that the management system of the mobile terminal undergoing test has the certificate management function, and only the certificate management and the rule configuration can be carried out by the system administrator, and only the application software with the compliant signature certificate is allowed to be installed and run, the result is conforming; otherwise it is non-conforming or partially conforming.

【L3-CES3-05/ L4-CES3-06 Interpretation and Description】

The targets of the evaluation index "should have the software whitelist function, and should be able to control the installation and operation of application software according to the whitelist" are the mobile terminal management system, etc.

The key points of evaluation implementation include: check whether the mobile terminal management system has the software whitelist function; check whether the configuration of the software whitelist can only be operated by the administrator and the user cannot modify it; check the validity of the software whitelist function. Especially check whether only the application software in the whitelist is allowed to be installed and run; whether the application software outside the whitelist cannot be installed and run, and whether relevant error information is given. Check whether the server records the log records of the above behaviors, and whether the details can be viewed.

If the results show that the management system of the mobile terminal undergoing test has the software white list function, and the software white list function can only be configured by the administrator, and the software white list rule is reasonable and effective, and the application software other than the software white list cannot be installed and run, the result is conforming; otherwise it is non-conforming or partially conforming.

【L4-CES3-07 Interpretation and Description】

The targets of the evaluation index "should be capable of receiving the mobile application software management strategy pushed by the mobile terminal management server and implementing management and control of the software according to the policy" are the mobile terminal, etc.

The key points of evaluation implementation include: check whether the mobile terminal management system has the function of pushing mobile application software to the mobile terminal, and whether the software configuration management policy is applied to the pushed mobile terminal; check whether the configuration of the management policy of mobile application software can only be operated by the administrator, and the user cannot modify it by himself; test and verify the effectiveness of the management policy of mobile application software. Focus on the test to verify whether the mobile terminal is only allowed to install and run mobile application software pushed by the system service end; whether the application software other than the system service end can not be installed and run, and whether relevant error information is given. Check whether the server records the log records of the above behaviors, and whether the details can be viewed.

If the results show that the mobile terminal management system undergoing test has the function of configuration management strategy for the pushed mobile application software, and the application software configuration management policy can only be configured by the administrator, and the application software configuration management strategy is reasonable and effective, and the application software other than the application software configuration management strategy cannot be installed and run, the result is conforming; otherwise it is non-conforming or partially conforming.

5.2.5 Security Management Center

Please refer to relevant chapters of this book for application interpretation of general security requirement evaluation.

5.2.6 Security Management System

Please refer to relevant chapters of this book for application interpretation of general security requirement evaluation.

According to the characteristics of the mobile interconnection system, this chapter lists the characterized interpretation of some General security requirement evaluation in the mobile interconnection environment.

Management System

[Standard Requirements]

Level III of the control point includes evaluation units L3-PSS1-02, L3-PSS1-03, L3-PSS1-04, and Level IV includes evaluation units L4-PSS1-02, L4-PSS1-03 and L4-PSS1-04.

[L3-PSS1-02/ L4-PSS1-02 Interpretation and Description]

The targets of the evaluation index "security management system should be established for various management contents in security management activities" are the mobile interconnection security management system documents.

The key points of evaluation implementation include: verify whether a special security management system has been established for the security management of the mobile Internet.

If the results show that the unit undergoing test has established a special security management system for the mobile interconnection security management, the result is conforming; otherwise it is non-conforming or partially conforming.

[L3-PSS1-03/ L4-PSS1-03 Interpretation and Description]

The targets of the evaluation index "the operation specification should be established for the daily management operation executed by the management personnel or operators" are the security operation procedure documents of the mobile Internet.

The key points of evaluation implementation include: verify whether special operation procedure documents have been established for the mobile connected security management.

If the results show that the unit undergoing test has established a special operation procedure document for the mobile interconnection security management, the result is conforming; otherwise it is non-conforming or partially conforming.

[L3-PSS1-04/ L4-PSS1-04 Interpretation and Description]

The targets of the evaluation index "should form a comprehensive security management system composed of security policy, management system, operation specification, record form, etc." are the mobile interconnection security management system documents.

The key points of evaluation implementation include: check whether the overall policy and strategy documents, management system, operation procedures and record forms are comprehensive, relevant and consistent; check whether the security management system contains the content related to mobile interconnection security management.

If the results show that the overall policy and strategy document, management system, operation procedure and record form of the tested unit are comprehensive, relevant and consistent, and the security management system system contains the content related to the security management of the mobile Internet, the result is conforming; otherwise it is non-conforming or partially conforming.

5.2.7 Security Management Organization

Please refer to relevant chapters of this book for application interpretation of General security requirement evaluation.

5.2.8 Security Management Personnel

Please refer to relevant chapters of this book for application interpretation of General security requirement evaluation.

5.2.9 Security Development Management

Please refer to relevant chapters of this book for application interpretation of General security requirement evaluation.

This chapter is the interpretation of the extended requirements for mobile interconnection security evaluation (Security Development Management).

1. Mobile Applications Procurement

[Standard Requirements]

Level Ⅲ of the control point includes evaluation units L3-CMS3-01 and L3-CMS3-02, and Level Ⅳ includes evaluation units L4-CMS3-01 and L4-CMS3-02.

[L3-CMS3-01/ L4-CMS3-01 Interpretation and Description]

The targets of the evaluation index "should ensure that the application software installed and operated by the mobile terminal comes from reliable distribution channel or is signed with reliable certificate" are mobile application software.

The key points of evaluation implementation include: verify whether the application software comes from a reliable distribution channel to reduce the risks caused by the installation of mobile application software; verify whether the application software installed and operated by the mobile terminal adopts certificate signature; if so, test the signature file, signature tool and development tool to check whether it is a reliable certificate signature and whether the integrity of the application software can be guaranteed.

If the results show that the application software installed and run by the mobile terminal undergoing test is from a reliable distribution channel or uses a reliable certificate signature, the result is conforming; otherwise it is non-conforming or partially conforming.

[L3-CMS3-02/ L4-CMS3-02 Interpretation and Description]

The targets of the evaluation index "should ensure that the application software installed and operated by the mobile terminal is developed by the designated developer" are mobile application software.

The key points of the evaluation include: check the system development and design documents and application developer information, and verify whether the mobile application software is developed by the designated developer.

If the results show that the application software installed and running by the mobile terminal undergoing test is developed by the designate developer, the result is conforming; otherwise it is non-conforming or partially conforming.

2. Development of Mobile Application Software

[Standard Requirements]

Level Ⅲ of the control point includes the evaluation units L3-CMS3-03 and L3-CMS3-04, and Level Ⅳ includes the evaluation units L4-CMS3-03 and L4-CMS3-04.

[L3-CMS3-03/ L4-CMS3-03 Interpretation and Description]

The targets of the evaluation index "qualification review should be conducted for mobile application software developers" are the person in charge of system construction.

The key points of evaluation implementation include: interview the person in charge of system construction, whether to conduct qualification examination on the developer, how to examine it and what measures to ensure the developers of mobile business application software can be controlled; interview the person in charge of system construction or other responsible persons to see whether the developer of mobile service application software has relevant qualification for software development; and check whether there are relevant review records.

If the results show that the unit undergoing test has formulated an examination system for the application software developer of the mobile service and strictly implements it, the result is conforming; otherwise it is non-conforming or partially conforming.

[L3-CMS3-04/ L4-CMS3-04 Interpretation and Description]

The targets of the evaluation index "the validity of the signature certificate for the development of mobile service application software" are the signature certificate of the software.

The key points of evaluation implementation include: check whether the signature certificate for developing mobile service application software is legal, whether the digital certificate used for electronic signature is issued by CA organization licensed by the Ministry of Industry and Information Technology, whether all the information of digital certificate is correct, and whether the signature technology is reliable.

If the results show that the signature certificate for developing the mobile service application software is legal, the result is conforming; otherwise it is non-conforming or partially conforming.

5.2.10 Security Operation and Maintenance Management

Please refer to relevant chapters of this book for application interpretation of General security requirement evaluation.

This chapter is the interpretation of theextended requirements for the security evaluation of mobile interconnection system (security operation and maintenance management).

1. Configuration Management

[Standard requirements]

Level Ⅲ of the control point includes the evaluation unit L3-MMS3-01, and Level Ⅳ includes the evaluation unit L4-MMS3-01.

[L3-MMS3-01/ L4-MMS3-01 Interpretation and Description]

The targets of the evaluation index that "the configuration library of legal wireless access device and legal mobile terminal should be established for identification of illegal wireless access equipment and illegal mobile terminal" are the record documents, mobile terminal management system or relevant components.

The key points of evaluation implementation include: check whether a trusted access mechanism for wireless access devices and mobile terminals is established, and whether there is a whitelist database of wireless access devices and mobile terminals; try to use unauthorized wireless access devices or mobile terminals for access test, and check whether there are identification measures.

If the results show that the unit undergoing test has established the configuration library of the wireless access equipment and the legal mobile terminal, and the illegal equipment is identified through the configuration library, the result is conforming; otherwise it is nonconforming or partially conforming.

Chapter 6 Application and Interpretation of the Extended Security Evaluation Requirements of IoT

6.1 Overview of IoT System

6.1.1 Characteristics of IoT System

The Internet of Things (IoT) is generally divided into three logical layers in terms of architecture, namely, the sensing layer, the network transport layer and the processing application layer. The sensing layer includes sensor node and sensor gateway node, or RFID tag and RFID reader-writer, as well as the short-distance communication (usually wireless) part between these sensing devices and sensor network gateway, RFID tag and reader; the network transport layer includes the network that transmits the sensing data to the processing center at a long distance, including the Internet, private network, mobile network, etc., as well as the integration of several different networks; the processing application layer includes the platform for storage and intelligent processing of the sensing data, and provides services for service application terminals. For large-scale, the processing application layer is generally a cloud computing platform and a service application terminal device.

6.1.2 Composition of IoT System

The composition diagram of the IoT is shown in Figure 6.1. The security protection for the IoT should include the sensing layer, the network transport layer and the processing application layer. Since the network transport layer and the processing application layer are usually composed of computer equipment, these two parts should be protected according to the requirements of the general security requirements. The extended security requirements of the IoT require special security requirements for the sensing layer, which together with the general security requirements constitute the complete security requirements for the IoT.

There are two development cases of the network transport layer in Figure 6.1.

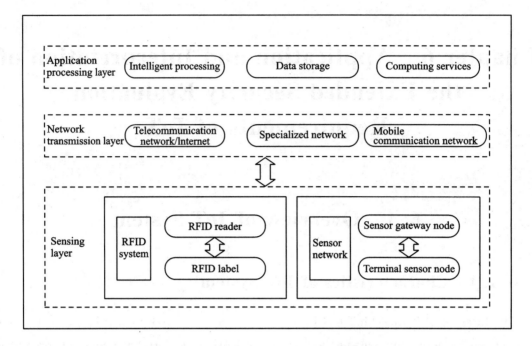

Figure 6.1 The Three Logical Layers of the IoT

① The special network such as special line and private network is used in the IoT system, and the network is controlled by the system itself, in this case, the network transport layer is protected by the operation and use unit of the IoT system, and the corresponding evaluation index of the network transport layer should be tested when evaluating the level of the IoT system;

② For the IoT system using the public networks such as mobile communication network and telecommunication network/Internet, the network transport layer is protected by the telecom operation and use unit, so it is not necessary to test the corresponding evaluation index of the network transport layer when evaluating the level of the IoT system.

The key to the security of the IoT system is the security protection of the IoT terminal (including the sensing gateway). The security requirements of the IoT system in different industries and application scenarios are quite different; the deployment environment and network structure of the IoT terminals are complex; and some scenarios are limited by cost and power consumption. According to the characteristics of IoT terminals, they can be divided into strong terminals and weak terminals, which face different threats and security needs, so they need to be distinguished in the evaluation.

Classification of IoT sensing terminal is shown in Table 6.1.

Chapter 6 Application and Interpretation of the Extended Security Evaluation Requirements of IoT

Table 6.1 Classification of IoT Sensing Terminal

Terminal type	Features	Typical product
Strong terminal	Strong computing capacity, large storage capacity, high network bandwidth, high data value, with operation system, etc.	AI camera, face recognition access control, smart car, sensing gateway, etc.
Weak terminal	Weak computing capacity, limited memory, low network bandwidth and sensitive to cost and power consumption	Intelligent water meter and electricity meter, intelligent gas meter, RFID, temperature, humidity and smoke detector, etc.

Due to the limitation of cost and power consumption, the computing power and storage capacity of the weak terminal are very limited, so it is difficult to run the complex security protection measures, and the protection ability is weak; meanwhile, the exposure surface of the terminal is limited, and the gain from the attack is relatively low. However, due to its high computing resources, network bandwidth resources and data resources, and the embedded general operation system, a strong terminal has a large protection area and high profits (such as stealing important data, hijacking a device to form abotnet, mining, etc.). Therefore, even in the same IoT system, the threats to the strong terminal and the weak terminal are not the same, so it is necessary to make specific analysis and select the corresponding evaluation index during the level evaluation. Take AI camera as an example, according to the model of IoT shown in Table 6.1, it can be regarded as the combination of sensing layer and processing application layer.

6.1.3 Overview of Extended Requirements for IoT Security

GB/T 22239—2019 *Baseline for Classified Protection of Cybersecurity* specifies that the General security requirement and the extended security requirements together constitute the security requirements for the objects to be protected at a level.

The extended security of the IoT requires that in the four security layers of Security physical environment, Security area boundary, Security computing environment and Security operation and maintenance management, the membership function between the relevant control points and the security layer is shown in Table 6.2.

Table 6.2 Security Layer of the Extended Security Requirements of the IoT

S/N	Security level	Control points (extended security requirements)
1	Security physical environment	Physical protection of sensor node devices
2	Security area boundary	Access control, intrusion prevention
3	Security computing environment	Sensor node device security, gateway node device security, anti-data playback, data fusion processing
4	Security operation and maintenance management	Sensor node management

On the basis of the general security requirements, the security requirements for physical protection, access control, intrusion prevention, security of sensory node devices, security of gateway node devices, anti-data playback, data fusion processing and sensor node management are added for IoT security.

The level-by-level change of control points required for extended security of the IoT in terms of the quantity of each requirement item is shown in Table 6.3.

Table 6.3　Change of the Number of Control Points Required for Extended Security of the IoT

S/N	Control point	Level I	Level II	Level III	Level IV
1	Physical protection of sensor node devices	2	2	4	4
2	Access control	1	1	1	1
3	Intrusion prevention	0	2	2	2
4	Security ofsensor node devices	0	0	3	3
5	Gateway node devices Security	0	0	4	4
6	Data replay resistant	0	0	2	2
7	Data fusion processing	0	0	1	2
8	Sensor node management	1	2	3	3

6.1.4　Basic Concepts

Internet of Things (IoT): It refers to the system formed by connecting the sensor node devices through the Internet and other networks.

Sensor node: Sensor node is a device that collects and/or performs operations on objects or environments and can communicate on a network. Sometimes, it is also referred to as a sensor terminal.

Sensor layer gateway: Gateway node is a device that collects data collected by the sensor node for summary, proper processing or data fusion, and transmits it. It is also called IoT gateway.

Security group: Security group is a kind of virtual firewall, which has the capability of state detection and packet filtering, and is used to divide the security domain in the cloud.

Timelines of data: ensure the time validity of data sending and receiving, and to ensure that the transmission of data is not played back.

Data freshness: a feature that identifies received historical data or data that is out of time.

Security function hardware: an independent processor and storage unit that can independently perform key generation, encryption and decryption calculation and random number generation, and can protect the Security storage of cryptographic materials such as keys and parameters.

Chapter 6 Application and Interpretation of the Extended Security Evaluation Requirements of IoT

6.2 Application and Interpretation of the Extended Security Evaluation Requirements of IoT at Level Ⅲ and Level Ⅳ

6.2.1 Security Physical Environment

This chapter interprets the extended security of "Security physical environment" of the security evaluation requirements of the IoT.

1. Physical Protection of Sensor node Devices

[Standard Requirements]

Level Ⅲ of the control point includes evaluation units L3-PES4-01, L3-PES4-02, L3-PES4-03, L3-PES4-04, and Level Ⅳ includes evaluation units L4-PES4-01, L4-PES4-02, L4-PES4-03 and L4-PES4-04.

[L3-PES4-01/ L4-PES4-01 Interpretation and Description]

The targets of the evaluation index "the physical environment in which the sensor node devices is located should not cause physical damage to the sensor node devices, such as squeezing and strong vibration" are the sensor node devices, such as video camera, LED display screen, etc.

The key points of evaluation implementation include: check the design or acceptance documents of the physical environment in which the sensor node devices is located, check the description of anti-squeezing and anti-strong vibration measures; on-site check whether the physical environment where the sensor node devices is located has taken anti-squeezing and anti-strong vibration protection measures.

If the results show that protective measure such as anti-squeezing and anti-strong vibration are taken for the sensor node devices, the result is conforming; otherwise it is non-conforming or partially conforming.

[L3-PES4-02/ L4-PES4-02 Interpretation and Description]

The targets of the evaluation index "the physical environment in which the sensor node devices is in the working state should correctly reflect the environmental state (for example, the temperature and humidity sensor cannot be installed in the direct sunlight area)" are sensor node devices, such as the camera, LED display, etc.

The key points of evaluation implementation include: check the design or acceptance documents of the physical environment in which the sensor node devices is located, whether there is a description of the physical environment in which the sensor node devices is in working state, and whether it is consistent with the actual situation; check whether the physical environment where the sensor node devices is located can correctly reflect the environmental conditions (for example, the temperature and humidity sensor cannot be installed in the direct sunlight area).

If the results show that the design or acceptance document of the physical environment in which the sensor node device is located has specified the requirements of the physical environment in the working state, and the physical environment in which the sensor node device is located can correctly reflect the environmental state, the result is conforming; otherwise it is non-conforming or partially conforming.

[L3-PES4-03/ L4-PES4-03 Interpretation and Description]

The targets of the evaluation index "the physical environment in which the sensor node devices is in the working state should not affect the normal operation of the sensor node devices, such as strong interference, blocking and shielding, etc." are sensor node devices, such as camera, LED display screen, etc.

The key points of evaluation implementation include: consult the design or acceptance documents of the sensor node devices, check the requirements for the physical environment of the sensor node devices in the working state, such as the requirements for strong interference prevention, blocking prevention and shielding, etc.; verify whether the physical environment of the sensor node devices has taken protective measures such as strong interference prevention and blocking prevention and shielding.

If the results show that the physical environment in which the sensor node devices is located adopts relevant protective measures to ensure that the regular operation of the sensor node devices is not affected, the result is conforming; otherwise it is non-conforming or partially conforming.

[L3-PES4-04/ L4-PES4-04 Interpretation and Description]

The targets of the evaluation index "the key sensor node devices should have the power supply for long-time operation (the key gateway node devices should have the lasting and stable power supply)" are sensor node devices, such as camera, LED display screen, etc.

The key points of evaluation implementation include: check whether the power supply design or acceptance documents of key sensor node devices (key gateway node devices) indicate the power supply requirements, and whether the power supply measures for ensuring long-time operation of key sensor node devices (lasting and stable power supply measures for key gateway node devices) are specified; check whether there are operation and maintenance records of relevant power supply measures and whether they are consistent with the power supply design.

If the test results show that the power supply requirements are indicated in the power supply design or acceptance documents of the critical sensor node devices, and the operation and maintenance records of the relevant power supply measures are consistent with the power supply design, the result is conforming; otherwise it is non-conforming or partially conforming.

6.2.2 Security Communication Network

This chapter interprets the extended security evaluation requirements of the "Security Communication Network" of IoT.

According to the characteristics of the IoT, this chapter gives an description of some General security requirement evaluation of IoT.

1. Network Architecture

[Standard requirements]

Level Ⅲ of the control point includes evaluation units L3-CNS1-01, L3-CNS1-02, L3-CNS1-03, L3-CNS1-04, L3-CNS1-05, and Level Ⅳ includes evaluation units L4-CNS1-01, L4-CNS1-02, L4-CNS1-03, L4-CNS1-04, L4-CNS1-05, L4-CNS1-06.

[L3-CNS1-01/ L4-CNS1-01 Interpretation and Description]

The targets of the evaluation index "should ensure that the service processing capacity of network equipment meets the needs of business peak" include routers, switches, firewalls, gateway nodes, etc.

The key points of evaluation implementation include: interview the management personnel to understand the business peak period; during the business peak period, check the usage of CPU and memory of the gateway node (or check the related usage conditions through the network management platform); check the alarm log of the gateway node or the running time of the equipment, and confirm whether there is equipment shutdown due to insufficient processing capacity of the equipment.

If the test results show that the peak utilization rate of CPU and memory of the relevant network equipment is not greater than 70%, and there is no device shutdown or abnormal restart alarm log in the gateway node, the result is conforming; otherwise it is non-conforming or partially conforming.

[L3-CNS1-02/ L4-CNS1-02 Interpretation and Description]

The targets of the evaluation index "should ensure that the bandwidth of each part of the network meets the needs of the business peak" include routers, switches, firewalls, gateway nodes, integrated network management system, etc.

The key points of evaluation implementation include: interviewing relevant management personnel to determine the business peak period. During the business peak period, check (or through the management system) the bandwidth occupation of relevant sensor gateway nodes; check the alarm log or traffic log of the sensor gateway node to confirm whether the equipment of the sensor node is not working normally due to insufficient bandwidth such as video cut-off, and picture quality reduction.

If the test results show that the peak bandwidth utilization rate of the relevant exit security equipment and network equipment in the peak period of service is not greater than 70%, and there is no alarm log such as service interruption and 404 error report, the result is conforming; otherwise it is non-conforming or partially conforming.

[L3-CNS1-03/ L4-CNS1-03 Interpretation and Description]

The targets of the evaluation index "should divide different network areas and allocate addresses for each network area according to the principle of convenient management and control" include routers, switches, firewalls, gateway nodes, etc.

The key points of evaluation implementation include: understand the division and importance of departments in the current organizational structure of the unit, check whether Vlan policy is configured in the main network equipment, or adopt VPC and security group, etc.; specifically check the Vlan information of main network equipment and the configuration information of VPC and security group, verify whether the divided network area corresponds to the current unit department one-to-one, and assign addresses.

If the test results show that the relevant aware gateway node has Vlan area division, and the division area is one-to-one corresponding to each area based on the importance of the deployment area of the sensor node, the result is conforming; otherwise it is non-conforming or partially conforming.

[L3-CNS1-04/ L4-CNS1-04 Interpretation and Description]

The targets of the evaluation index "avoid the deployment of important network areas at the boundary, and adopt reliable technical isolation between the important network areas and other network areas" include the topological graph, router, switch, firewall, gateway node, etc.

The key points of evaluation implementation include: check the connection status of equipment in the machine room according to the latest network topology provided by the current customer, and confirm whether the network topology diagram is consistent with the actual network operation environment; analyze the network topology structure and check whether there are security protection measures such as sensory node access between the boundaries of important areas and the external connection access boundary; check the isolation measures at the boundary, such as the sensory node access equipment, sensory node gateway and firewall, and confirm whether there are effective isolation measures, such as access equipment, protocol, port and address filtering; check whether the boundary isolation measures are effective, and verify whether the isolation measures are effective.

If the results show that the topological graph is consistent with the actual operation situation, reliable technical isolation means are deployed at the boundary of each area and the strategy configuration is reasonable, and only the port, protocol and address to be used are open, the result is conforming; otherwise it is non-conforming or partially conforming.

[L3-CNS1-05/ L4-CNS1-05 Interpretation and Description]

The targets of the evaluation index "hardware redundancy of communication line, key network equipment, and key computing equipment should be provided to ensure the availability of the system" includes topological graph, router, switch, firewall, gateway node, data processing system, etc.

The key points of evaluation implementation include interview with network administrator or service administrator to verify that whether the site provides hardware redundancy and communication line redundancy of key network equipment and security equipment. Interview with network administrator/ business administrator to verify whether hardware redundancy (active/ standby or dual-active) and communication line redundancy of key computing equipment are provided according to site conditions.

If the test results show that the related communication line, key network equipment, and key computing equipment have hardware redundancy and can be recovered timely in case of system failure, the result is conforming; otherwise it is non-conforming or partially conforming.

【L4-CNS1-06 Interpretation and Description】

The targets of the evaluation index "bandwidth should be allocated according to the importance of business services, and priority should be given to important services" include topology diagram, router, switch, network management system, etc.

The points of the evaluation include: interview with the network administrator or service administrator to understand the various services carried by the system and their importance degree; according to the network topology diagram, understand the regional division, communication and bandwidth allocation of various services, check whether the bandwidth can be allocated according to the business importance degree, and give priority to guarantee the communication demand of the important business.

If the results show that the system undergoing test defines or defines the importance degree of the service, and reasonably allocates the bandwidth according to the service importance degree, the result is conforming; otherwise it is non-conforming or partially conforming.

2. Communication Transmission

【Standard requirements】

Level Ⅲ of the control point includes evaluation units L3-CNS1-06 and L3-CNS1-07, and Level Ⅳ includes evaluation units L4-CNS1-07, L4-CNS1-08, L4-CNS1-09 and L4-CNS1-10.

【L3-CNS1-06/ L4-CNS1-07 Interpretation and Description】

The targets of the evaluation index "data integrity in the communication process should be ensured by using verification technology or cryptography" include the sensor node, the gateway node and the data processing system, etc.

The key points of evaluation implementation include: check the current evaluation target and test whether the verification technology or cryptography technology is used in the data transmission process to ensure the integrity of the data; if it is not possible to confirm whether the integrity verification function is available in the transmission process, the verification test can be carried out by means of the tool grabbing the package and modifying the package contents and then uploading the package to check whether the integrity guarantee measures are available.

If the results show that the data transmission lay of the sensor node of the relevant terminal has the integrity check function and the function verification is valid, the result is conforming; otherwise it is non-conforming or partially conforming.

[L3-CNS1-07/ L4-CNS1-08 Interpretation and Description]

The targets of the evaluation index "the encryption technology should be adopted to ensure the confidentiality of data in the communication process" include VPN, encryption machine or other transmission encryption components.

The key points of the evaluation include: check whether the current evaluation target uses the cryptography technology to ensure the confidentiality of the data in the data transmission process; using the tool to retrieve the data package to test whether the information is in the ciphertext mode.

If the results show that the communication protocol of the relevant sensor node has the privacy function and the function verification is valid, the result is conforming; otherwise it is non-conforming or partially conforming.

[L4-CNS1-09 Interpretation and Description]

The targets of the evaluation index "verify/authenticate the both parties to the communication based on cryptography prior to communication" include VPN, encryptor, or other transport encryption component.

The key points of evaluation implementation include: check whether the current evaluation target uses cryptographic techniques such as digital certificate to verify or authenticate both parties of communication before communication, and understand the specific technical measures; simulate the communication process, and test whether the measures to verify or authenticate the identities of both parties based on cryptography technology are adopted before communication.

If the results show that the communication protocol of the relevant sensor node has the capability of authentication or authentication for both communication parties, and the function is validated to be valid, the result is conforming; otherwise it is non-conforming or partially conforming.

[L4-CNS1-10 Interpretation and Description]

The targets of the evaluation index "cryptographic operation and key management for important communication processes based on hardware cryptographic module" includes VPN, encryptor or other hardware cryptographic modules.

The key points of the evaluation include: check whether the current evaluation target generates the key based on the hardware cryptography module and calculate the cryptography; check whether the relevant products have obtained the valid test report or cryptography product model certificate specified by the national cryptography management department.

If the results show that the relevant sensor node is a hardware cryptographic device or bear a hardware cryptographic module, a valid detection report or a cryptography product model certificate specified by the national cryptographic management authority is obtained, and a key is generated and cryptography calculation is performed on the basis of the hardware cryptographic module, the result is conforming; otherwise it is non-conforming or partially conforming.

3. Trusted Verification

[Standard Requirements]

Level III of the control point includes the evaluation unit L3-CNS1-08, and Level IV includes the evaluation unit L4-CNS1-11.

[L3-CNS1-08/ L4-CNS1-11 Interpretation and Description]

The targets of the evaluation index "can conduct trusted verification on the system boot program, system program, important configuration parameters and communication application program of communication equipment based on the trusted root, and conduct dynamic trusted verification in the key execution links of the application program, give an alarm after the credibility is detected to be damaged, and send the verification result into audit record to the security management center" include the trusted verification platform, the trusted root, and the security management center.

The key points of evaluation implementation include: check whether the gateway node of the current sensing layer and whether similar trusted root measures are deployed for the communication and access, and check the description of the trusted verification device or the deployment record; when there are trusted verification measures, test and verify whether to give an alarm when the main identity identifier of the sensor node and the gateway or between them is changed, i. e. the credibility is damaged; check and verify whether the alarm results are reported to the security management center in the form of audit records.

If the results show that the relevant gateway node devices and the sensor node devices have reliable verification measures and can give an alarm after the credibility is damaged, and the alarm content is sent to the security management center for storage, the result is conforming; otherwise it is non-conforming or partially conforming.

6.2.3 Security Area Boundary

This chapter is the interpretation of the extended security evaluation requirement of "security area boundary" of IoT.

According to the characteristics of the IoT, this chapter also lists the characterized interpretation of some General security requirement evaluation of IoT.

1. Boundary Protection

[Standard Requirements]

The Level Ⅲ of the control point includes the evaluation units L3-ABS1-01, L3-ABS1-02, and the Level Ⅳ includes the evaluation units L4-ABS1-01, L4-ABS1-02, L4-ABS1-05, L4-ABS1-06.

[L3-ABS1-01/ L4-ABS1-01 Interpretation and Description]

The target of the evaluation indicator "should ensure that the access across the boundary and the data flow communicate through the controlled interface provided by the boundary equipment" are gateway node devices, gateway, firewall, router, switch, wireless access gateway and other devices or relevant components providing access control function.

The key points of the implementation include: check whether the topology diagram is consistent with the actual network link, whether the gateway node device or access device is deployed at the boundary of the perception layer, and confirm that the link access port is correct; check whether the gateway node devices communicates with the access sensing node device using the controlled port; check whether the security policy is configured and enabled on the development port; check or test whether there are no other uncontrolled ports for cross-boundary network communication through other technical means (such as wireless network device positioning, check device configuration information, etc.).

If the results show that the gateway node device or the admission device is deploy at the boundary of the sensing layer, the network communication across the boundary is carried out for the access sensing node device using the controlled port, the security strategy is applied to the network communication port across the boundary by the gateway node device or the admission device to the access sensing node device, and other technical means are used to verify or test that there are no other uncontrolled ports for network communication across the boundary, the result is conforming; otherwise it is non-conforming or partially conforming.

[L3-ABS1-02/L4-ABS1-02 Interpretation and Description]

The targets of the evaluation indicator "should be able to check or limit the unauthorized equipment connected to the internal network without permission" are gateway node devices, IoT security access control system, etc.

The key points of implementation include: interview with network administrator, ask what technical means to control the unauthorized device connecting to internal network without permission, and verify its effectiveness; check whether all idle ports of related equipment such as routers and switches have been closed; If the access control is realized by deploying the IoT admission control system, check whether the IoT sensing node devices is included in the unified management of the access control system; if IP-MAC binding is adopted to realize admission control, check whether the access layer network equipment is configured with IP/ MAC address binding and other measures.

If the test results show that the gateway node devices and the IoT security access control system have the function of preventing unauthorized devices from accessing, the sensing node devices of the IoT are uniformly managed; if that idle port of the gateway node devices and the network equipment are both in the close state, the result is conforming; otherwise it is non-conforming or partially conforming.

【L4-ABS1-05 Interpretation and Description】

The targets of the evaluation index "should be able to effectively block unauthorized devices or internal users in case of unauthorized connection to the internal network" are gateway node devices, IoT security access control system, business terminals, management terminals, etc.

The key points of evaluation implementation include: check whether the technical measures can effectively block the access of unauthorized devices to the internal network, try to use unauthorized devices to connect to the internal network privately, especially check whether the technical measures can control the USB and wireless network card of the terminal, and verify and test the effectiveness of relevant technical control measures with the cooperation of the network administrator, such as inserting the USB wireless network card on the terminal.

If the results show that the gateway node device, the security access control system of the IoT and the like have the function of preventing the unauthorized devices from accessing, and the sensor node devices of the IoT are uniformly included in the management, the result is conforming; otherwise it is non-conforming or partially conforming.

【L4-ABS1-06 Interpretation and Description】

The targets of the evaluation index that "trusted verification mechanism should be adopted for the equipment connected to the network, so as to ensure the authenticity and credibility of the equipment connected to the network" are sensor node devices.

The key points of evaluation implementation include: check whether the equipment of the sensor node has built-in trusted root; checkwhether the trusted verification mechanism is used to conduct the trusted verification for the equipment connected to the network; test and verify whether the authentication can be performed for the equipment connected to the internal network.

If the results show that the sensor node device has a built-in trusted root; when the sensor node device accesses the network and the firmware is updated, the trusted root is need to be used for verification, the result is conforming; otherwise it is non-conforming or partially conforming.

2. Access Control

【Standard Requirements】

Level Ⅲ of the control point includes evaluation units L3-ABS1-05, L3-ABS1-06, L3-ABS1-07, and Level Ⅳ includes evaluation units L4-ABS1-07, L4-ABS1-08, L4-ABS1-09.

[L3-ABS1-05/ L4-ABS1-07 Interpretation and Description]

The targets of the evaluation index "set access control rules according to the access control policy between network boundaries or areas, and reject all communication through the controlled interface except for the allowed communication by default" are gateway node devices, gateway, firewall, router, switch, wireless access gateway and other devices or relevant components providing access control function.

The key points of evaluation implementation include: check whether gateway node access device is deployed between network boundaries or areas and whether the access control policy is enabled; check whether the last access control policy of the device is to prohibit all network communication.

If the results show that the gateway node access device is deployed at the network boundary, the gateway node device and the access control device adopt the white list mechanism, and the last strategy is to prohibit all network communication, the result is conforming; otherwise it is non-conforming or partially conforming.

[L3-ABS1-06/ L4-ABS1-08 Interpretation and Description]

The targets of the evaluation index "delete redundant or invalid access control rules, and optimize the access control list, and minimize the number of access control rules" should be the evaluation targets of the gateway node devices, as well as the equipment or relevant components providing access control function such as the sensor node access device, gateway, firewall, router, switch and wireless access gateway.

The key points of evaluation implementation include: check whether there are redundant or invalid access control policies in the access control device according to the actual business requirements and security policies of the IoT system, and analyze whether the policies are effective in combination with the number of policy hits; check whether the access control policies prohibit the all-pass policy, whether the redundant ports and address limits are too large; check whether the logical relationship between different access control policies and the sequence of preceding and following are reasonable.

If the test results show that, according to the actual service requirements and security policies of the system, there is no redundant or invalid access control policy in the relevant access control device, and the policy is valid; the all-pass policy is prohibited in the access control policy, the redundant port and address limit range is reasonable; the logical relationship and sequence of different access control policies are reasonable, the result is conforming; otherwise it is non-conforming or partially conforming.

[L3-ABS1-07/ L4-ABS1-09 Interpretation and Description]

The targets of the evaluation index "check the source address, destination address, source port, destination port and protocol to allow/reject data packets entering and leaving" are gateway node devices, gateway, firewall, router, switch, wireless access gateway and other devices or relevant components providing access control function.

The key points of evaluation implementation include: check whether the source address, destination address, source port, destination port and protocol and other configuration parameters are clearly set in the access control policy of gateway node and other devices according to the actual business requirements and security policy of the IoT system; verify the effectiveness and control granularity of the access control policy through tool test.

If the results show that, according to the actual service requirements and security policies of the system, configuration parameters such as the source address, the destination address, the source port, the destination port and the protocol are clearly set in the access control policy of the relevant access control device; the access control policy and the control granularity of the deployed access control system of the video device are validated through the test, the result is conforming; otherwise it is non-conforming or partially conforming.

3. Intrusion Prevention

[Standard Requirements]

Level III of the control point includes evaluation units L3-ABS1-12 and L3-ABS1-13, and Level IV includes evaluation units L4-ABS1-14 and L4-ABS1-15.

[L3-ABS1-12/ L4-ABS1-14 Interpretation and Description]

The targets of the evaluation index, "take technical measures to analyze the network behaviors, especially the new network attacks" is the APT analysis platform of the IoT.

The key points of evaluation implementation include: check whether relevant systems or components are deployed for detection and analysis of new network attacks; test and verify whether to analyze network behaviors to realize detection and analysis of network attacks, especially unknown new network attacks; check whether the rule base version of relevant systems or equipment has been updated to the latest version.

If the results show that the traffic data collection device is deployed at the key nodes of the network; the APT analysis platform will continuously perform big data analysis on the data collected by each traffic collection probe, and the APT analysis platform can identify the network attack behavior through the traffic data generated by the simulation tool through the APT simulation tool verification, the result is conforming; otherwise it is non-conforming or partially conforming.

[L3-ABS1-13/ L4-ABS1-15 Interpretation and Description]

The targets of the evaluation index "When an attack is detected, record the attack source IP, attack type, attack target, attack time, and provide alarm in case of serious intrusion event" are APT analysis platform of IoT, security access equipment, intrusion protection equipment, etc.

The key points of evaluation implementation: check whether the records of relevant systems or components include relevant contents such as attack source IP, attack type, attack target, attack time, etc.; test and verify whether the alarm policies of relevant systems or components are effective.

If the results show that the system deploys the safe access equipment, APT analysis platform and the like; the security access equipment or APT analysis platform can identify the network attack behavior through the traffic data generated by the simulation tool; the network attack recorded by the security access equipment and the like include the attack source IP, attack type, attack target, attack practice, etc.; and after the APT alarm policy is adjusted, the security access device or APT analysis platform triggers the subsequent network attack alarm to execute according to the new policy, the result is conforming; otherwise it is non-conforming or partially conforming.

4. Security Audit

[Standard Requirements]

Level III of the control point includes evaluation units L3-ABS1-16, L3-ABS1-17, L3-ABS1-18, and Level IV includes evaluation units L4-ABS1-18, L4-ABS1-19, L4-ABS1-20.

[L3-ABS1-16/ L4-ABS1-18 Interpretation and Description]

The targets of the evaluation index "conduct security audit at the network boundary and important network nodes, and cover each user, and audit the important user behavior and important security events" are the gateway node devices and the integrated security audit system.

The key points of evaluation implementation include: interview with the network administrator and view the network topology diagram, understand the overall network condition of the system undergoing test, sort out and analyze the network boundary and important network nodes of the system undergoing test; check whether the integrated security audit system or the system platform with similar functions is deployed, and check whether the equipment deployment location is reasonable; check whether the security audit scope covers each user, and whether important user behaviors and important security events are audited.

If the results show that an integrated security audit system is deployed at the boundary of the sensing layer or the gateway node has an audit function; the audit scope of the gateway node device covers all users log in the gateway node device; the gateway node device audits important events such as user login, configuration change, gateway firmware update, etc.; the result is conforming; otherwise it is non-conforming or partially conforming.

[L3-ABS1-17/ L4-ABS1-19 Interpretation and Description]

The targets of the evaluation index "the audit record should include the date and time of the event, user, event type, success of the event and other audit-related information" are the gateway node devices and the security audit system of the IoT.

The key points of evaluation implementation include: check the system platform with similar functions such as the integrated security audit system and network audit system, and check whether the audit record information includes the date and time of the event, user, event type, success of the event and other audit-related information, and extra attention should be paid to the accuracy and clock synchronization of the date and time.

If the test results show that the audit information of the security audit system of the IoT includes the audit-related information such as the date and time of the event, the user, the type of the event, whether the event is successful, etc., the result is conforming; otherwise it is non-conforming or partially conforming.

[L3-ABS1-18/ L4-ABS1-20 Interpretation and Description]

The targets of the evaluation index "protect and back up the audit records regularly to avoid unexpected deletion, modification or coverage" are gateway node devices and integrated security audit system.

The key points of the evaluation include: check whether the technical measures are taken to protect the audit records; check whether the technical measures are taken to back up the audit records regularly, and check the backup strategy.

If the results show that the local audit record of the gateway node device is encrypted and stored and uploaded to the integrated security audit system or the cloud management platform at the first time; the integrated security audit system or the cloud management platform is set with a reasonable regular backup policy, and the unauthorized user has no right to delete, modify and cover the audit record; the integrated security audit system or the cloud management platform regularly backs up the audit records according to the backup policy, the result is conforming; otherwise it is non-conforming or partially conforming.

5. Trusted Verification

[Standard Requirements]

Level III of the control point includes the evaluation unit L3-ABS1-20, and Level IV includes the evaluation unit L4-ABS1-21.

[L3-ABS1-20/ L4-ABS1-21 Interpretation and Description]

The targets of the evaluation index "conduct trusted verification on the system boot program, system program, important configuration parameters and boundary protection application program of the boundary device based on the trusted root, and conduct dynamic trusted verification in the key execution links of the application program, and give an alarm after the credibility is detected to be damaged, and sendthe audit records to the security management center" are gateway node devices.

The key points of evaluation implementation include: check whether the system boot program, system program, important configuration parameters and boundary protection application of boundary equipment are authenticated based on the trusted root; check whether dynamic trusted verification is conducted in the key execution links of the application program; test and verify whether the alarm will be given when the credibility of the boundary equipment is detected to be damaged; test whether the verification results are sent to the security management center in the form of audit records.

If the results show that the relevant sensor node device has a built-in trusted root; the access network and firmware update need to use the trusted root for verification; and the relevant event information of the sensor node device is correctly recorded into the audit database by use of the trusted root verification, the result is conforming; otherwise it is non-conforming or partially conforming.

6. Access control

[Standard Requirements]

Level III of the control point includes the evaluation unit L3-ABS4-01, and Level IV includes the evaluation unit L4-ABS4-01.

[L3-ABS4-01/ L4-ABS4-01 Interpretation and Description]

The targets of the evaluation index "only authorized sensor nodes can be accessed" include sensor nodes, gateway nodes and other devices.

The key points of evaluation implementation include: check whether the design document of the access mechanism of the sensor node includes the mechanism for preventing the illegal device of the sensor node from accessing the network and the description of the identity authentication mechanism; check whether the access control measures (such as disabling the idle port, setting the access control policy, deploying the security management system, etc.) are adopted to manage the access of the device; perform penetration test on the network of the boundary and the sensing layer, and test whether there is no method to bypass the white list or relevant access control measures and the identity authentication mechanism.

If the results show that the relevant sensor node device and the gateway node device have unique identifications, and the device identity authentication mechanism is adopted to control the access of the sensor node; an access security management system is deployed in the sensing layer to bind the brand, model, IP and MAC of the front-end device, and only the devices that have passed the authentication are allowed to access; the penetration test is performed on the boundary and the sensing layer network, and there is no method to bypass the white list or the related access control measures and the identity authentication mechanism, the result is conforming; otherwise it is non-conforming or partially conforming.

7. Intrusion Prevention

[Standard Requirements]

Level III of the control point includes the evaluation units L3-ABS4-02 and L3-ABS4-03, and Level IV includes the evaluation units L4-ABS4-02 and L3-ABS4-03.

[L3-ABS4-02/ L4-ABS4-02 Interpretation and Description]

The targets of the evaluation index "should be able to limit the target address communicating with the sensor node to avoid the attack on the unfamiliar address" include the sensor node, gateway node and other devices.

The key points of evaluation implementation include: check the security design document of the sensing layer and whether there are control measures for the communication target address of the sensor node; check whether the sensor node device is configured with control measures for the communication target address of the sensor node and whether the relevant parameter configuration meets the design requirements; perform penetration test on the sensor node device to test whether it can restrict the sensor node device from accessing or attacking the communication target address violating the access control policy.

If the results show that the system undergoing test can limit the communication target address range of the front-end equipment such as the sensor node device, and can configure the communication destination address and port through the network communication management and control policy, and the penetration test verify that there is no method for the front-end equipment to bypass the network communication management and control policy, the result is conforming; otherwise it is non-conforming or partially conforming.

[L3-ABS4-03/ L4-ABS4-03 Interpretation and Description]

The targets of the evaluation index "should be able to restrict the target address of communication with the gateway node to avoid the attack on the unknown address" are gateway node devices.

The key points of evaluation implementation include: check the security design document of the sensing layer and whether there are control measures for the communication target address of the gateway node; check whether the gateway node devices is configured with control measures for the communication target address of the gateway node and whether the relevant parameter configuration conforms to the design requirements; perform penetration test on the sensor node devices to test whether it can restrict the gateway node devices from accessing or attacking the communication target address violating the access control policy.

If the results show that the system undergoing test has the capability of restricting network communication management and control between the gateway node device and the strange address device, can configure the communication destination IP address through the network communication management and control policy, the network communication management and control policy can prevent the communication of the strange IP address of the gateway node device according to the network communication management and control policy, and the penetration test verify that there is no method for the gateway node device to bypass the network communication management and control policy, the result is conforming; otherwise it is non-conforming or partially conforming.

6.2.4 Security Computing Environment

This chapter is the interpretation of the extended security evaluation requirement of "Security computing environment" of the IoT.

According to the characteristics of the IoT, this chapter also lists the interpretation of some General security requirement evaluation of the IoT.

1. Identification

[Standard Requirements]

Level Ⅲ of the control point includes evaluation units L3-CES1-01, L3-CES1-02, L3-CES1-03, L3-CES1-04; Level Ⅳ includes evaluation units L4-CES1-01, L4-CES1-02, L4-CES1-03, L4-CES1-04.

[L3-CES1-01/ L4-CES1-01 Interpretation and Description]

The targets of the evaluation index of "identity identification and authentication should be carried out for the logged-in user, the identity identification should be unique, and the identity authentication information should be subject to the complexity requirements and should be replaced regularly" are the sensor node devices and business application system.

The key points of evaluation implementation include: check whether the user adopts identity authentication measures when logging in; check the user list to confirm whether the user ID is unique; check the user configuration information or test to verify whether there is no empty cryptography user; check whether the user authentication information has complexity requirements and replace it regularly.

If the results show that the sensor node device or the front-end device adopts the identity identification and authentication measures when logging in, the user ID of the user list of the front-end device is unique, and the user with the same user ID cannot be added; there is no empty cryptography user in the user configuration information; the user login cryptography of the front-end device has the complexity requirement and the cryptography is changed periodically, the result is conforming; otherwise it is non-conforming or partially conforming.

[L3-CES1-02/ L4-CES1-02 Interpretation and Description]

The targets of the evaluation index "should have the login failure processing function, and should be configured and enabled to end the session, limit the number of illegal login times, and automatically exit when the login connection is overtime" are the sensor node devices, and business application system.

The key points of evaluation implementation include: check whether the login failure processing function is configured and enabled; check whether the function of restricting illegal login is configured and enabled; take specific actions after the illegal login reaches a certain number of times, such as account locking; check whether the login connection timeout and auto-logout function are configured and enabled.

If the results show that the sensor node device is configured and enabled with the login failure processing function, the end session and the login connection timeout automatic exit function, the illegal login function is configured and enable and the illegal login function is enabled, and a specific action is taken after the illegal login reaches a certain number of times, and the relevant measures such as ending the session, limiting the number of illegal login times and automatically exiting when the login connection timeout are validated by the test, the result is conforming; otherwise it is non-conforming or partially conforming.

[L3-CES1-03/ L4-CES1-03 Interpretation and Description]

The targets of the evaluation index that "necessary measures should be taken to prevent the authentication information from being eavesdropped during the network transmission" are sensor node devices.

The key points of evaluation implementation include: verify whether the system is remotely managed in a Security manner such as encryption, so as to prevent the authentication information from being eavesdropped in the process of network transmission.

If the results show that the remote maintenance login mode of the sensor node devices adopts the encryption technology such as HTTPS for communication protection, the result is conforming; otherwise it is non-conforming or partially conforming.

[L3-CES1-04/ L4-CES1-04 Interpretation and Description]

The targets of the evaluation index "two or more identification technologies such as cryptography, cryptography technology and biological technology should be adopted to authenticate the user's identity, and at least one of the authentication technologies should be realized by cryptography technology" are the sensor node devices and business application system.

The key points of evaluation implementation include: check that there are two or more combination authentication technologies in the high security scenario of the system, for example, whether two or more combination authentication technologies such as cryptography, digital certificate, biotechnology and device fingerprint are used to authenticate the user's identity. Check whether one of the authentication is implemented by using cryptographic techniques.

If the results show that the login of the relevant system management platform adopt two or more than two combination technologies such as cryptography, digital certificate, biotechnology and device fingerprint to authenticate the user identity, and one of the authentication technologies is realized by using the cryptography technology, the result is conforming; otherwise it is non-conforming or partially conforming.

2. Access Control

[Standard Requirements]

Level Ⅲ of the control point includes evaluation units L3-CES1-05, L3-CES1-06, L3-CES1-07, L3-CES1-08, L3-CES-09 and Level Ⅳ includes evaluation units L4-CES1-05, L4-CES1-06 and L4-CES1-07, L4-CES1-08, L4-CES1-09.

[L3-CES1-05/ L4-CES1-05 Interpretation and Description]

The targets of the evaluation index "account and authority should be allocated to the logged-in user" are the IoT business application system and the sensor node devices.

The key implementation points include: check whether users have been assigned accounts and permissions and related settings; verify that access to anonymous, default accounts has been disabled or restricted.

If the test result shows that the sensor node in the relevant system can set different account for different login users and allocate corresponding permissions; the sensor node device in the system cannot log in with the anonymous account or the default account, the result is conforming; otherwise it is non-conforming or partially conforming.

【L3-CES1-06/ L4-CES1-06 Interpretation and Description】

The targets of the evaluation index "should rename or delete the default account, and modify the default cryptography of the default account" are the sensory node device.

The key points of the evaluation include: check whether the default account has been renamed or deleted; check whether the default cryptography of the default account has been modified.

If the results show that the relevant sensor node device has rename or deleted the default account, the result is conforming; otherwise it is non-conforming or partially conforming.

【L3-CES1-07/ L4-CES1-07 Interpretation and Description】

The targets of the evaluation index "delete or disable redundant and expired accounts in time to avoid the existence of shared accounts" are sensor node devices and business application systems.

The key points of the evaluation include: check whether there is no redundant or expired account, and whether there is a one-to-one correspondence between the administrator user and the account; test and verify whether the redundant or expired account has been deleted or deactivated.

If the results show that there is no redundant or expired account in the system or management platform, and the account and the user are in a one-to-one correspondence; the redundant and expired accounts have been deleted or disabled, the result is conforming; otherwise it is non-conforming or partially conforming.

【L3-CES1-08/ L4-CES1-08 Interpretation and Description】

The targets of the evaluation index "grant the minimum authority required by the management user to realize the authority separation of the management user" are the IoT business application system and the sensor node.

The key points for evaluation implementation include: check whether role division is performed; check whether the permissions of management users have been separated; and check whether the permissions of management users are the minimum permissions required for their work tasks.

Check whether the management user is divided into account administrator, security administrator, auditor, etc.; check whether the authority of the management user is reasonably separated; check whether the authority of the management user is the minimum authority required to complete the normal work tasks.

If the results show that the relevant sensor node divide the roles of the management user, the authority of the management user is reasonably separated, and the authority distribution of the management user of different roles is based on the minimum authority distribution required to complete the normal work task, the result is conforming; otherwise it is non-conforming or partially conforming.

【L3-CES1-09/ L4-CES1-09 Interpretation and Description】

The targets of the evaluation index "the authorized subject should configure the access control policy, and the access control policy specifies the access rules of the subject to the object" are the sensor node, the access control device, etc.

The key points of evaluation implementation include: check whether the authorization subject (such as the management user) is responsible for configuring the access control policy; check whether the authorization subject has configured the access rules of the subject to the object according to the security policy; and test whether the authentication user has the situation of unauthorized access.

If the test results show that only the account with access device management authority can log in the access device management platform to configure the access policy; the admission device management platform is configured with an access policy for the device of the sensor node to access the network, the result is conforming; otherwise it is non-conforming or partially conforming.

3. Security Audit

【Standard Requirements】

Level III of the control point includes evaluation units L3-CES1-12, L3-CES1-13, L3-CES1-14, L3-CES1-15, and Level IV includes evaluation units L4-CES1-12, L4-CES1-13, L4-CES1-14 and L4-CES1-15.

【L3-CES1-12/ L4-CES1-12 Interpretation and Description】

The targets of the evaluation index "able the security audit function, the audit should cover each user, and audit the important user behavior and important security events" are the IoT business application system.

The key points of evaluation implementation include: verify whether the security audit function is provided and enabled; verify whether the security audit scope covers each user; and verify whether important user behaviors and important security events are audited.

If the results show that the integrated security audit system is deployed and accessed to the audit information of the IoT application system, or the IoT application system itself has the audit function; the audit covers all the important users, including the critical user of the application system and the privileged account number; and the behaviors of the user on the management platform, including setting the configuration information of the accessed IoT device, setting the group information of the access device, modifying the policy configuration, developing the cryptography, creating the user, changing the user authority, etc., are audited and can be backtracked, the result is conforming; otherwise it is non-conforming or partially conforming.

[L3-CES1-13 Interpretation and Description]

The targets of the evaluation index "the audit record should include the date and time of the event, user, event type, success of the event and other audit-related information" are the IoT business application system.

The key points of the evaluation include: check whether the audit record information includes the date and time of the event, the user, the type of the event, the success of the event and other audit-related information.

If the results show that the information record by the audit function of the management platform includes the audit-related information such as the date and time of the event, the user, the type of the event, the success of the event and the like, the result is conforming; otherwise it is non-conforming or partially conforming.

[L4-CES1-13 Interpretation and Description]

The targets of the evaluation index "the audit record should include the date and time of the event, event type, subject identification, object identification and result" are the IoT business application system.

The implementation points of the evaluation include: check whether the audit record information includes the date and time of the event, the subject identification, the object identification, the event type, the success of the event and other audit-related information.

If the results show that the information record by the audit function of the system undergoing test management platform includes the audit-related information such as the date and time of the event, the subject identification, the object identification, the event type, the success of the event and the like, the result is conforming; otherwise it is non-conforming or partially conforming.

[L3-CES1-14/ L4-CES1-14 Interpretation and Description]

The targets of the evaluation index "protect and back up the audit records regularly to avoid unexpected deletion, modification or coverage" are the evaluation targets of the IoT business application system.

The key points of the evaluation include: check whether the technical measures are taken to protect the audit records; check whether the technical measures are taken to back up the audit records regularly, and check the backup strategy.

If the results show that the system adopt such technical measures as encryption storage and record anti-tampering for the audit data, the audit data backup strategy is reasonable, and the audit data is backed up regularly according to the backup strategy, the result is conforming; otherwise it is non-conforming or partially conforming.

[L3-CES1-15/ L4-CES1-15 Interpretation and Description]

The targets of the evaluation index "audit process should be protected to prevent unauthorized interruption" are the IoT business application system.

The key points of evaluation implementation include: test to verify that the audit process is interrupted through other accounts other than the audit administrator, and verify that the audit process is protected.

If the test results show that the audit function of the system undergoing test cannot be closed by using the non-audit administrator account and the audit process cannot be terminated through the process management tool, the result is conforming; otherwise it is non-conforming or partially conforming.

4. Intrusion Prevention

[Standard Requirements]

Level Ⅲ of the control point includes evaluation units L3-CES1-17, L3-CES1-18, L3-CES1-19, L3-CES1-20, L3-CES1-21, L3-CES1-22, and Level Ⅳ includes evaluation units L4-CES1-16, L4-CES1-17, L4-CES1-18, L4-CES1-19, L4-CES1-20, L4-CES1-21.

[L3-CES1-17/ L4-CES1-16 Interpretation and Description]

The targets of the evaluation index "follow the principle of minimum installation, and install the required components and applications only" are the components and applications installed by the sensor node device.

The key evaluation implementation points include: verify that the minimum installation principle is followed; and verify that non-essential components and applications are not installed.

If the results show that the relevant sensor node devices follow the minimum installation principle, only the necessary components are deployed, and there are no other non-essential components or application program except for the normal operation of the guarantee system, the result is conforming; otherwise it is non-conforming or partially conforming.

[L3-CES1-18/ L4-CES1-17 Interpretation and Description]

The targets of the evaluation index "close unneeded system services, default sharing and high-risk ports" are sensor node devices device.

The key points of evaluation implementation include: check whether unnecessary system services and default shares are closed; check whether there are no unnecessary high-risk ports.

If the results show that the non-essential system service such as FTP, Telnet and the default share are closed by the relevant sensor node device, and there is no unnecessary high-risk port, such as 445,21,23 network port, the result is conforming; otherwise it is non-conforming or partially conforming.

【L3-CES1-19/ L4-CES1-18 Interpretation and Description】

The targets of the evaluation index "restrict the management terminal managed through the network by setting the terminal access mode or network address range" are sensor node devices.

The key points of evaluation implementation include check whether the configuration file or parameter limits the access range of the terminal.

If the results show that the policy configuration of the relevant admission device can restrict the access mode of the terminal or the IP address range of the network, the result is conforming; otherwise it is non-conforming or partially conforming.

【L3-CES1-20/ L4-CES1-19 Interpretation and Description】

The targets of the evaluation index that "the data validity test function should be provided to ensure that the content input through the man-machine interface or the communication interface conforms to the system setting requirements" are sensor node devices.

The key points of evaluation implementation include: check whether the content of system design document includes the content or module of data validity inspection function; test and verify whether the validity of the input content of man-machine interface or communication interface is conducted.

If the test results show that the relevant management platform has the function of check the validity of the input content of the interface and the module description; after inputting the illegal content such as SQL injection or invalid content in the device search condition input interface, the management platform will prompt that the input content is illegal and refuse to execute, then the result is conforming; otherwise it is non-conforming or partially conforming.

【L3-CES1-21/ L4-CES1-20 Interpretation and Description】

The targets of the evaluation index that "it should be able to find possible known vulnerabilities and repair them in time after full test and evaluation" are sensor node devices.

The key points of evaluation implementation include: check whether there is no high-risk vulnerability through vulnerability scanning and penetration test; verify whether the vulnerability is repaired in a timely manner after adequate test and evaluation.

If the results show that the firmware of the relevant sensor node device does not have a high-risk vulnerability detected by the firmware analysis tool; the management platform of the sensor node device has the function of updating the online firmware of the access sensor node device; the sensor node device can obtain the leak patch from the management platform and repair the loophole in time, the result is conforming; otherwise it is non-conforming or partially conforming.

【L3-CES1-22/ L4-CES1-21 Interpretation and Description】

The targets of the evaluation index "should be able to detect the behavior of invading important nodes and provide alarm in case of serious intrusion event" are the situation awareness platform of IoT, threat intelligence monitoring system, etc.

The key points of evaluation implementation include: interview and check whether there are intrusion detection measures; check whether the alarm is provided in case of serious intrusion event.

If the test results show that the relevant situation-aware platform has detection measures for network attack, malware spread, APT and other intrusion behaviors, and the situation-aware platform can identify and generate an alarm after using APT simulation tool to generate threat traffic, the result is conforming; otherwise it is non-conforming or partially conforming.

5. Trusted verification

[Standard Requirements]

Level III of the control point includes the evaluation unit L3-CES1-24, and Level IV includes the evaluation unit L4-CES1-23.

[L3-CES1-24 Interpretation and Description]

The targets of the evaluation index "conduct trusted verification on the system boot program, system program, important configuration parameters and application program of the computing device based on the trusted root, conduct dynamic trusted verification in the key execution links of the application program, give an alarm after the credibility is detected to be damaged, and send the verification result into audit record and send to the security management center" are sensor node devices.

The key points of evaluation implementation include: check whether the system boot program, system program, important configuration parameters and application program of the computing device are authenticated based on the trusted root; check whether the dynamic trusted verification is performed in the key execution links of the application program; test and verify whether the alarm is given when the credibility of the computing device is detected to be damaged; and test whether the verification results are sent to the security management center in the form of audit records.

If the results show that the boot program perform trusted verification on the boot firmware by means of a trusted key or a trusted hash in the startup process of the relevant sensor node device; the firmware is trusted verify in the key links such as system boot and firmware update; the tampered or incomplete firmware is used for updating, the sensor node device alarms during boot; and the related operation and alarm information is sent to the management platform or other security management center in the form of audit records; the result is conforming; otherwise it is non-conforming or partially conforming.

[L4-CES1-23 Interpretation and Description]

The targets of the evaluation index "conduct trusted verification on the system boot program, system program, important configuration parameter and application program of the computing device based on the trusted root, and conduct dynamic trusted verification in all the execution links of the application program, and give an alarm after the credibility is detected to be damaged, and the verification result should be sent to the security management center in the form of audit record, and the dynamic association perception should be carried out" are sensor node devices.

The key points of evaluation implementation include: check whether the system boot program, system program, important configuration parameters and application program of the computing device are authenticated based on the trusted root; check whether the dynamic trusted verification is performed in all the execution links of the application program; test and verify whether the alarm is given when the credibility of the computing device is detected to be damaged; test whether the verification result is sent to the security management center in the form of audit record; and check whether the dynamic association perception can be performed.

If the results show that the boot program perform trusted verification on the boot firmware by means of a trusted key or a trusted hash in the startup process of the relevant sensor node devices machine; the firmware or application is trusted verify in all execution links such as system boot, firmware update, application upgrade, etc. ; the tampered or incomplete firmware is used for updating and an alarm is given during boot; the related operation and alarm information is sent to the management platform or other security management center in the form of audit record; and the audit record formed by the trusted verification result is dynamically associated with the situation awareness platform and the threat information center, the result is conforming; otherwise it is non-conforming or partially conforming.

6. Data Integrity

[Standard Requirements]

Level Ⅲ of the control point includes evaluation units L3-CES1-25 and L3-CES1-26, and Level Ⅳ includes evaluation units L4-CES1-24 and L4-CES1-25.

[L3-CES1-25/ L4-CES1-24 Interpretation and Description]

The targets of the evaluation index "adopt verification technology or cryptography technology to ensure the integrity of important data during transmission, including but not limited to authentication data, important business data, important audit data, important configuration data, important video data and important personal information" are sensor node devices.

The key points of evaluation implementation include: interview the system administrator to check whether the system design documents, authentication data, important business data, important audit data, important configuration data, important video data and important personal information have adopted verification technology or cryptography technology in the transmission process to ensure the integrity; use tools to tamper the authentication data, important business data, important audit data, important configuration data, important video data and important personal information in the communication message, and check whether the integrity of the data in the transmission process can be detected and recovered in time.

If the test results show that the designing document of the relevant sensor node device includes the verification technology or the cryptographic techniques to protect the integrity of the important data in the transmission process, and can detect the tampering of the important data during the transmission process, the result is conforming; otherwise it is non-conforming or partially conforming.

[L3-CES1-26/ L4-CES1-25 Interpretation and Description]

The targets of the evaluation index of "verification technology or cryptography technology should be adopted to ensure the integrity of important data in the storage process, including but not limited to authentication data, important business data, important audit data, important configuration data, important video data and important personal information" are sensor node devices.

The key points of evaluation implementation include: ask the system administrator whether the authentication data, important business data, important audit data, important configuration data, important video data and important personal information of the system are stored with verification technology or cryptography technology to ensure the integrity; use tools to tamper the authentication data, important business data, important audit data, important configuration data, important video data and important personal information in the communication message, and check whether the integrity of the data in the storage process can be detected and recovered in time.

If the test results show that the relevant sensor node devices adopts the verification technology or the cryptographic techniques to protect the integrity of the important data in the storage process, and can detect the tampering of the stored important data and recover in time, the result is conforming; otherwise it is non-conforming or partially conforming.

7. Data Confidentiality

[Standard Requirements]

Level Ⅲ of the control point includes evaluation units L3-CES1-27 and L3-CES1-28, and Level Ⅳ includes evaluation units L4-CES1-27 and L4-CES1-28.

[L3-CES1-27/ L4-CES1-27 Interpretation and Description]

The targets of the evaluation index "adopt cryptography technology to ensure the confidentiality of important data in the transmission process, including but not limited to authentication data, important business data and important personal information" are sensor node devices.

The key points of evaluation implementation include: ask the system administrator whether cryptography technology is used to ensure the confidentiality of authentication data, important business data and important personal information in the transmission process; catch the data package in the transmission process by sniffing, and whether the authentication data, important business data and important personal information are encrypted during the transmission process.

If the results show that the sensor node devices undergoing test adopt the cryptography technology to ensure the confidentiality of the relevant data in the transmission process, and the relevant data transmission is encrypted through the sniffing network data packet verification, the result is conforming; otherwise it is non-conforming or partially conforming.

[L3-CES1-28/ L4-CES1-28 Interpretation and Description]

The targets of the evaluation index "cryptography technology should be adopted to ensure the confidentiality of important data in the storage process, including but not limited to authentication data, important business data and important personal information" are sensor node devices.

The key points of the evaluation include: inquire the system administrator whether the cryptography technology is adopted to ensure the confidentiality of authentication data, important business data and important personal information in the storage process; check the relevant fields in the database or configuration file, and check whether the authentication data, important business data and important personal information are encrypted and stored.

If the results show that the sensor node device undergoing test adopts the cryptography technology to ensure the confidentiality of the relevant data in the storage process, and the relevant data is encrypted and stored, the result is conforming; otherwise it is non-conforming or partially conforming.

8. Sensor Node Devices Security

[Standard Requirements]

Level III of the control point includes evaluation units L3-CES4-01, L3-CES4-02, L3-CES4-03, and Level IV includes evaluation units L4-CES4-01, L4-CES4-02 and L4-CES4-03.

[L3-CES4-01/ L4-CES4-01 Interpretation and Description]

The targets of the evaluation index that "only authorized users can configure or change the software application on the sensor node device" are sensor node devices.

The key points of evaluation implementation include: check whether the sensor node device adopts certain technical means to prevent unauthorized users from configuring or changing the software application on the device; try to access and control the sensor network to access the unauthorized resources, and test and verify whether the access control measures of the sensor node device are effective in controlling the behavior of illegally accessing and illegally using the resources of the sensor node device.

If the results show that the management configuration interface can be entered only after successful login, only the authorized user can configure or change it, and the sensor node resource cannot be accessed and used illegally by means of counterfeit authorized network address, phishing authorized user, phishing authorized device, etc., then the result is conforming; otherwise it is non-conforming or partially conforming.

[L3-CES4-02/ L4-CES4-02 Interpretation and Description]

The targets of the evaluation index "should be able to identify and authenticate the connected gateway node devices (including the card reader)" are gateway node devices (including the card reader).

The key points of evaluation implementation include: check whether the connected gateway node devices (including the card reader) are identified and authenticated, and whether the parameters conforming to the security policy are configured; test and verify whether there is no method to bypass the identity identification and authentication functions.

If the results show that when the gateway node devices access the network, the gateway node devices is identified and authenticated; only the access gateway node devices conforms to the identity authentication strategy and is allowed to access the network after passing the identity identification and authentication with other equipment; and there is no method to bypass the identity identification and authentication function of the gateway node devices through the penetration test, the result is conforming; otherwise it is non-conforming or partially conforming.

[L3-CES4-03/ L4-CES4-03 Interpretation and Description]

The targets of the evaluation index "should be able to identify and authenticate the connected other sensor node devices (including routing nodes)" are other sensor node devices (including routing nodes).

The key points of evaluation implementation include: check whether the identity identification and authentication of other connected sensor node devices (including routing node) devices are performed, and whether the parameters conforming to the security policy are configured; and whether there is no method to bypass the identity identification and authentication functions should be tested.

If the results show that the other sensor node devices, such as the routing node devices, conduct identity identification and authentication to the node devices when accessing the network; only theconnected routing node devices conform to the identity authentication policy, and are allowed to access the network after passing the identity identification and authentication with other equipment; and the penetration test verifies that there is no method to bypass the identity identification and authentication function of the gateway node devices, the result is conforming; otherwise it is non-conforming or partially conforming.

9. Gateway Node Devices Security

[Standard Requirements]

Level Ⅲ of the control point includes evaluation units L3-CES4-05, L3-CES4-06, L3-CES4-07, L3-CES4-08, and Level Ⅳ includes evaluation units L4-CES4-05, L4-CES4-06,

L4-CES4-07 and L4-CES4-08.

【L3-CES4-05/ L4-CES4-05 Interpretation and Description】

The targets of the evaluation index "should be able to identify the legally connected equipment (including terminal node, routing node and data processing center)" are gateway node devices.

The key points of evaluation implementation include: check whether the gateway node devices can identify the connected equipment (including terminal node, routing node and data processing center) and configure the authentication function; test and verify whether there is no method to bypass the identity identification and authentication function.

If the results show that the gateway node can identify the connected device, and can use the identity authentication information of the connected device to configure the authentication function and perform the device authentication; the gateway access node device only allows the legal device that has undergone the identity authentication to access; and it is confirmed through the penetration test that there is no method to bypass the identity identification and authentication function, the result is conforming; otherwise it is non-conforming or partially conforming.

【L3-CES4-06/ L4-CES4-06 Interpretation and Description】

The targets of the evaluation index "should be able to filter the data sent by illegal nodes and forged nodes" are gateway node devices.

The key points of evaluation implementation include: check whether it has the function of filtering the data sent by illegal nodes and forged nodes; test and verify whether it can filter the data sent by illegal nodes and forged nodes.

If the results show that the gateway node devices undergoing test has the function of identifying the illegal node and the counterfeit node and filtering the data sent by the gateway node; through the test verification, the gateway node can filter the data sent by the illegal node and the forged node, the result is conforming; otherwise it is non-conforming or partially conforming.

【L3-CES4-07/ L4-CES4-07 Interpretation and Description】

The targets of the evaluation index "the authorized user should be able to update the critical key online during the use of the device" are gateway node devices.

The key points of evaluation implementation include: check whether the gateway node devices update its critical keys online.

If the results show that the relevant IOT security gateway use the key, and the critical key can be updated online during the operation of the IOT security gateway; the result is conforming; otherwise it is non-conforming or partially conforming.

【L3-CES4-08/ L4-CES4-08 Interpretation and Description】

The targets of the evaluation index "the authorized user should be able to update the key configuration parameters online during the use of the equipment" are gateway node devices.

The key points of evaluation implementation include: check whether online updating of key configuration parameters is supported and whether online updating method is effective.

If the results show that the authorize user can update the key configuration parameters such as the MAC address white list and the communication port white list of the access-sensor node device through the online management platform or the remote configuration mode, and the related gateway equipment work correctly according to the new parameters after the configuration parameters are updated, the result is conforming; otherwise it is non-conforming or partially conforming.

10. Data Replay Resistant

[Standard Requirements]

Level Ⅲ of the control point includes evaluation units L3-CES4-09 and L3-CES4-10, and Level Ⅳ includes evaluation units L4-CES4-09 and L4-CES4-10.

[L3-CES4-09/ L4-CES4-09 Interpretation and Description]

The targets of the evaluation index "should be able to identify the freshness of data and avoid replay attack of historical data" are sensor node devices device.

The key points of evaluation implementation include: check whether the measures for data freshness identification of sensor node devices can avoid the replay of historical data; check whether the application system of IoT processing application layer can recognize the replay of historical data of sensor node and discard the replay data; conduct replay test on historical data of sensor node devices to verify whether the protective measures are effective.

If the results show that the measure such as time stamp or counting mark are adopted by the device of the tested sensor node to the data to avoid the replay of the historical data; the corresponding management platform can identify the historical replay data of the sensor node and discard the replay data, the result is conforming; otherwise it is non-conforming or partially conforming.

[L3-CES4-10/ L4-CES4-10 Interpretation and Description]

The targets of the evaluation index "should be able to identify the illegal modification of historical data and avoid the modification and replay attack of data" are sensor node devices device.

The key points of evaluation implementation include: check whether the sensing layer is equipped with measures to detect the illegal tampering of the historical data of the sensor node devices, and whether necessary recovery measures can be taken when the modification is detected; test and verify whether the modification replay attack of data can be avoided.

If the results show that the relevant management platform has the measure of detecting that the historical data of the sensor node device is illegally tampered; when the sensor node data is detect to be tampered, necessary recovery measures can be taken; and the protective measures are effective by tampering with the historical data of the sensor node device, the result is conforming; otherwise it is non-conforming or partially conforming.

11. Data Fusion Processing

[Standard Requirements]

Level III of the control point includes the evaluation units L3-CES4-11, and Level IV includes the evaluation units L4-CES4-11 and L4-CES4-12.

[L3-CES4-11/ L4-CES4-11 Interpretation and Description]

The targets of the evaluation index "data from sensor network should be data fusion processing, so that different types of data can be used on the same platform" are the application systems of IoT.

The key evaluation points include: check whether the function of data fusion processing for data from sensor network is provided; test and verify whether the data fusion processing function can process different kinds of data.

If the results show that the gateway node devices or the application system has the function of fusion process for multiple data, and the gateway node devices or the application system can correctly perform fusion processing on the multi-data of the sensing network, the result is conforming; otherwise it is non-conforming or partially conforming.

[L4-CES4-12 Interpretation and Description]

The targets of the evaluation index "the dependence and restriction relationship between different data should be intelligently processed, for example, when one type of data reaches a certain threshold, it can affect the management instruction to another data acquisition terminal" is the IoT application system.

The key points of the evaluation include: check whether it is able to handle the dependency and constraint relationship between different data intelligently.

If the results show that the IoT system management platform can carry out intelligent processing according to the dependence relation and restriction relation between different perceived data, and when some data reach the threshold, it can affect the management instruction to another type of data acquisition terminal, then the result is conforming; otherwise it is non-conforming or partially conforming.

6.2.5 Security Management Center

According to the characteristics of the IoT, this chapter interprets several general security requirement evaluation of the IoT.

1. System Management

[Standard Requirements]

Level III of the control point includes the evaluation units L3-SMC1-01 and L3-SMC1-02, and Level IV includes the evaluation units L4-SMC1-01 and L4-SMC1-02.

[L3-SMC1-01/ L4-SMC1-01 Interpretation and Description]

The targets of the evaluation index "identify system administrators should be identified, only allowed to conduct system management operations through specific command or operation interface, and audit these operations" mainly include the unified management system of sensor nodes, the unified management system of network equipment, and the system providing centralized management of security-related matters such as security policy, malicious code, patch upgrade, etc., such as SOC operation and maintenance platform, fortress machine, security management system, log audit platform, etc.

The key points of evaluation implementation include: check whether the system of unified management system of sensor node, unified management system of network equipment, SOC operation and maintenance platform, fortress machine, security management system and log audit platform conduct identity authentication for administrator login; check whether system administrator is only allowed to conduct system management operation through specific command or operation interface; check whether to audit system management operation, and the audit content covers all management operations.

If the test results show that the centralized management system authenticates the administrator's identity, allows the administrator to perform system management operation only through specific command or operation interface, and audits its operation through the system itself or the centralized log audit platform, the result is conforming; otherwise it is non-conforming or partially conforming.

[L3-SMC1-02/ L4-SMC1-02 Interpretation and Description]

The targets of the evaluation index "configure, control and manage the system resources and operation through the system administrator, including user identity, resource allocation, system loading and startup, exception handling of system operation, data and equipment backup and recovery, etc." include the unified management system of sensor node, the unified management system of network equipment, and the system providing centralized management for security-related matters such as security policy, malicious code, patch upgrade, etc., such as SOC operation and maintenance platform, fortress machine, security management system, log audit platform, etc.

The key points of evaluation implementation include: interview the system administrator to understand the configuration, control and management of system resources and operation, including user identity, resource allocation, system loading and startup, exception handling of system operation, backup and recovery of data and equipment and other specific operation modes and roles of operators; check whether the operators and authorities contained in the system providing these management functions have the administrative authority of the system administrator, and whether other non-administrators use the administrator account for operation; check the specific log of these operations to see whether all operation contents are recorded and whether they are implemented by the system administrator.

If the test results show that the configuration, control and management of system resources and operation, including user identity, resource configuration, system loading and startup, exception handling of system operation, backup and recovery of data and equipment, etc., are only implemented by the system administrator, and comprehensive audit function is provided, the result is conforming; otherwise it is non-conforming or partially conforming.

2. Audit Management

[Standard Requirements]

Level Ⅲ of the control point includes the evaluation units L3-SMC1-03 and L3-SMC1-04, and Level Ⅳ includes the evaluation units L4-SMC1-03 and L4-SMC1-04.

[L3-SMC1-03/ L4-SMC1-03 Interpretation and Description]

The targets of the evaluation index that "identify audit administrator and only allow to conduct security audit operations through specific command or operation interface, and audit these operations" include the unified management system of sensor node, the unified management system of network equipment, and the system providing centralized management of security-related matters such as security policy, malicious code, patch upgrade, etc., such as SOC operation and maintenance platform, fortress machine, security management system, and audit function.

The key points of evaluation implementation include: verify whether the unified management system of sensor node, unified management system of network equipment, SOC operation and maintenance platform, fortress machine and security management system provide audit function, whether special audit administrator is set, and whether the identity of audit administrator is identified; test and verify whether the audit function of equipment management system, SOC operation and maintenance platform, fortress machine and security management system is only authorized by audit administrator to view and manage, and whether audit administrator is only allowed to conduct security audit operation through specific command or operation interface; test and verify whether audit administrator has relevant operation records after logging in the audit system and performing operations such as view, backup and clearing (if any).

If the test results show that the relevant management system is equipped with a special audit administrator to authenticate the identity of the audit administrator, only the audit administrator has the authority to view and manage the audit function, and only the audit administrator is allowed to perform the security audit operation through specific command or operation interface, and relevant operation records are available, the result is conforming; otherwise it is non-conforming or partially conforming.

[L3-SMC1-04/ L4-SMC1-04 Interpretation and Description]

The targets of the evaluation index "analyze and process the audit records through audit administrator according to the analysis results, including the storage, management and query of audit records according to the security audit strategy" include the unified management system of sensor nodes, the unified management system of network equipment, and the system providing centralized management of security-related matters such as security policy, malicious code, patch upgrade, etc., such as the audit records generated by SOC operation and maintenance platform, fortress machine and security management system.

The key points of evaluation implementation include: check whether the management system of sensor node, unified management system of network equipment, SOC operation and maintenance platform, fortress machine and security management system provide audit function, and whether special audit administrator is set; whether the audit administrator analyzes and processes the audit records, and whether the audit records are processed according to the analysis results, including storing, managing and querying audit records according to the security audit policy.

If the results show that the relevant management system provide the audit function and a special audit administrator is set, and the audit administrator analyzes and processes the audit record, including storing, managing and querying the audit record according to the security audit policy, the result is conforming; otherwise it is non-conforming or partially conforming.

3. Security Management

[Standard Requirements]

Level III of the control point includes evaluation units L3-SMC1-05 and L3-SMC1-06, and Level IV includes evaluation units L4-SMC1-05 and L4-SMC1-06.

[L3-SMC1-05/ L4-SMC1-05 Interpretation and Description]

The targets of the evaluation index that "should identify the security administrators, and only allows to conduct security management operations through specific commands or operation interfaces, and audit these operations" include the unified management system of sensor nodes, the unified management system of network equipment, and the system providing centralized management of security-related matters such as security policy, malicious code, patch upgrade, etc., such as SOC operation and maintenance platform, fortress machine, security management system, etc.

The key points of evaluation implementation include: check whether the systems providing centralized management function, such as the unified management system of sensor node, unified management system of network equipment, SOC operation and maintenance platform, fortress machine and security management system, are equipped with security administrator and identify the security administrator login; for these systems, check whether the security administrator is only allowed to perform security management operation through specific command or operation interface, such as security policy setting, patch distribution, malicious code library upgrade, etc., and there is no other method to bypass the operation mode; for these systems, check whether to audit the security management operation, and the audit content covers security policy setting, patch distribution, malicious code base upgrade and other events.

If the results show that the relevant management system provide the identity authentication function of the security administrator, only allows the security administrator to carry out the security management operation through the operation interface of the management system, there is no other operation mode to bypass the system, and the system security management operation is provided with a comprehensive audit function, the result is conforming; otherwise it is non-conforming or partially conforming.

[L3-SMC1-06 Interpretation and Description]

The targets of the evaluation index "configure the security strategy in the system through security administrator, including the setting of security parameters, unified security marking for the subject and object, authorization for the subject, configuration of trusted verification strategy, etc." include the unified management system of sensor node, the unified management system of network equipment, and the system providing centralized management of security-related matters such as security policy, malicious code, patch upgrade, etc., such as SOC operation and maintenance platform, fortress machine, security management system, log audit platform, etc.

The key points of evaluation implementation include: check whether the security policy in the system is configured by the security administrator of the systems providing centralized management function, such as the unified management system of sensor node, SOC operation and maintenance platform, fortress machine and security management system, including setting of security parameters, unified security marking of subject and object, authorization of subject, and configuration of trusted verification strategy, etc.

If the test results show that the security parameters are set, the subject and the object are uniformly marked, the subject is authorized, and the trusted verification policy is configured by the security administrator, the result is conforming; otherwise it is non-conforming or partially conforming.

[L4-SMC1-06 Interpretation and Description]

The targets of the evaluation index "configure the security strategy in the system through security administrator, including the setting of security parameters, unified security marking for the subject and object, authorization for the subject, configuration of trusted verification strategy, etc." include the unified management system of sensor node, the unified management system of network equipment, and the system providing centralized management of security-related matters such as security policy, malicious code, patch upgrade, etc., such as SOC operation and maintenance platform, fortress machine, security management system, log audit platform, etc.

The key points of evaluation implementation include: check whether the system providing centralized management function such as the unified management system of sensor node, SOC operation and maintenance platform, fortress machine and security management system has set up a special security administrator, and understand the specific implementation mode of the operations including setting of security parameters, unified security marking for the subject and object, authorizing the main body, and configuring the trusted verification policy, etc.; check whether the established operators and authorities in the system providing these security management functions have the security policy configuration authority, and whether other non-security administrators use the security administrator account for operation; check the specific log of these operations, whether all the operation contents are recorded, and whether they are implemented by the security administrator.

If the test results show that the security parameter setting, unified security marking for the subject and object, authorization for the subject, configuration of the trusted verification policy and other operations are only implemented by the security administrator, and the comprehensive audit function is provided, the result is conforming; otherwise it is non-conforming or partially conforming.

4. Centralized Management and Control

[Standard Requirements]

Level III of the control point includes evaluation units L3-SMC1-07, L3-SMC1-08, L3-SMC1-09, L3-SMC1-10, L3-SMC1-11, L3-SMC1-12, and Level IV includes evaluation units L4-SMC1-07, L4-SMC1-08, L4-SMC1-09, L4-SMC1-10, L4-SMC1-11, L4-SMC1-12, L4-SMC1-13.

[L3-SMC1-07/ L4-SMC1-07 Interpretation and Description]

The targets of the evaluation index "designate special management area manage and control the security devices or security components distributed in the network" are the overall network. In the implementation of the evaluation, the overall situation of the network topology should be checked according to the actual situation of the IoT system, including the conventional security devices such as firewall, IDS and fortress, as well as the security protection equipment of near-field communication sensor nodes and edge computing nodes.

The key points of evaluation implementation include: for the system background, check whether separate network areas are divided for deployment of firewall, IDS, fortress machine, network access system, log audit system, etc.; for each edge node gateway, check whether isolation area and security devices are set through the gateway; check whether firewall, IDS, fortress machine, network access system and log audit system are deployed in a centralized way in a separate network area, and check whether the security devices of edge nodes are deployed in the designated isolation area.

If the results show that the back end of the system divide a separate network area, and the edge area is also set with a special area for centralized deployment of the security device or the security component, the result is conforming; otherwise it is non-conforming or partially conforming.

[L3-SMC1-08/ L4-SMC1-08 Interpretation and Description]

The targets of the evaluation index "should be able to establish a Security information transmission path and manage the security devices or components in the network" include routers, switches, firewalls, gateway nodes, etc.

The key points of evaluation implementation include: interview the security administrator to check whether a Security information transmission path (such as SSH, HTTPS, IPSec, etc.) can be established for the security devices or security components in the network, such as firewall, IDS, fortress machine, network access system and log audit system, etc.; check whether there are other ways to manage the security devices by bypassing the Security information transmission path.

If the test results show that a Security information transmission path is established through a Security connection mode, the security devices or components in the network are managed, and a separate out-of-band management network is set up, the result is conforming; otherwise it is non-conforming or partially conforming.

[L3-SMC1-09/ L4-SMC1-09 Interpretation and Description]

The targets of the evaluation index "the operation status of network link, security equipment, network equipment and server should be monitored intensively" are the integrated network management system, server monitoring system and sensor node management system, etc.

The key points of evaluation implementation include: interview the system administrator, whether the system or equipment with the operation status monitoring function is deployed, and can conduct centralized monitoring on the operation status of the network links, routers, switches, firewalls, gateway nodes, sensing devices and servers, etc., such as whether the traffic monitoring system is configured to monitor the running conditions of network equipment and network traffic, whether to monitor the CPU and memory of the network equipment, security devices and servers, and whether to configure the threshold limit alarm function. For the sensor device, the gateway node and the edge computing node, whether the operation status monitoring function is deployed or not, the operation status of the sensing device and the edge computing node can be monitored. Test and verify whether the operation condition monitoring system alarms according to the working state of the network link, security equipment, network equipment, gateway node, sensor device and server, etc. according to the set threshold, such as alarm by SMS, email, etc., and whether the threshold value is set reasonably.

If the results show that the system or equipment with the function of running state monitoring for the back end and the edge computing node of the system can centrally monitor the running state of the network link, the security device, the network equipment, the sensing device, the edge computing node and the server, and can give a real-time alarm according to the state or the set threshold value, the result is conforming; otherwise it is

non-conforming or partially conforming.

[L3-SMC1-10/ L4-SMC1-10 Interpretation and Description]

The targets of the evaluation index "the audit data scattered on each equipment should be collected, summarized and intensively analyzed, and the retention time of audit records should be ensured to meet the requirements of laws and regulations" include the comprehensive security audit system, database audit system and fortress machine, etc.

The key points of evaluation implementation include: interview the administrator to understand whether the unified log audit system is configured and collect the log data of routers, switches, firewalls, gateway nodes and servers; check whether the network equipment, security equipment, server and database are configured and enabled with relevant policies, and send the audit data to the security audit system independent of the equipment itself; check whether the comprehensive security audit system, database audit system and fortress machine are configured and enabled with relevant policies, collect and store the equipment logs in a unified manner, and conduct centralized analysis and processing of the audit data on a regular basis; check whether the audit data of the comprehensive security audit system, database audit system, fortress machine and other log audit systems have a retention time of at least 6 months, whether the logs are backed up, and whether the backup frequency and method can ensure that the data will not be lost.

If the test results show that the audit data scattered on each equipment can be collected, summarized and analyzed in a centralized manner through the integrated security audit system, database audit system, fortress machine and sensory equipment management system, and sufficient backup measures are taken to ensure that the audit records are kept for more than 6 months, the unit will judge the result as conforming; otherwise the unit is deemed as non-conforming or partially conforming.

[L3-SMC1-11/ L4-SMC1-11 Interpretation and Description]

The targets of the evaluation target "manage such security-related matters as security policy, malicious code and patch upgrading in an integrated manner" include the unified management system of sensor node, the unified management system of network equipment, and the system providing centralized management for security-related matters such as security policy, malicious code, patch upgrade, etc., such as SOC operation and maintenance platform, fortress machine, security management system, etc.

The key points of evaluation implementation include: interview the security administrator to know whether the security policy (firewall access control policy, intrusion protection system prevention policy, and WAF security protection policy, etc.) is managed in a centralized manner. For example, the security administrator should set it uniformly through the security equipment management system, and verify it by viewing the policy settings and log records of the security management system; interview the security administrator to understand whether the anti-malicious code system of the operation system and the network malicious code protection equipment are managed in a centralized manner, check whether the malicious code check and killing system installed on the host supports the unified upgrade function, and check whether the anti-malicious code system of the operation system and the network malicious code protection equipment are configured centrally by the security administrator; interview the security administrator whether to conduct centralized management on the patch upgrade of each operation system, network equipment and sensor node, the specific products and methods to be used, whether the patch to be upgraded is checked regularly, and whether the security evaluation is conducted before the upgrade; Check whether the security administrator manages the patch upgrade of the equipment through the unified SOC operation and maintenance platform.

If the results show that the security-related matters such as security policy, malicious code and patch upgrade can be managed centrally through the unified management system of the sensor node, the unified management system of the network equipment, etc., the result is conforming; otherwise it is non-conforming or partially conforming.

【L3-SMC1-12/ L4-SMC1-12 Interpretation and Description】

The targets of the evaluation index "should be able to identify, alarm and analyze various security events occurring in the network" include the unified management system of sensor node and the unified management system of network equipment.

The key points of evaluation implementation include: interview system administrator, whether the system or equipment with security event analysis function is deployed, such as situation awareness platform, threat analysis platform, Web firewall, etc., which can identify CC attack, Web attack, crawler, machine swipe and other security events intermingled in the request, and give an alarm through sound, light and short message; interview the security administrator and check relevant processing records, and check whether the security event alarm is analyzed and handled in time, whether there is relevant processing record and security event analysis report; understand the data flow of the system, check the situation awareness platform, threat analysis platform, Web firewall and other security management equipment, whether it can cover all the key data flow, such as Internet data collection path, VPN or other private line data uploading path, enterprise subnet data transmission path, and the transmission path of each sensor node data upload and instruction issue.

If the test results show that the back-end and edge computing nodes of the system can identify various security events occurring in the network, provide real-time alarm function, timely analyze malicious traffic or take blocking measures, the result is conforming; otherwise it is non-conforming or partially conforming.

【L4-SMC1-13 Interpretation and Description】

The targets of the evaluation index "the time within the system should be generated by a unique determined clock to ensure the consistency of management and analysis of various data in time" include the unified management system of sensor nodes, the unified management system of network equipment, and the system providing centralized management of security-related matters such as security policy, malicious code, patch upgrade, etc., such as SOC operation and maintenance platform, fortress machine, security management system, log audit platform, etc.

The key points of evaluation implementation include: check whether the unified management system of sensor node, unified management system of network equipment, SOC operation and maintenance platform, fortress machine, security management system, log audit platform, etc. providing centralized management function, and check whether all relevant devices under centralized management and control use the unique determined clock source, such as sensor node devices, gateway node devices, network equipment, server, etc.

If the results show that both the centralized management system and the relevant devices under the centralized management use the unique determined clock source, the result is conforming; otherwise it is non-conforming or partially conforming.

6.2.6 Security Management System

Please refer to relevant chapters of this book for the application and interpretation of the general requirements of this security evaluation.

6.2.7 Security Management Organization

Please refer to relevant chapters of this book for the application and interpretation of the general requirements of this security evaluation.

6.2.8 Security Management Personnel

Please refer to relevant chapters of this book for the application and interpretation of the general requirements of this security evaluation.

6.2.9 Security Development Management

Please refer to relevant chapters of this book for the application and interpretation of the general requirements of this security evaluation.

6.2.10 Security Operation and Maintenance Management

This chapter interprets the extended security evaluation requirement of "Security operation and maintenance management" of IoT.

1. Sensor Node Management

[Standard Requirements]

Level III of the control point includes evaluation units L3-MMS4-01, L3-MMS4-02, L3-MMS4-03, and Level IV includes evaluation units L4-MMS4-01, L4-MMS4-02 and L4-

MMS4-03.

[L3-MMS4-01/ L4-MMS4-01 Interpretation and Description]

The targets of the evaluation index "designate personnel should regularly check the deployment environment of the sensor node devices and gateway node devices, record and maintain the environmental abnormalities that may affect the normal operation of the sensor node devices and gateway node devices" mainly include the check inspection, environmental anomaly, and maintenance records of the sensor node devices and gateway node devices. The typical objects include the relevant pages of regular inspection, abnormality, fault and maintenance records in the asset management system of the sensing equipment.

The key points of evaluation implementation include: interview the person in charge of the system to find out whether there is regular maintenance for the sensor node devices and gateway node devices, and determine the maintenance period and responsible person; check the maintenance records of the sensor node devices and gateway node devices in the management information system of the sensing equipment to confirm that the records contain the maintenance date, maintenance person, maintenance equipment information, failure causes, maintenance results and other key contents; check whether the maintenance date interval in the maintenance record conforms to the maintenance period known by the person in charge of the system.

If the results show that the system responsible person can provide the information of regular maintenance and maintenance period for the sensor node devices and the maintenance responsible person, the maintenance record contain the maintenance date, maintenance person and maintenance equipment information, and the maintenance record is consistent with the information provided by the system responsible person, the result is conforming; otherwise it is non-conforming or partially conforming.

[L3-MMS4-02/ L4-MMS4-02 Interpretation and Description]

The targets of the evaluation index "specify the rules for the process of warehousing, storage, deployment, carrying, maintenance, loss and scrapping of the sensor node devices and gateway node devices, and conduct the whole process management" include that the security management documents of the sensor node devices and the gateway node devices are the evaluation targets, and the typical evaluation targets include the relevant parts of the sensor node devices of the IoT equipment or the fixed assets management method.

The key points for evaluation implementation include: check whether the verification document clearly covers the warehousing, storage, deployment, carrying, maintenance, loss and scrapping of the sensor node devices and gateway node devices; whether the verification document contains the relevant clear process description on the warehousing, storage, deployment, carrying, maintenance, loss and scrapping of the sensor node devices and gateway node devices; in the verification document, check whether it required to designate a responsible person in the management process of the sensor node devices and gateway node

Chapter 6 Application and Interpretation of the Extended Security Evaluation Requirements of IoT

devices; for the remote maintenance equipment, the remote maintenance security specification should be provided.

If the test results show that the asset management document describes the management regulations for the warehousing, storage, deployment, carrying (transportation), maintenance, loss and scrapping of the sensor node devices and the gateway node devices; the specific processes of the receipt, storage, deployment, carrying (transportation), maintenance, loss and scrapping of the sensor node devices and the gateway node devices are described and required to be recorded; if the life cycle of the sensor node devices and the gateway node devices can be traced completely after consulting and comparing the actual records, and the relevant link record information contains the specific responsible person information, the result is conforming; otherwise it is non-conforming or partially conforming.

[L3-MMS4-03/ L4-MMS4-03 Interpretation and Description]

The targets of the evaluation index of "strengthen the confidentiality of the deployment environment of sensor node devices and gateway node devices, including that the personnel responsible for inspection and maintenance should immediately return relevant inspection tools and inspection and maintenance records when they are transferred from their posts" include the management system documents of sensory node devices and gateway node devices.

The key points of evaluation implementation include: check whether the management documents of the sensor node devices and gateway node devices include the contents of the verification and maintenance personnel who are transferred from their posts to immediately return the relevant verification tools and check the maintenance records, etc. ; check whether the management documents of the sensor node devices and the gateway node devices have relevant confidentiality management requirements for the equipment deployment environment; check whether the records related to the deployment environment of the sensor node devices and the gateway node devices meet the confidentiality management requirements.

If the test results show that the management system has been formulated and it is specified that the personnel responsible for the verification and maintenance of the sensor node devices and gateway node devices should immediately return relevant verification tools and verification and maintenance records when they are transferred from their posts; if the confidentiality management requirements for the deployment environment of the sensor node devices and gateway node devices are specified and verified by the actual inspection, the result is conforming; otherwise it is non-conforming or partially conforming.

Chapter 7 Application and Interpretation of the Extended Security Evaluation Requirements of Industrial Control System

7.1 Overview of Industrial Control System

7.1.1 Characteristics of Industrial Control System

Industrial control system (ICS) is a generic term for several types of control systems, including supervisory control and data acquisition (SCADA), distributed control systems (DCS), and other control systems, such as programmable logic controllers (PLC), which are frequently used in industrial sectors and critical infrastructure. Industrial control systems are commonly used in industries such as power, water and sewage treatment, oil and gas, chemicals, transportation, pharmaceuticals, pulp and paper, food and beverages, and discrete manufacturing (e.g., automotive, aerospace, and durable goods).

Industrial control system is mainly composed of process level (field control layer and field equipment layer), operation level (process monitoring layer) and communication network between and inside each level. For large-scale control system, it also includes management level (production management layer and enterprise resource layer). Process level includes controlled object, field control equipment and measuring instrument, etc., operation level includes engineer and operator station, man-machine interface and configuration software, control server, etc., management level includes production management system and sometimes enterprise resource system, and communication network includes commercial Ethernet, industrial Ethernet, field bus, etc.

Industrial control system is the core of industrial infrastructure, which has the characteristics of high availability and real-time requirement, long system life cycle and high business continuity. There are many differences between the industrial control system and the traditional information system, and the main differences are shown in Table 7.1.

Chapter 7 Application and Interpretation of the Extended Security Evaluation Requirements of Industrial Control System

Table 7.1 Differences Between Industrial Control System and Traditional Information System

Classification	Traditional Information System	Industrial Control System
Performance Requirements	• Non-real-time • High Throughput • High latency and jitter allowed • Response Consistency	• Real-time • Moderately low throughput • Low latency and/or jitter allowed • Response Urgency
Availability Requirements	• Restart or interrupt acceptable • Usability defects can be tolerated, but depending on the operation requirements of the system	• Unable to accept restart or interrupt, redundant system may be required • Interrupts must be executed as planned • Requires detailed pre-deployment testing
Management Requirements	• Data confidentiality and integrity are the most important • Fault tolerance is less important, temporary downtime is not a major risk • Main risk impact is significant delays in business operations	• Personal security is the most important, followed by process protection • Fault tolerance is essential and even momentary downtime may not be acceptable • Main risk impact is non-compliance, environmental impact, loss of life, equipment or production
System Architecture Security Focus	• The focus is on protecting IT assets and the information stored on them and transferred to each other • Central servers may require more protection	• The primary goal is to protect edge clients (e.g., field devices, such as process controllers) • Central server protection is also important
Time-critical Interaction	• Emergency interaction is less important • Tightly restricted access control can be implemented based on the necessary level of security	• Response to people and other emergency interactions is critical • Access to the ICS should be strictly controlled, but should not interfere with the human-computer interaction
System Operation	• Using a typical operation system • Easy system upgrade	• Specific operation system, often without built-in security features • Software changes must be made with care, usually by the software supplier
Resource constraints	• System has sufficient resources to support additional third-party applications	• Systems are designed to support anticipated industrial processes and may not have sufficient resources to support additional security functions

Classification	Traditional Information System	Industrial Control System
Communication	• Standard communication protocol • Mainly wired with a little bit of localized wireless functionality • Typical IT Networking Practices	• Many proprietary and standard communication protocols • Multiple types of media are used, including dedicated wired and wireless (radio and satellite) • Networks are complex and sometimes require the expertise of a control engineer
Change Management	• Software changes are applied quickly and automatically when good security policies and procedures are in place	• Software changes must be thoroughly tested and deployed incrementally throughout the system to ensure the integrity of the control system. There must be a break plan, scheduled in advance (days/weeks). There is an operation system that is no longer supported by the manufacturer
Management Support	• Allow diverse support models	• Often relies on a single supplier
Component Lifecycle	• 3—5 years	• 15—20 years
Component Access	• Components are usually local and easily accessible	• Components can be isolated, remote, and often require significant resources to gain access to them

Note: This comparison table is referenced from NIST SP800-82.

7.1.2 Functional Hierarchy Model of Industrial Control System

The functional hierarchy model is shown in Figure 7.1. The hierarchical model is divided into five levels from top to bottom, which are enterprise resource layer, production management layer, process monitoring layer, field control layer and field equipment layer. The real-time requirements of different levels are different. Enterprise resource layer mainly includes ERP system function unit, which is used to provide decision-making operation means for enterprise decision-making staff; production management layer mainly includes MES system function unit, which is used to manage production process, such as manufacturing data management and production scheduling management; process monitoring layer mainly includes monitoring server and HMI system function unit, which are used to collect and monitor production process data and realize human-computer interaction through HMI system; site control layer mainly includes various controller units, such as PLC and DCS control unit, etc., which are used to control each execution equipment; field device layer mainly includes various process sensing equipment and

Chapter 7 Application and Interpretation of the Extended Security Evaluation Requirements of Industrial Control System

execution equipment units, which are used to sense and operate the production process.

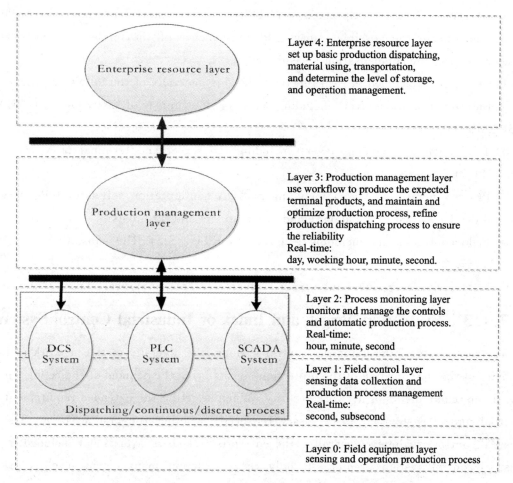

Figure. 7.1 Functional Hierarchy Model of Industrial Control System

Note: The figure is a classical hierarchical model of industrial control system, which is originated from the international standard IEC 62264-1. However, with the development of Industry 4.0 and information physics system, it is not fully applicable. Therefore, for the actual development of different industrial enterprises, it is allowed to combine partial levels.

The industrial control system is usually a hierarchical protection object with high requirements on availability. Some devices in the industrial control system may terminate their continuous operation if they implement specific type of security measures. In principle, the security measures should not adversely affect the basic functions of the high-availability industrial control system. For example, the account used for basic functions should not be locked, even for a short time; the deployment of security measures should not significantly increase the delay and affect the system response time; for high-availability control systems, the failure of security measures should not interrupt the basic functions, etc.

283

When relevant clauses of security level protection requirements cannot be implemented due to assessment of great impact on availability, security statement should be made to analyze and explain the possible impact and consequence of the implementation of this clause, as well as the compensation measures used.

The equipment included in each functional level is generally of the following types:

① Enterprise resource layer: including servers and terminals of ERP, OA, NC, EAS, PMS, LMS and other systems;

② Production management layer: including the server and terminal of MES, SIS, EMS, APC and other systems;

③ Process monitoring layer: including SCADA configuration software, OPC server, historical database, real-time database and other servers and terminals;

④ Field control layer: including PLC, DCS, RTU, SIS, RIO substation and other control units;

⑤ Field device layer: including sensor, actuator, robot, numerical control center, etc.

7.1.3 Evaluation Target and Index of Industrial Control System

The complexity of the industrial control system, the diversity of the network and the flexibility of the classification of the protection objects bring the demand of choice for the use of the basic requirements of network security. When selecting the extended requirements of industrial control system, it is necessary to specify whether the grading target has the industrial control attribute. Generally, the enterprise resource layer is not included in the industrial control system grading. If only the production management level is graded separately, the extended index of industrial control system should not be considered, such as MES system of manufacturing industry and SIS system of power plant (without control and optimization function).

See Table 7.2 for typical evaluation targets in industrial control system.

Table 7.2 Typical Evaluation Targets

Security Class or Level	Evaluation Target
Security Physical Environment	Physical places such as system machine room, centralized control room and unattended monitoring room[1]
Security Communication Network	• Network devices such as switches and routers[2] • Security devices such as firewalls, gateways, encryption devices, etc.
Security Zone Boundary	• Network equipment such as data transmission radio station and wireless gateway • Security devices such as gateway, firewall, IDS, IPS, anti-virus detection, security audit, etc.

Security Class or Level	Evaluation Target
Security Computing Environment	• Server, operation terminal • Storage devices such as disk array, etc. • Control equipment, intelligent instrument, remote substation with Ethernet communication • MES system, EMS system, APC system and other production management software[3] • SCADA software, DCS monitoring software, OPC communication software, real-time/ historical database software, network management software, etc. • PLC programming software, DCS configuration software, SIS programming software, communication configuration software, firmware upgrade software, etc. • Operation system, anti-malicious code software, etc.
Security Management Center	Security operation center, situation awareness platform, audit system, etc.

(1) The outdoor control box installed in the field also needs to meet the extended requirements of the industrial control system;

(2) Non-management switch is not regarded as evaluation target;

(3) Production management level software that is graded separately, or production management software cannot meet the extended requirements of the industrial control system due to the lack of the controlling function.

According to the hierarchical content of industrial control system in Annex G of GB/T22239—2019 *Baseline for Classified Protection of Cybersecurity*, the mapping relationship between each level and relevant technical requirements is given in Table 7.3 according to the function hierarchy model and unit mapping model of each layer.

Table 7.3 Mapping Relation of Basic Requirements of Each Level and Basic Requirements of Classified Protection

Functional Hierarchy	Technical Requirements
Enterprise Resource Layer	General security requirement (Security physical environment)
	General security requirement (Security communication network)
	General security requirement (Security computing environment)
	General security requirement (Security area boundary)
	General security requirement (Security management center)

(Continued)

Functional Hierarchy	Technical Requirements
Production Management Layer	General security requirement (Security physical environment)
	General security requirement (Security communication network) Extended security requirement (Security communication network)
	General security requirement (Security computing environment)
	General security requirement (Security area boundary) Extended security requirement (Security area boundary)
	General security requirement (Security management center)
Process Monitoring Layer	General security requirement (Security physical environment)
	General security requirement (Security communication network) Extended security requirement (Security communication network)
	General security requirement (Security computing environment)
	General security requirement (Security area boundary) Extended security requirement (Security area boundary)
	General security requirement (Security management center)
Field Control Layer	General security requirement (Security physical environment) Extended security requirement (Security physical environment)
	General security requirement (Security communication network) Extended security requirement (Security communication network)
	General security requirement (Security computing environment) Extended security requirement (Security computing environment)
	General security requirement (Security area boundary) Extended security requirement (Security area boundary)
Field Equipment Layer	General security requirement (Security physical environment) Extended security requirement (Security physical environment)
	General security requirement (Security communication network) Extended security requirement (Security communication network)
	General security requirement (Security computing environment) Extended security requirement (Security computing environment)
	General security requirement (Security area boundary) Extended security requirement (Security area boundary)

In the industrial control system scenario, the software and hardware equipment and application system in the industrial control system should be checked. The corresponding evaluation targets should be selected according to Appendix A of GB/T22239—2019, and other general information systems should be evaluated according to the general security evaluation requirements.

7.1.4 Typical Industrial Control System

1. SCADA System

SCADA system is the data acquisition and supervisory control system. Widely used in electric power, metallurgy, petroleum, chemical industry, gas, railway and other fields of data acquisition and monitoring control and process control and many other fields.

The C/S architecture is usually adopted in SCADA system. The hardware equipment usually includes client, server, controller, measurement and control unit and field equipment. The network networking and communication modes of SCADA system are various. The measurement and control unit or controller can be connected to the server through serial bus, Ethernet or wireless mode, as well as the Ethernet network. The bus connection usually adopts RS485 and RS422, and the point-to-point connection usually adopts RS232. Typical network structure of SCADA system is shown in Figure 7.2.

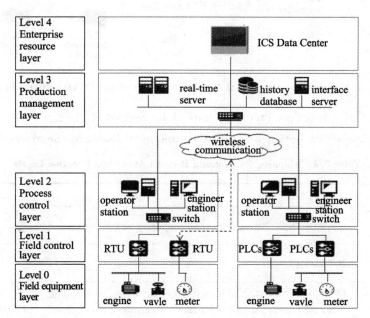

Figure 7.2 Typical SCADA System Network Structure

Taking the urban gas SCADA system shown in Figure 7.3 as an example, the system is generally divided into three areas: central control room, two gate stations and acquisition area. The operator station in the central control room is responsible for monitoring the gas pipeline pressure, collecting information such as user gas consumption, and controlling the gas pipeline pressure and valve on-off. The gate station realizes local control, and usually deploys the local operator station, and deploys the controller in the door station machine room. The acquisition area includes residents and enterprises, and the remote collected gas consumption is sent to the remote central control room through the RTU acquisition terminal through 3G or 4G communication.

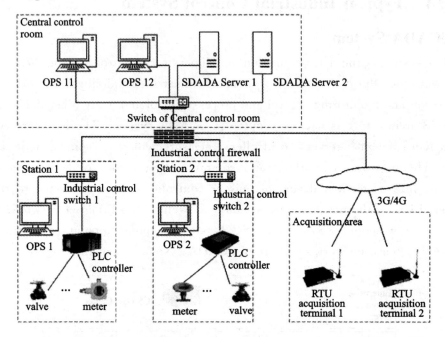

Figure. 7.3　Schematic Diagram of Network Structure of Gas SCADA System in a City

According to the function hierarchy model in Section 4.1.2, Table 7.4 shows the corresponding relationship between assets and functional levels in the above figure.

Table 7.4　Mapping Relationship Between Assets and Function Levels

Functional Hierarchy	Assets
Enterprise Resource Layer	Not involved
Production Management	Not involved
Process Monitoring Layer	OPS 11, OPS 12, SERVER 1, SERVER 2, central control room switch, industrial switch 1, industrial control switch 2, industrial firewall, monitoring software and database software
Field Control Layer	OPS 1, OPS 2, PLC controller, RTU controller, monitoring software and database software
Field Equipment Layer	Valve, instrument, RTU acquisition terminal 1 set terminal list, acquisition terminal n

2. DCS System

DCS system is distributed control system. Widely used in electric power, metallurgy, oil refining, chemical industry, water treatment, pharmacy, heating and other fields, with the characteristics of decentralized control, centralized operation, hierarchical management, flexible configuration and convenient configuration.

DCS system usually adopts the network structure of double bus, ring or double star to ensure the real-time and reliability of the network. Hardware equipment generally includes engineer station, operator station, master station, IO control station and field equipment, etc. Typical network structure of DCS system is shown in Figure 7.4.

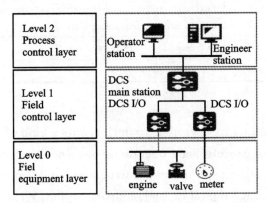

Figure. 7.4 Typical DCS System Network Structure

Taking the DCS system of a unit in the power plant shown in Figure 7.5 as an example, the system is generally divided into three physical areas: main control room, electronic room and engineer station machine room. The operator station in the main control room is responsible for monitoring the operation status of the unit and real-time controlling the on-site equipment such as valves. The engineer station is responsible for the engineering configuration and also has the function of the operator station. The electronics room is used to deploy the control cabinet. Optical fiber, twisted pair or hardwired connection is usually used between the control cabinet and the field instrument.

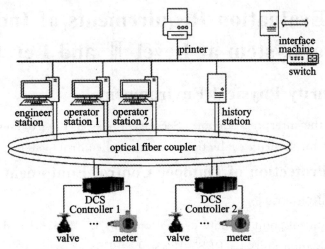

Figure. 7.5 Schematic Diagram of DCS System Network Structure of One Unit in a Power Plant

According to the function hierarchy model in Section 4.1.2, Table 7.5 shows the corresponding relationship between assets and functional levels in Figure 7.4.

Table 7.5 Corresponding Relationship Between Assets and Function Levels

Functional Hierarchy	Assets
Enterprise Resource Layer	Not involved
Production Management Layer	Not involved
Process Monitoring Layer	Engineer station, operator station 1, operator station 2, history station, switch, printer, monitoring software
Field Control Floor	DCS controller 1, DCS controller 2
Field Equipment Layer	Valves and instruments

3. PLC System

PLC (Programmable Logic Controller), which adopts a kind of programmable memory, is used to store program, execute logic operation, sequence control, timing, counting and arithmetic operation and other user-oriented instructions, and control all kinds of machinery or production process through digital or analog input/ output. Widely used in electric power, petroleum, chemical industry, steel, building materials, machinery manufacturing, automobile, textile, transportation and other industries. It has the characteristics of flexible configuration, convenient operation, quick response and accurate action.

PLC system is similar to DCS system in network structure. The hardware equipment is usually composed of engineer station, operator station, controller and field equipment, which will not be described in detail here.

7.2 Application and Interpretation of the Extended Security Evaluation Requirements of Industrial Control System at Level III and Level IV

7.2.1 Security Physical Environment

This chapter is the interpretation of "Security Physical Environment", which is an extended requirement for security evaluation of industrial control systems.

1. Physical Protection of Outdoor Control Equipment

[Standard Requirements]

Level III of the control point includes evaluation units L3-PES5-01 and L3-PES5-02, and Level IV includes evaluation units L4-PES5-01 and L4-PES5-02.

[L3-PES5-01/ L4-PES5-01 Interpretation and Description]

The evaluation index "outdoor control equipment should be placed in the box or device made of iron plate or other fireproof materials and fastened; the box or device should have

Chapter 7 Application and Interpretation of the Extended Security Evaluation Requirements of Industrial Control System

the capability of ventilation, heat dissipation, theft prevention, rain protection and fire prevention" should be the outdoor or local control equipment of the industrial control system.

The key points of evaluation implementation include: check whether it is placed in the box or device made of iron plate or other fireproof materials and fastened in the box or device; check whether the box or device has the capability of ventilation, heat dissipation, anti-theft, rain-proof and fire prevention; check the proof that the box or device is made of iron plate or other fireproof materials; check whether the box or device is provided with ventilation and heat dissipation port, cooling hole or exhaust device, which can provide ventilation and heat dissipation, or whether the ambient temperature is within the normal working range of the control equipment, whether the box or device has anti-theft measures, and whether there is any rain trace in the box or device.

If the test results show that the box in which the outdoor control equipment is located is installed Securityly, and the box is capable of ventilation, heat dissipation, anti-theft, rain-proof and fire-proof, the result is conforming; otherwise it is non-conforming or partially conforming.

[L3-PES5-02/ L4-PES5-02 Interpretation and Description]

The targets of the evaluation index "Outdoor control equipment should be placed far away from strong electromagnetic interference, strong heat source and other environments, and emergency treatment and maintenance should be conducted in time to ensure normal operation of the equipment if it is unavoidable" are the outdoor or local control equipment of industrial control system.

The key points of evaluation implementation include: check whether outdoor control equipment is far away from strong electromagnetic interference and strong heat source, such as strong electromagnetic interference environment such as thunder and lightning, sandstorm, high-power start-stop equipment, high-voltage transmission line, and strong heat source environment such as heating furnace and steam; for outdoor control equipment that cannot be far away from strong electromagnetic interference and strong heat source, check whether it has emergency treatment and maintenance records.

If the test results show that there is no strong electromagnetic or strong heat source near the outdoor control equipment, or there is strong electromagnetic or strong heat source near the outdoor control equipment, emergency disposal and recent maintenance records should be checked, and the equipment operates normally, the result is conforming; otherwise it is non-conforming or partially conforming.

7.2.2 Security Communication Network

This chapter is the interpretation of "Security Communication Network", which is an extended requirement for security evaluation of industrial control systems.

According to the characteristics of the industrial control system, this chapter also lists the characterized interpretation contents of some General security requirement evaluation under the industrial control system environment.

1. Network Architecture

[Standard Requirements]

Level Ⅲ of the control point includes evaluation units L3-CNS1-01, L3-CNS1-02, L3-CNS1-03, L3-CNS1-04, L3-CNS1-05, L3-CNS5-01, L3-CNS5-02, L3-CNS5-03, and Level Ⅳ includes evaluation units L4-CNS1-01, L4-CNS1-02, L4-CNS1-03, L4-CNS1-04, L4-CNS1-05, L4-CNS1-06, L4-CNS5-01, L4-CNS5-02, L4-CNS5-03.

[L3-CNS1-01/ L4-CNS1-01 Interpretation and Description]

The targets of the evaluation index "network processing capacity of network equipment should meet the demand of business peak" are the equipment or relevant components providing network communication function, such as router, switch, wireless access gateway, encryption gateway, interface machine, protocol conversion device, network isolation product, industrial control firewall, etc.

The key points of evaluation implementation include: determine whether there is a business peak. If there is a business peak, check the CPU utilization rate and memory utilization rate of the key network node and security node devices. The CPU and memory utilization rate should not exceed 70% (the limit may be adjusted according to the system operation condition); check the running time of the equipment and its log alarm information, and confirm whether there is any communication failure caused by the performance problem of the equipment.

If the test results show that the CPU and memory utilization rate of the tested device meet the requirements and the device has not experienced shutdown and restart event, the result is conforming; otherwise it is non-conforming or partially conforming.

[L3-CNS1-02/ L4-CNS1-02 Interpretation and Description]

The targets of the evaluation index "should ensure that the bandwidth of each part of the network meets the demand of the business peak" are the equipment or relevant components providing network communication function such as integrated network management system, switch, router, wireless access gateway, encryption gateway, interface machine, protocol conversion device, network isolation product and firewall.

The key points of evaluation implementation include: firstly, master the service flow of the industrial control system during the business peak period, interview the network administrator whether to deploy the integrated network management system or flow control equipment, or query whether the QoS function is enabled in the relevant equipment; if the equipment does not have the function, check the technical documents of the equipment, inquire whether the supporting bandwidth of the equipment meets the demand of peak traffic, and ask the system administrator whether there is any system crash or other security events caused by insufficient network bandwidth.

If the results show that the test equipment can allocate the bandwidth and no security event caused by insufficient bandwidth has occurred, the unit judges that the result is conform; otherwise it is inconsistent or partially conforming.

[L3-CNS1-03/ L4-CNS1-03 Interpretation and Description]

The targets of the evaluation index "Different network areas should be divided and addresses should be allocated for each network area according to the principle of convenient management and control". are the network devices with VLAN function such as layer 3 switch, router, wireless access gateway, wireless network controller, industrial control firewall, etc.

The key points for evaluation implementation include: check whether different network area divisions are conducted on important network equipment according to address planning and security area protection requirements; check configuration information of relevant network equipment, and verify whether the divided network area is consistent with the division principle.

If the results show that the unit undergoing test can divide the VLAN according to the function of the department, the importance degree of the grade protection object, and the level and service of the application system, and the configuration is reasonable, the result is conforming; otherwise it is non-conforming or partially conforming.

[L3-CNS1-04/ L4-CNS1-04 Interpretation and Description]

The targets of the evaluation index "the deployment of important network areas at the boundary should be avoided, and reliable technical isolation measures should be adopted between the important network areas and other network areas" are the network topology, industrial control firewall and other devices or components providing access control functions.

The key points of evaluation implementation include: check whether the important network area is deployed at the boundary of the Internet or other network areas; if it is deployed at the network boundary, check whether reliable technical isolation measures are adopted; check whether the important network area and other network area boundaries adopt reliable technical isolation measures, such as Security access area or deployment of network isolation products, firewalls and devices with access control (such as ACL list) functions.

If the results show that the important service area is not directly deploy at the boundary of the Internet or other network area, and the security isolation equipment is provided, and the correct physical interface and the safe and effective isolation policy are configured, the result is conforming; otherwise it is non-conforming or partially conforming.

[L3-CNS1-05/ L4-CNS1-05 Interpretation and Description]

The evaluation index "hardware redundancy of communication line, key network equipment and key computing equipment should be provided to ensure the availability of the system" mainly includes network topology, switch, router, security device and server, etc.

The key points of evaluation implementation include: organize the network structure of the industrial control system by interview and verification, check whether the main network links of the system, key network equipment (exit router, core switch) and key computing equipment (important server and control equipment) are hardware redundancy; interview the network administrator whether the key equipment of the system is in the master-standby or double-active mode, and check whether the configuration mode can effectively avoid single point of failure.

If the results show that the system undergoing test is equipped with hardware redundancy such as communication link, core switch, important server, controller, etc., the result is conforming; otherwise it is non-conforming or partially conforming.

[L4-CNS1-06 Interpretation and Description]

The targets of the evaluation index "Bandwidth should be allocated according to the importance of business services, and priority should be given to the guarantee of important services" are switches, routers, firewalls, flow control equipment and other devices or relevant components providing bandwidth control functions.

The key points of evaluation implementation include: interview the system administrator to understand the system business conditions, determine the importance of each business to the system, distinguish the key business from the general business, interview the network administrator to determine whether the bandwidth is allocated according to the importance of the business, and verify whether the bandwidth control equipment is configured according to the importance of the business service and enable the bandwidth strategy.

If the results show that the equipment undergoing test can allocate the bandwidth according to the service importance degree, the result is conforming; otherwise it is non-conforming or partially conforming.

[L3-CNS5-01 Interpretation and Description]

The evaluation index "The industrial control system and other systems of the enterprise should be divided into two areas, and the one-way technical isolation means should be adopted between the areas" are mainly the firewall, industrial control firewall, network isolation products, wireless access gateway, encrypted gateway, router providing access control function, layer-3 switch and other equipment or components.

The key points of evaluation implementation include: check whether the industrial control system boundary communicates with other systems of the enterprise and which communication mode is adopted; which access control equipment or technical means are used to realize unidirectional isolation; check whether the one-way isolation policy is effective; if the industrial control system of wireless communication is used, check whether the wireless communication of the industrial control system adopts effective one-way isolation measures.

If the results show that the overall network architecture of the unit undergoing test is divided into two areas: the industrial control system and the other system of the enterprise, and the one-way isolation device is deployed between the areas, and the device strategy configuration is reasonable and effective, the result is conforming; otherwise it is non-conforming or partially conforming.

【L4-CNS5-01 Interpretation and Description】

The evaluation index "industrial control system and other systems of the enterprise should be divided into two areas, and one-way technical isolation should be adopted between the areas" are firewall, network isolation products, wireless access gateway, encrypted gateway, router providing access control functed, layer-3 switch, and other equipment or components. If the production area is connected wirelessly, a wireless gateway could be used to check this situation.

The key points of evaluation implementation include: on-site inspection of the boundary of industrial control system, whether there is communication with other systems of the enterprise, and which communication mode; whether unidirectional technical isolation measures are adopted between the industrial control system and other systems to ensure unidirectional transmission of data flow; check whether the one-way security isolation equipment meets the professional product requirements specified by the state or the industry.

If the results show that the one-way isolation device is deployed between the industrial control system and other systems of the enterprise, the IP, MAC, port and other contents of the device isolation strategy configuration are valid, and the device is tested and authenticated by the national designated department, the result is conforming; otherwise it is non-conforming or partially conforming.

【L3-CNS5-02/ L4-CNS5-02 Interpretation and Description】

The evaluation index "The interior of industrial control system should be divided into different security domains according to business characteristics, and technical isolation means should be adopted between security domains" are the equipment or components providing access control function, such as firewall, industrial control firewall, network isolation products, routers and layer-3 switches.

The key points of evaluation implementation include: firstly, analyze the network topology to understand the service characteristics and the division principle of security domain; check whether different security domains are divided within the industrial control system according to the service characteristics; check whether the devices with access control function are used between different security domains and various levels, and check the effectiveness of the access control policies.

If the results show that the overall network architecture of the system undergoing test is divided into different security domains, and the isolation devices such as industrial control firewall are deployed between the different security domains, and the device policy configuration is reasonable and effective, the result is conforming; otherwise it is non-conforming or partially conforming.

[L3-CNS5-03/ L4-CNS5-03 Interpretation and Description]

The evaluation index "industrial control system involving real-time control and data transmission should use independent network equipment networking to realize safe isolation from other data networks and external public information network on the physical layer" are network topologies.

The key points for evaluation implementation include: first, understand whether the tested industrial control system involves real-time control and data transmission; check whether the industrial control system involving real-time control and data transmission is independently networked on the physical layer, and there is no common equipment with other systems or non-real-time services.

If the test results show that the system involving the real-time control and data transmission service is isolated from other data networks and external public information networks, and the system network equipment, communication lines, etc. are physically independently networked, the result is conforming; otherwise it is non-conforming or partially conforming.

2. Communication transmission

[Standard Requirements]

Level III of the control point includes evaluation units L3-CNS1-06, L3-CNS1-07, L3-CNS5-04, and Level IV includes evaluation units L4-CNS1-07, L4-CNS1-08, L4-CNS1-09, L4-CNS1-10, L4-CNS5-04.

[L3-CNS1-06 Interpretation and Description]

The targets of the evaluation index "data integrity in the communication process should be ensured by using verification technology or cryptography technology" are the equipment or components providing verification technology or cryptography, and the industrial control system supporting communication verification. If I/O interface analog communication is used between the on-site equipment layer and the on-site control layer of the system undergoing test, this evaluation index is not available.

The key points of evaluation implementation include: check whether the equipment or components using verification technology (CRC check, parity check, etc.) or cryptographic techniques (hash check, etc.) in the transmission process guarantee the data integrity of authentication data, important business data, important audit data, important configuration data, important video data and important personal information.

If the results show that the important business data transmission of the equipment or system undergoing test has a verification mechanism to ensure the data integrity in the communication process, the result is conforming; otherwise it is non-conforming or partially conforming.

Chapter 7　Application and Interpretation of the Extended Security Evaluation Requirements of Industrial Control System

【L4-CNS1-07 Interpretation and Description】

The targets of the evaluation index "adopt cryptographic technology to ensure the integrity of data in the communication process" are the equipment or components providing cryptographic technology and the industrial control system supporting communication verification. If analog communication of I/O interface is adopted between the field device layer of the tested system and the field control layer, this evaluation index does not applies to this level.

The implementation points of the evaluation include: check whether the equipment or components that use cryptographic technology (hash check, etc.) for authentication data, important business data, important audit data, important configuration data, important video data and important personal information during transmission ensure the integrity of data during communication.

If the results show that the system under test adopts the password verification mechanism to ensure the integrity of the communication data in the data transmission process, the result is conforming; otherwise it is non-conforming or partially conforming.

【L3-CNS1-07/ L4-CNS1-08】

The targets of the evaluation index "adopt cryptography technology to ensure the confidentiality of data in the communication process" are encryption machine, encryption authentication equipment, encryption module and other devices or components providing cryptographic technology.

The implementation points of evaluation include: check whether encryption protection measures are taken during communication, specific technical measures such as encryption machine, encryption module, VPN equipment, encryption protocol (IPsec protocol) or encryption algorithm (AES, DES, etc.) adopted; when conditions permit, test to verify whether the sensitive information field or the entire message is encrypted during communication; if conditions are not available, check the system design document to confirm whether the sensitive field or the entire message is encrypted during communication.

If the results show that the system undergoing test adopts cryptographic technology or equipment to encrypt the communication data, and the data transmission is ciphertext, the result is conforming; otherwise it is non-conforming or partially conforming.

【L4-CNS1-09 Interpretation and Description】

The targets of the evaluation index "the two sides of the communication should be verified or authenticated based on cryptographic techniques before communication" are the equipment or components that provide cryptographic techniques, such as encryption machine, encryption authentication equipment, encryption module, etc.

The key points of the evaluation implementation include: check whether the session initialization verification or authentication is conducted by using the cryptography technology before the communication parties establish the connection, and the specific technical measures adopted.

If the test results show that the system undergoing test adopts the cryptography to verify both sides of the communication, the unit determines that the result is consistent; otherwise it is non-conformance or partial conformity.

[L4-CNS1-10 Interpretation and Description]

The target of the evaluation index "cryptographic operation and key management for important communication processes based on the hardware cryptographic module" is the equipment or components providing cryptographic techniques, such as encryption machine, encryption authentication equipment, encryption module, etc.

The key points of evaluation implementation include: check whether the communication process in the industrial control system design document adopts the hardware cryptography module; check whether the relevant products have obtained the valid test report or cryptography product model certificate specified by the national cryptography management department.

If the test results show that the instruction manual of the tested equipment has the description of the hardware cryptography module and the test report or cryptography product model certificate issued by the relevant department of national cryptography management, the result is conforming; otherwise it is non-conforming or partially conforming.

[L3-CNS5-04/ L4-CNS5-04 Interpretation and Description]

The evaluation index "encryption authentication, access control and data encryption transmission should be adopted for the control instruction or relevant data exchange using WAN in the industrial control system". This does not apply if the WAN is not used.

The key points of evaluation implementation include: interviewing network administrators to understand whether the industrial control system has control instructions with WAN or relevant data exchange requirements, and identifying whether there are control commands transmitted through WAN; verify whether the industrial control system and WAN use encryption authentication equipment and whether the equipment is configured with encryption strategy; testing and verify the effectiveness of the encryption and authentication techniques used. If conditions permit (such as shutdown for maintenance), technical means can be used to verify whether the data is transmitted in ciphertext.

If the results show that the encrypt authentication device is used for identity authentication, access control and data encryption transmission when the industrial control system is communicate with the WAN, and the policy configuration is reasonable and effective, the result is conforming; otherwise it is non-conforming or partially conforming.

3. Trusted Verification

[Standard Requirements]

Level III of the control point includes the evaluation unit L3-CNS1-08, and Level IV includes the evaluation unit L4-CNS1-11.

[L3-CNS1-08 Interpretation and Description]

The evaluation index "the system boot program, system program, important configuration parameters and communication application program of the communication equipment can be trusted to verify based on the root of trust, and the dynamic trusted verification can be carried out in the key execution links of the application program, the alarm will be given after the credibility is detected to be damaged, and the verification result will be formed into audit record and sent to the security management center". The evaluation targets are the switch, router or other communication equipment and other devices or components that provide trusted verification and provide the system with centralized audit function.

The key points of evaluation implementation include: check whether the system boot program, system program, important configuration parameters and communication application program of the equipment adopt the trusted verification technology, and whether the dynamic trusted verification is carried out in the key execution links of the application program; test and verify whether the alarm is given when the credibility of the equipment is detected to be damaged, and whether the verification results are sent to the security management center in the form of audit records.

If the test results show that the tested equipment or system adopt the trusted chip, the key execution links of the application program are subject to dynamic trusted verification, and an alarm is sent to the security management center after the credibility is damaged, the result is conforming; otherwise it is non-conforming or partially conforming.

[L4-CNS1-11 Interpretation and Description]

The evaluation index "can conduct trusted verification on the system boot program, system program, important configuration parameter and communication application program of the communication equipment based on the trusted root, and conduct dynamic trusted verification in the key execution links of the application program, and give an alarm after the credibility is detected to be damaged, and the verification result should form an audit record and send it to the security management center, and conduct dynamic association perception". The evaluation targets is the equipment or component providing the trusted verification such as the switch, router or other communication equipment, which provides the centralized audit function.

The key points of evaluation implementation include: check whether the system boot program, system program, important configuration parameters and communication application program of the equipment adopt the trusted verification technology, and whether the dynamic trusted verification is conducted in the key execution links of the application program; test and verify whether the alarm is given when the credibility of the equipment is detected to be damaged, and whether the verification result is sent to the security management center in the form of audit record; check whether the trusted verification mechanism can dynamically associate and perceive the audit records.

If the test results show that the tested equipment or system adopts the trusted chip, the key execution link of the application program carries out dynamic trusted verification, and at the same time, an alarm is sent to the security management center and the situation-aware equipment after the credibility is damaged, the result is conforming; otherwise it is non-conforming or partially conforming.

7.2.3 Security Area Boundary

This chapter is the interpretation of the expanded requirements for security evaluation of industrial control systems, "Security area boundary".

According to the characteristics of the industrial control system, this chapter also lists the characterized interpretation contents of some General security requirement evaluation under the industrial control system environment.

1. Boundary Protection

[Standard Requirements]

Level Ⅲ of the control point includes evaluation units L3-ABS1-01, L3-ABS1-02, L3-ABS1-03, L3-ABS1-04, and Level Ⅳ includes evaluation units L4-ABS1-01, L4-ABS1-02, L4-ABS1-03, L4-ABS1-04, L4-ABS1-05, L4-ABS1-06.

[L3-ABS1-01/ L4-ABS1-01 Interpretation and Description]

The targets of the evaluation index "should ensure that the access and data flow crossing the boundary communicate through the controlled interface provided by the boundary equipment" include the routers, firewalls, network isolation products, wireless access gateway, encryption gateway, layer-3 switch, interface machine, protocol conversion device and other devices or components that can provide access control functions among levels, security domains, wired and wireless networks, and between all boundaries between the system undergoing test and other systems.

The key points of evaluation implementation include: whether the network security administrator system has the latest network topology; check the physical wiring on site to confirm that the network topology is consistent with the physical connection, and confirm whether there is any physical link between the areas isolated by the access control device; according to the system deployment, analyze the security area of the system, the level of each area and whether the access control device or component (firewall, network isolation product, encryption gateway, router, etc.) is deployed at the network boundary of different levels. If the access control device is not deployed, this item is judged as non-compliant; if the switch without management function is deployed, this item is judged as non-conformance; check the access control device on site and check the access control policy. If the firewall is deployed, whether the firewall is configured to restrict the source address and destination address of the device, whether the port is designated for cross-boundary network communication, and whether the specified port is configured and enabled with security policy; if the vertical encryption device is deployed, whether the encryption device restricts the source address and destination address, whether the specified port is designated for cross-boundary network communication, and whether the data communication is ciphertext transmission; if the network isolation product is used, whether the source address and the destination address are restricted, check whether the access control port is specified; if the router device is used, it is necessary to check the configuration information of the router device to confirm whether to restrict the physical port and only allow the specified IP address to communicate; other technical means (such as illegal wireless network device location, device configuration information verification, and out-of-area address access attempt) can be used to verify whether there is any network communication across the boundary by other ports that are not controlled.

If the test results indicate that the cross-boundary access and the data flow communicate through the controlled interface provided by the boundary device, the result is conforming; otherwise it is non-conforming or partially conforming.

【L3-ABS1-02/ L4-ABS1-02 Interpretation and Description】

The targets of the evaluation index "should be able to check or restrict the behavior of unauthorized equipment connecting to the internal network" are inter-level, inter-security domain, between wired network and wireless network, and between all boundaries between the system undergoing test and other systems, special access control equipment, terminal management system with access control function, router, switch, industrial control network audit equipment, etc.

The key points of evaluation implementation include: check whether access layer network equipment is configured with measures such as IP/ MAC address binding; check whether to deploy special access control equipment or intranet access control system, and check whether it has sales license for special network security products and whether it is configured with strict access control or inspection measures; if there is no technical measure, interview network administrator, and whether there are other auxiliary measures, such as strict physical access control measures, all-round video monitoring measures, management measures, etc.

If the results show that the test device can check or restrict the behavior of the unauthorized device connecting to the internal network, the unit determine that the result is compliant; otherwise it is non-compliant or partially conforming.

【L3-ABS1-03/ L4-ABS1-03 Interpretation and Description】

The targets of the evaluation index "should be able to check or restrict the unauthorized connection of internal users to the external network" are the intranet security management system, special access control equipment, terminal management system with access control function, router, switch, industrial control network audit equipment, terminal computer, server, etc. between all boundaries between the system undergoing test and other systems.

The key points of the evaluation include: interview whether the network security administrator adopts technical measures to monitor or restrict the unauthorized outreach behavior of internal users, such as adopting technical measures, and conducting on-the-spot inspection for them with different technical measures; check the special access control equipment on site, and check whether it has the sales license for special network security products and whether it is equipped with strict access control or inspection measures; check the intranet security management system on site to confirm whether all terminal devices have been accessed and managed, and enable relevant policies, such as disabling dual network cards, USB interface, Modem, wireless network, etc.; check the computer configuration information on site, such as restricting the user authority of the terminal computer (installation program or driver is prohibited), closing the peripheral interface (such as USB interface, wireless network card, etc.), prohibiting the modification of IP address and gateway, etc., and analyzing whether it is sufficient to monitor or restrict unauthorized external activities; if there are no technical measures, interview the network administrator to see whether there are other auxiliary measures, such as regular inspection of operation system access records, all-round video monitoring measures and management measures.

If the results show that the device undergoing test can check or restrict the behavior of unauthorized connection of the internal user to the external network, the unit determine that the result is conforming; otherwise it is non-conforming or partially conforming.

【L3-ABS1-04/ L4-ABS1-04 Interpretation and Description】

The targets of the evaluation index "the use of wireless network should be restricted to ensure that the wireless network is connected to the internal network through the controlled boundary device" are the wireless access gateway, wireless network controller, network topology diagram, wireless network access control equipment, boundary access control equipment for wireless network access to wired network, etc.

The key points of evaluation implementation include: interview with network security administrator, whether the system adopts wireless network, such as wifi-Fi, 4G/ 5G mobile communication, etc.; if the wireless network is not used, this item is not applicable; on-site check whether the wireless network is connected to the internal wired network through the controlled boundary protection equipment; if that security access gateway device is adopt, whether the security access gateway restrict the access IP address and whether the encryption device performs encrypted transmission management on the data, which limit the port; on-site check whether the control measures for the unauthorized wireless equipment are deployed in the network, such as wireless sniffer, wireless intrusion detection system, handheld wireless signal detection system, etc.

If the test results indicate that the wireless network accesses the internal network through the controlled boundary device, the result is conforming; otherwise it is non-conforming or partially conforming.

【L4-ABS1-05 Interpretation and Description】

The targets of the evaluation index "should be able to effectively block the behavior of unauthorized equipment or unauthorized connection of internal users to the external network" are the internal network security management system, special access control equipment and terminal management system with access control function among levels, security domains, wired network and wireless network, and all boundaries between the system undergoing test and other systems.

The key points of evaluation implementation include: especially check the special access control equipment, check whether it has the sales license for special network security products and whether it is configured with strict access control or inspection measures; on-site verification of the intranet security management system to confirm whether all terminal devices have been accessed and whether relevant policies are enabled to block the unauthorized external and inline behaviors; if there are no technical measures, interview with the network administrator to see whether there are other auxiliary measures, such as regular inspection of operation system access records, comprehensive video monitoring measures and management measures, etc.

If the results show that the equipment can effectively block the behavior of the unauthorized device connecting to the internal network or the behavior of the internal user connecting to the external network without authorization, the result is conforming; otherwise it is non-conforming or partially conforming.

[L4-ABS1-06 Interpretation and Description]

The evaluation targets of "trusted verification mechanism should be adopted for the equipment in the access network to ensure the authenticity and credibility of the equipment connected to the network" are terminal management system, router, switch, firewall, network isolation product, security access gateway, industrial control network audit equipment, terminal computer, server, etc.

The key points of evaluation implementation include: interview the network security administrator, whether the system adopts the trusted verification mechanism, and conducting the trusted verification for the equipment connected to the system; on-site verification whether the system boot program, system program, important configuration parameters and boundary protection application program of the boundary equipment are authenticated based on the trusted root.

If the test results show that the device in the access network can be authenticated by using the trusted verification mechanism, the result is conforming; otherwise it is non-conforming or partially conforming.

2. Access Control

[Standard Requirements]

Level Ⅲ of the control point includes evaluation units L3-ABS1-05, L3-ABS1-06, L3-ABS1-07, L3-ABS1-08, L3-ABS1-09, L3-ABS5-01, L3-ABS5-02, and Level Ⅳ includes evaluation units L4-ABS1-07, L4-ABS1-08, L4-ABS1-09, L4-ABS1-10, L4-ABS1-11, L4-

ABS5-01, L4-ABS5-02.

[L3-ABS1-05/ L4-ABS1-07 Interpretation and Description]

The evaluation index "Access control rules should be set according to the access control policy between network boundaries or areas, and all communication should be denied by the controlled interface except for allowed communication by default" include the routers, firewalls, network isolation products, wireless access gateway, encryption gateway, layer-3 switch, interface machine, protocol conversion device and other devices or components that can provide access control function between levels, security domains, wired network and wireless network, and between all boundaries between the system undergoing test and other systems.

The key points of evaluation implementation include: check whether access control devices (firewalls, isolation devices, encryption devices, etc.) are deployed between network boundaries or areas; on-site check whether access control policies are enabled by access control devices; check whether rules are configured in the way of default prohibition, such as deploying firewalls, configuring firewalls, restricting device source and destination addresses, specifying ports for cross-boundary network communication, and whether designated ports are configured and enabled with security policies; if deploying vertical encryption devices, whether encrypting devices restrict source address and destination addresses, and whether designated ports are used for cross-boundary network communication, and whether data communication is ciphertext transmission, such as using isolation devices, limiting source addresses and destination addresses, and check whether access control ports are specified; besides external network boundaries, it is also necessary to check access control rules between different areas and layers within the system undergoing test.

If the results show that the access control policy set between the network boundary or areas only allow the controlled interface to communicate, the result is conforming; otherwise it is non-conforming or partially conforming.

[L3-ABS1-06/ L4-ABS1-08 Interpretation and Description]

The "redundant or invalid access control rules should be deleted, the access control list should be optimized, and the number of access control rules should be minimized" should be the evaluation targets of firewall, industrial control firewall, network isolation products, wireless access gateway, encryption gateway, router providing access control function, layer-3 switch and other equipment or components between levels, security domains, wired network and wireless network, and all boundaries between the system undergoing test and other systems.

The key points of evaluation implementation include: interview the network security administrator whether the access control device (firewall, switch, encryption device) has set access control rules and open the access control list; on-site check whether the logical relationship between different access control policies and the sequence before and after are reasonable. For the access control rule of the equipment, interview the network security administrator one by one to ask whether the rule is useful and consistent with the requirements. If not, delete it. For example, whether the firewall device enables the all-pass policy, the access control port is not restricted, the access address range is too large, and whether the address isolation device and the encryption device store the redundant access control policy. If the access control device has a rule matching count, check whether there is a rule with matching number of 0. A rule with matching number of 0 is generally considered as redundant rule.

If the results show that the number of access control rules of the device undergoing test is minimized, the result is conforming; otherwise it is non-conforming or partially conforming.

【L3-ABS1-07/ L4-ABS1-09 Interpretation and Description】

The targets of the evaluation index "source address, destination address, source port, destination port and protocol should be checked to allow/ deny data packets entering and leaving" are routers, firewalls, network isolation products, wireless access gateway, encryption gateway, layer-3 switch, interface machine, protocol conversion device and other devices or components that can provide access control function between levels, security domains, wired and wireless networks, and between all boundaries between the system undergoing test and other systems.

The key points of evaluation implementation include: whether the network administrator has set access control rules for the access control equipment (switch, firewall, isolation device, etc.); on-site verification that the access control policy of the equipment is configured according to relevant parameters such as source address, destination address, destination port and protocol; in principle, access control is not allowed in the way of network segment; whether the source address, destination address and protocol are restricted in the firewall access control policy. Whether the port is restricted in the isolation device. For the special access control device that allows only certain protocol or service to pass through, if the source destination address is restricted, this item can be judged as conforming.

If the results show that the source address, the destination address, the source port, the destination port and the protocol are effectively restrict by the device under test, the unit determines that the result is accord; otherwise it is inconsistent or partially conforming.

【L3-ABS1-08/ L4-ABS1-10 Interpretation and Description】

The targets of the evaluation index "the ability to provide clear permission/ denial of access for the incoming and outgoing data flow according to the session state information" are the routers, firewalls, wireless access gateways, network isolation products, encryption devices, etc. among levels, security domains, wired and wireless networks, and all boundaries between the system undergoing test and other systems.

The key points of evaluation implementation include: interviewing the network security administrator whether to provide clear permission/ denial of access for the incoming and outgoing data flow by using mechanisms such as session authentication; on-site check whether the source address, destination address and protocol are restricted in the firewall security policy of the access control device.

If the results show that the device undergoing test has respect to the source address, the destination address, the source port, the destination port and the protocol, the result is conforming; otherwise it is non-conforming or partially conforming.

[L3-ABS1-09 Interpretation and Description]

Evaluation index: Access control based on application protocol and application content should be implemented for the data flow in and out of the network.

Selection of evaluation targets: the evaluation targets of this evaluation index are routers, firewalls, wireless access gateways, network isolation products and encryption devices among levels, security domains, wired and wireless networks, and all boundaries between the system undergoing test and other systems.

The key points of evaluation implementation include: interview the network security administrator whether to deploy the access control device of the next generation firewall or security component and enable the access control policy; on-site check whether the access control policy of the access control device can be configured for the application protocol type (non-port); on-site check whether the access control policy of the application content is available, such as disabling some specific control parameters and location address of ModBusTCP.

If the results show that the equipment or system undergoing test is configured with relevant policies, access control is performed on the application protocol and application content, and the validity of the policy is verified, the result is conforming; otherwise it is non-conforming or partially conforming.

[L4-ABS1-11 Interpretation and Description]

The targets of the evaluation index "data exchange should be carried out through communication protocol conversion or communication protocol isolation at the network boundary" are routers, firewalls, wireless access gateways, network isolation products, encryption equipment, etc. among levels, security domains, wired and wireless networks, and all boundaries between the system undergoing test and other systems.

The key points of evaluation implementation include: interview the network security administrator whether to deploy the access control device by the next generation firewall or security component and enable the access control policy; on-site check whether the access control policy of the access control device can be configured based on the communication protocol conversion; and on-site check whether the access control policy of the application content is available.

If the results show that the device undergoing test is configured with a relevant policy, the communication protocol is access control, and the validity of the policy is verify, the result is conforming; otherwise it is non-conforming or partially conforming.

[L3-ABS5-01/ L4-ABS5-01 Interpretation and Description]

The evaluation index "general network services such as E-mail, Web, Telnet, Rlogin, FTP, etc. that should be deployed between the industrial control system and other systems of the enterprise, configured with access control policy and prohibited from crossing the boundary of the area" are the routers, firewalls, network isolation products, wireless access gateways, encryption gateways, layer-3 switch, interface machine, protocol conversion device and other device or components that can provide access control functions between the system undergoing test and other systems.

The key points of evaluation implementation include: interview the network security administrator whether the access control device (firewall, isolation device) is deployed between the industrial control system and other systems of the enterprise, and whether the access control device enables the access control policy; on-site check the access control policy of the access control device and whether the general network services such as E-mail, Web, Telnet, Rlogin and FTP are prohibited.

If the test results show that the tested equipment is configured with an access policy to prohibit any common network services such as E-mail, Web, Telnet, Rlogin, FTP, etc. crossing the area boundary, the result is conforming; otherwise it is non-conforming or partially conforming.

[L3-ABS5-02/ L4-ABS5-02 Interpretation and Description]

The targets of the evaluation index "should give an alarm in time when the boundary protection mechanism between the security domain and the security domain in the industrial control system fails" are the network isolation product between the security domain and the security domain boundary, industrial firewall, router and switch, intranet security management system and other devices that provide access control function.

The key points of evaluation implementation include: interview the network security administrator whether to deploy the network monitoring and early warning system or other relevant modules, and give an alarm when the boundary protection mechanism fails; the on-site verification is to monitor whether the relevant functions of the early warning system or relevant modules are enabled, and give an alarm in time when the boundary protection mechanism fails.

If the test results show that the access control device between the security domains is configured with an access policy, and an alarm is given in time when the boundary protection mechanism fails, the result is conforming; otherwise it is non-conforming or partially conforming.

3. Intrusion Prevention

[Standard Requirements]

Level Ⅲ of the control point includes evaluation units L3-ABS1-10, L3-ABS1-11, L3-ABS1-12, L3-ABS1-13, and Level Ⅳ includes evaluation units L4-ABS1-12, L4-ABS1-13, L4-ABS1-14 and L4-ABS1-15.

[L3-ABS1-10/ L4-ABS1-12 Interpretation and Description]

The targets of the evaluation index "network attack initiated from outside should be detected, prevented or restricted at key network nodes" are IDS, IPS, UTM, firewall, etc. among levels, security domains, wired network and wireless network, and all boundaries between the system undergoing test and other systems.

The key points of evaluation implementation include: verify whether to deploy special equipment or equipment with intrusion detection/defense function module, etc.; on-site check whether the configuration information or security policy of relevant equipment can cover all network traffic of the boundary; on-site check whether the version of rule base or threat intelligence database of relevant equipment has been updated to the latest version; on-site check whether relevant equipment has enabled intrusion prevention function or whitelist mechanism, and whether it can detect network attack initiated from outside.

If the results show that the test equipment contains the intrusion prevention module, the version of the rule base has been updated to the latest version, and the security policy is to cover all the key nodes of the network, the result is conforming; otherwise it is non-conforming or partially conforming.

[L3-ABS1-11/ L4-ABS1-13 Interpretation and Description]

The targets of the evaluation index of "detecting, preventing or limiting the network attack initiated from inside at the key network nodes" are IDS, IPS, multi-functional security gateway UTM including intrusion prevention module, etc. among levels, security domains, wired and wireless networks, and all boundaries between the system undergoing test and other systems.

The key points of evaluation implementation include: interview the network security administrator, whether the system deploys special equipment or equipment with intrusion detection/ defense function module, etc.; on-site check whether the relevant equipment has enabled the intrusion prevention function or the whitelist mechanism and whether it is able to detect the internal network attack behavior; on-site check whether the rule base version or threat intelligence database of relevant equipment has been updated to the latest version; on-site check whether the configuration information or security policy of relevant equipment can cover all key nodes of the network.

If the results show that the equipment undergoing test contains the intrusion prevention module, the version of the rule base has been updated to the latest version, and the security policy is to cover all the key nodes of the network, the result is conforming; otherwise it is

non-conforming or partially conforming.

[L3-ABS1-12/ L4-ABS1-14 Interpretation and Description]

The evaluation index "technical measures should be taken to analyze the network behavior and realize the analysis of the network attack, especially the new network attack behavior" is mainly the anti-APT attack system, the network backtracking system, the industrial control audit system, the situation awareness system, the comprehensive log analysis system, etc., and all boundaries between the system undergoing test and other systems.

The key points of evaluation implementation include: interview whether the security network security administrator system is deployed with anti-APT attack system, network backtracking system, industrial control audit system, situation awareness system, comprehensive log analysis system and other systems to detect and analyze new network attacks; relevant network security products or systems should adopt network behavior analysis technology and comprehensive log analysis technology to identify new network attacks instead of identifying existing network attacks based on existing knowledge base; on-site check whether the network backtracking system or APT attack and other systems enable access control rules and whether the rule base is updated to the latest version.

If the results show that the device or system undergoing test is capable of analyzing the network behavior, the unit determines that the result is conform; otherwise it is non-compliant or partially conforming.

[L3-ABS1-13/ L4-ABS1-15 Interpretation and Description]

"When an attack is detected, record the attack source IP, attack type, attack target, attack time, and provide alarm in case of serious intrusion event." The evaluation targets are the anti-APT attack system, network backtracking system, industrial control audit system, situation awareness system, comprehensive log analysis system, etc., and all boundaries between the system undergoing test and other systems.

The key points of evaluation implementation include: interview the network security administrator system whether to deploy the equipment with intrusion function module; on-site check whether the logging function is enabled for relevant systems or components. Whether the log records include the attack source IP, attack type, attack target, attack time, etc.; test and verify whether the alarm strategy of the relevant system or component is valid, the alarm function has been enabled and is in normal use status, and the alarm information should be able to timely notify relevant personnel by effective means, such as SMS, large monitoring screen, etc.

If the results show that the device or system undergoing test contains an intrusion prevention module, enable the control policy, and records the attack IP, attack type, attack purpose, attack time and other information, the result is conforming; otherwise it is non-conforming or partially conforming.

4. Malicious Code and Spam Prevention

[Standard Requirements]

Level Ⅲ of the control point includes evaluation units L3-ABS1-14 and L3-ABS1-15, and Level Ⅳ includes evaluation units L4-ABS1-16 and L4-ABS1-17.

[L3-ABS1-14/ L4-ABS1-16 Interpretation and Description]

"Malicious code should be detected and cleared at key network nodes, and upgrade and update of malicious code protection mechanism should be maintained". The evaluation targets are anti-virus gateway, traffic malicious code detection product, multi-function security gateway UTM including anti-virus module, etc. among security domains, and all boundaries between the system undergoing test and other systems.

The key points of evaluation implementation include: check whether anti-malicious code products or malicious code detection products based on network traffic are deployed at key network nodes; if traffic-based malicious code detection products are deployed, check whether there is a special production license and the products cannot be blocked, then this judgment is partially met; interview the network security administrator, ask whether to upgrade the feature base of the anti-malicious code product and the specific upgrade method, log in to view the upgrade of the feature base, and check whether the current version is the latest version. In principle, the malicious code feature library is within the latest month.

If the results show that the equipment or system undergoing test contains an anti-virus module, and the version of the rule base has been updated to the latest version, the unit judges that the result is conform; otherwise it is non-conforming or partially conforming.

[L3-ABS1-15/ L4-ABS1-17 Interpretation and Description]

The evaluation index of "detecting and protecting spam at key network nodes, and maintaining the upgrade and update of spam protection mechanism" is the anti-spam product between security domains and all boundaries between the system undergoing test and other systems. This indicator is generally not applicable to industrial control systems, unless the user has the need for the application.

The key points of the evaluation implementation include: check whether the system uses the email protocol and whether the email server is deployed; if the email protocol is used in the system, interview the network security administrator whether the anti-spam product is deployed at the key network node and other technical measures; on-site check whether the anti-spam product operates normally, and whether the anti-spam rule base has been updated to the latest.

If the results show that the system undergoing test has enabled the anti-spam product and the version of the rule base has been updated to the latest version, the unit determines that the result is conform; otherwise it is non-conforming or partially conforming.

5. Security Audit

[Standard Requirements]

Level Ⅲ of the control point includes evaluation units L3-ABS1-16, L3-ABS1-17, L3-ABS1-18, L3-ABS1-19, and Level Ⅳ includes evaluation units L4-ABS1-18, L4-ABS1-19, L4-ABS1-20.

[L3-ABS1-16/ L4-ABS1-18 Interpretation and Description]

The evaluation targets "Security audit should be carried out at network boundaries and important network nodes. The audit should cover each user and audit important user behaviors and important security events" include comprehensive audit system, industrial control audit products, internal network security management platform of industry special network security platform, network security platform of power monitoring system, router, switch, firewall, network isolation products and other access control equipment, etc. among levels, security domains, wired network and wireless network, and all boundaries between the system undergoing test and other systems.

The key points of evaluation implementation include: interview the network security administrator whether the integrated security audit system or the system platform with similar functions is deployed or whether the access control equipment has enabled the traffic communication audit function; on-site check whether the deployed industrial control audit system or the system platform with similar functions has enabled the security audit function, and whether the security audit scope covers all the communication traffic at the boundary; on-site check whether the important user behavior and important security events are audited; if there is no system platform with audit system or similar functions, it can check whether the audit function of the boundary equipment and important network node devices is enabled.

If the results show that the device or system undergoing test has an audit function, and the audit content can cover each user and contain important user behavior and important security events, the result is conforming; otherwise it is non-conforming or partially conforming.

[L3-ABS1-17/ L4-ABS1-19 Interpretation and Description]

The targets of the evaluation index "the audit record should include the date and time of the event, the user, the type of the event, the success of the event and other audit-related information" are the integrated audit system, industrial control audit products, industry-specific network security platform, intranet security management platform, power monitoring system network security platform, routers, switches, firewall network isolation products and other access control devices between levels, security domains, wired and wireless networks, and all boundaries between the system undergoing test and other systems.

The key points of evaluation implementation include: interview whether the traffic audit function is enabled in the access control policy of the security device (firewall, isolation device and encryption device) in the network security administrator system, and whether the audit record contains the date and time, user and event type of the event; on-site check whether the log function in the access control policy of the security device (firewall, isolation device and encryption device) is enabled; the audit function is enabled by default for common switch routers; the audit record includes the date and time of the event, the user, the type of the event, the success of the event and other information related to the audit.

If the results show that the equipment or system undergoing test has an audit function, and the audit record information includes the information of the date and time of the event, the user, the type of the event, the success of the event and the like, the result is conforming; otherwise it is non-conforming or partially conforming.

【L3-ABS1-18/ L4-ABS1-20 Interpretation and Description】

The targets of the evaluation index "audit records should be protected, backed up regularly to avoid unexpected deletion, modification or coverage" mainly include integrated audit systems, industrial control audit products, industry-specific network security platform intranet security management platform, power monitoring system network security platform, routers, switches, firewalls, network isolation products and other access control devices between levels, security domains, wired and wireless networks, and all boundaries between the system undergoing test and other systems.

The key points of evaluation implementation include: check whether protective measures are taken to protect audit records, including two levels: one is to set up special administrator to manage audit records (including local devices and backup logs), and unauthorized users are not authorized to delete, modify or cover audit records; second, there are log storage rules and synchronous outgoing backup mechanism to prevent accidental loss of original audit logs or erasure by attackers; check whether audit records are backed up regularly and verify log storage time.

If the results show that the equipment or system undergoing test has the audit function, the audit record is protected and the backup is made regularly, the unit determines that the result is conform; otherwise it is non-conformance or partial compliance.

【L3-ABS1-19 Interpretation and Description】

The targets of the evaluation index "should be able to independently conduct behavior audit and data analysis on the user's behavior of remote access and Internet access" are the comprehensive audit system between levels, security domains, wired network and wireless network, and all boundaries between the system undergoing test and other systems, industrial control audit products, industry-specific network security platform intranet security management platform, power monitoring system network security platform, routers, switches, firewalls, network isolation products and other access control devices. If the user in the system cannot access the Internet and cannot access remotely, the judgment of this item is not applicable.

The key points of evaluation implementation include: interview the network security administrator and whether there are remote access users and Internet access users; if there are remote access users and Internet access users, on-site check whether the comprehensive security audit system or the system platform with similar functions separately audits the user behaviors and whether there are corresponding audit records.

If the results show that the equipment or system undergoing test has an audit function, and the remote access user and the internet access user are separately audited, the result is conforming; otherwise it is non-conforming or partially conforming.

6. Trusted Verification

[Standard Requirements]

Level III of the control point includes the evaluation unit L3-ABS1-20, and Level IV includes the evaluation unit L4-ABS1-21.

[L3-ABS1-20 Interpretation and Description]

The evaluation index "can conduct trusted verification on the system boot program, system program, important configuration parameters and boundary protection application program of boundary equipment based on the trusted root, and conduct dynamic trusted verification in the key execution links of the application program, and give an alarm after the credibility is detected to be damaged, and the verification result should form an audit record and send to the security management center". The evaluation targets are the routers, switches, firewalls and network isolation products, and all the boundaries between the system undergoing test and other systems.

The key points of evaluation implementation include: check whether the system boot program, system program, important configuration parameters and boundary protection application of boundary equipment are authenticated based on the trusted root; check whether dynamic trusted verification is conducted in the key execution links of the application program; test and verify whether the alarm will be given when the credibility of the boundary equipment is detected to be damaged; test whether the verification results are sent to the security management center in the form of audit records.

If the results show that the boundary device undergoing test has a trusted verification function based on the trusted root, an alarm is given after the credibility is detected to be damaged, and the dynamic trusted verification is performed in the key execution link of the application program, and the verification result is sent to the security management center as an audit record, the result is conforming; otherwise it is non-conforming or partially conforming.

[L4-ABS1-21 Interpretation and Description]

The evaluation index "can conduct trusted verification on the system boot program, system program, important configuration parameters and boundary protection application program of boundary equipment based on the trusted root, and conduct dynamic trusted verification in all execution links of the application program, and give an alarm after the credibility is detected to be damaged, and the verification result should form an audit record and send to the security management center for dynamic association perception". The evaluation targets are the routers, switches, firewalls and network isolation products between all the boundaries between the system undergoing test and other systems.

The key points of evaluation implementation include: check whether the system boot program, system program, important configuration parameters and boundary protection application of the boundary equipment are authenticated based on the trusted root; verify whether the dynamic trust verification is performed in all the execution links of the application program; test and verify whether the alarm is given when the credibility of the boundary equipment is detected to be damaged; and test whether the verification results are sent to the security management center in the form of audit records and performing dynamic association perception.

If the results show that the boundary device undergoing test has a trusted verification function based on the trusted root, an alarm is given after the credibility is detected to be damaged, and dynamic trusted verification is carried out in all the execution links of the application program, and the verification result is sent to the security management center as an audit record, and the dynamic association perception is carried out, the result is conforming; otherwise it is non-conforming or partially conforming.

7. Dial Use Control

[Standard Requirements]

Level Ⅲ of the control point includes the evaluation units L3-ABS5-03 and L3-ABS5-04, and Level Ⅳ includes the evaluation units L4-ABS5-03, L4-ABS5-04 and L4-ABS5-05.

[L3-ABS5-03/ L4-ABS5-03 Interpretation and Description]

"If the industrial control system needs to use dial-up access service, the number of users with dial-up access authority should be limited, and user identification and access control measures should be taken. The evaluation targets are dial-up server, client, VPN, etc."This clause does not apply when dial-up access service is not used.

The key points of evaluation implementation include: on-site verification of whether the dial-up equipment restricts the number of users with dial-up access rights, whether the dial-up server and client use identity authentication methods such as account/ cryptography, and whether access control measures such as controlling account authority are adopted; if VPN access is used for dial-up access, check whether the access control measures are taken.

If the results show that the device or system undergoing test adopts dial-up service, and the security policy is configured to restrict the number of users with dial-up access rights, and measures such as user identity authentication and access control are taken, the result is conforming; otherwise it is non-conforming or partially conforming.

[L3-ABS5-04/ L4-ABS5-04 Interpretation and Description]

The targets of the evaluation index "both the dial-up server and the client should use the operation system with security reinforcement, and take the measures of digital certificate authentication, transmission encryption and access control" are the dial-up server and the client. This clause does not apply when dial-up access service is not used.

The key points of evaluation implementation include: interview the network security administrator whether the security reinforced operation system is adopted for the dialing server and the client, whether the communication between the server and the client adopts encryption measures and whether the digital certificate authentication method is adopted; on-site verify that the Security and reinforced operation system is used by the dial-up server and the client, such as closing the unnecessary ports and services, enabling the login failure processing function, setting the cryptography policy, setting the access control list, adopting the data certificate authentication method for data communication, deploying the encryption authentication device or the server cipher machine for encrypted communication during the communication process between the server and the client.

If the test results show that the tested equipment or system adopts the operation system with security reinforcement, and the security protection measures such as encryption, digital certificate authentication and access control are taken, the result is conforming; otherwise it is non-conforming or partially conforming.

[L4-ABS5-05 Interpretation and Description]

The main evaluation target of "the industrial control system involv real-time control and data transmission prohibits the use of dial-up access service" is dial-up server, client, etc. When the system does not involve real-time control and data transmission, this clause does not apply.

The key points of evaluation implementation include: interview the network security administrator to control whether the industrial control system for real-time control and data transmission adopts dial-up access service; on-site check whether there is dial-up access service equipment.

If the results show that the industrial control system undergoing test does not use the dial-up access service, the unit determines that the result is conform; otherwise it is either non-compliant or partially conforming.

8. Wireless Usage Control

[Standard Requirements]

Level III of the control point includes evaluation units L3-ABS5-05, L3-ABS5-06, L3-ABS5-07, L3-ABS5-08, and Level IV includes evaluation units L4-ABS5-06, L4-ABS5-07, L4-ABS5-08, L4-ABS5-09.

[L3-ABS5-05/ L4-ABS5-06 Interpretation and Description]

The evaluation index "should provide unique identification and authentication for all users (personnel, software process or equipment) participating in wireless communication" mainly includes wireless router, wireless access gateway, wireless terminal, etc. This clause does not apply when the system does not use wireless communication.

The key points of evaluation implementation include: on-site verify of whether the wireless communication user adopts identity authentication measures when logging in, such as account/cryptography, biometric identification, etc., communication equipment should be identified by unique identification, such as MAC address, equipment serial number, etc.; on-site verify of whether the user identification is unique.

If the results show that the device undergoing test provides unique identification and authentication for all users (personnel, software process or equipment) participating in the wireless communication, the unit determines that the result is conform; otherwise it is non-conformance or partial conformity.

[L3-ABS5-06/ L4-ABS5-07 Interpretation and Description]

The evaluation index "all users (personnel, software process or equipment) participating in wireless communication should be authorized and limited in execution and use" are mainly the wireless router, wireless access gateway, wireless terminal, etc. When the system does not use wireless communication, this evaluation index is not applicable.

The key points of evaluation implementation include: check whether the user is authorized during the wireless communication process, check whether the specific authority is reasonable, and check whether unauthorized use can be found and give an alarm.

If the results show that the device undergoing test authorize all the users (personnel, software process or equipment) participating in the wireless communication and restricts the execution and use, the result is conforming; otherwise it is non-conforming or partially conforming.

[L3-ABS5-07/ L4-ABS5-08 Interpretation and Description]

The targets of the evaluation index "the security measures of transmission encryption should be taken for wireless communication to realize the confidentiality protection of the transmitted message" are the encryption authentication equipment. When the system does not use wireless communication, this evaluation index is not applicable.

The key points of evaluation implementation include: interview the network security administrator whether to adopt encryption transmission measures in wireless communication transmission and the type of wireless communication technology adopted in the system; on-site check whether encryption authentication equipment or encryption module is deployed in wireless communication transmission for encryption transmission and data encryption processing, so as to ensure the confidentiality of the transmitted message; if 4G/5G wireless communication technology is adopted, the default is compliance; if WIFI communication is adopted, on-site check the AP configuration and whether the encryption transmission mode is enabled, such as WPA2, etc.

If the results show that the wireless device undergoing test adopts the encryption module to realize the encrypted transmission, the unit judge the result as conforming; otherwise it is non-conforming or partially conforming.

[L3-ABS5-08/ L4-ABS5-09 Interpretation and Description]

"For the industrial control system controlled by wireless communication technology, it should be able to identify the unauthorized wireless equipment transmitted in its physical environment, and report the unauthorized attempt to access or interfere with the control system". The main test objects are wireless sniffer, wireless intrusion detection/ defense system, handheld wireless signal detection system, etc. When the system does not use wireless communication, this evaluation index clause does not apply.

The key points of evaluation implementation include: interview the network security administrator whether the industrial control system can monitor the unauthorized wireless devices transmitted in its physical environment in real time; on-site check whether the industrial control system can monitor the unauthorized wireless devices transmitted in its physical environment in real time, and whether the unauthorized wireless devices can give an alarm in time and shield the wireless devices attempting to be connected.

If the results show that the device or the system undergoing test can monitor the unauthorized wireless equipment transmit in its physical environment in real time, give an alarm in time when the unauthorized wireless equipment is found, and can shield the wireless equipment attempting to access, the result is conforming; otherwise it is non-conforming or partially conforming.

7.2.4 Security Computing Environment

This chapter is the interpretation of "Security computing environment", which is an extended requirement for industrial control system security evaluation.

According to the characteristics of the industrial control system, this chapter also lists the characterized interpretation contents of some General security requirement evaluation under the industrial control system environment.

1. Identification

[Standard Requirements]

Level Ⅲ of the control point includes evaluation units L3-CES1-01, L3-CES1-02, L3-CES1-03, L3-CES1-04, and Level Ⅳ includes evaluation units L4-CES1-01, L4-CES1-02, L4-CES1-03 and L4-CES1-04.

[L3-CES1-01/ L4-CES1-01 Interpretation and Description]

The evaluation targets of "identity identification and authentication should be carried out for logged-in users, identity identification should be unique, and identity authentication information should be subject to complexity requirements and periodic replacement" should include network equipment (including virtualized network equipment), security equipment (including virtualized security equipment), operation system in workstation and server equipment (including host computer and virtual machine operation system), control equipment, control equipment management system, business application system, database system, middleware, system management software and system design documents, etc.

The key points of evaluation implementation include: check whether there is no null cryptography user in the user configuration information; if the evaluation target has null cryptography or weak cryptography, further check whether double-factor authentication or other equivalent management and control means are used for protection; check whether the user authentication information has complexity requirements and replace it regularly; if the above requirements cannot be realized due to technical constraints, check whether the same function is realized through management means; if the evaluation target is controller equipment, check whether the programming cryptography is configured and whether the programming cryptography has complexity requirements and replace it regularly; if the evaluation target is controller equipment, if the control equipment is deployed in a complete set with the upper unit software, and the configuration software is used to issue the controller configuration, the configuration software of the upper computer should be checked for the configuration of the controller: whether identity authentication measures are adopted, whether the user identification is unique, and whether the authentication information has complexity requirements and is replaced regularly.

If the test results show that the tested equipment or system carries out identity identification and authentication for the logged-in user, and the identity identifier is unique, and the authentication information has complexity requirements and is replaced regularly, the unit determines that the result is consistent; otherwise it is not conforming or partially conforming.

[L3-CES1-02/ L4-CES1-02 Interpretation and Description]

The targets of the evaluation index "should have the login failure processing function, configure and enable the end of session, limit the number of illegal login and automatically exit when the login connection is overtime". The evaluation targets include the network equipment (including the virtualized network equipment), the security equipment (including the virtualized security equipment), the operation system in the workstation and server equipment (including host computer and virtual machine operation system), the control equipment, the control equipment management system, the business application system, the database system, the middleware, the system management software and the system design document, etc.

The implementation points of the evaluation include: check whether the login failure processing function is equipped and enabled; check whether the illegal login restriction function is equipped and enabled, and take specific actions after the illegal login reaches a certain number of times, such as account locking, etc. ; check whether the login connection timeout and automatic logout function are configured and enabled; if the evaluation target is a controller device, if the control device is deployed in a complete set with its host group software and the configuration software is used to issue the configuration of the controller, it is necessary to check the configuration software of the upper computer for the controller: whether the illegal login restriction function is equipped and enabled, and whether the login connection timeout and automatic logout function are configured and enabled.

If the results show that the system undergoing test is configured with a login failure handling strategy and the login timeout exit function is configured, the result is conforming; otherwise it is non-conforming or partially conforming.

[L3-CES1-03/ L4-CES1-03 Interpretation and Description]

When conducting remote management, necessary measures should be taken to prevent authentication information from being eavesdropped during network transmission. The evaluation targets include network equipment (including virtualized network equipment), security equipment (including virtualized security equipment), operation system in workstation and server equipment (including host computer and virtual machine operation system), control equipment, control equipment management system, business application system, database system, middleware, system management software and system design document. This clause does not apply when only local administration exists.

The key points of evaluation implementation include: check whether encryption and other security methods are used for remote management of the system to prevent authentication information from being eavesdropped in the process of network transmission; if the evaluation target is controller equipment, if the control device is deployed in a complete set with its upper unit state software and the configuration software is used to issue the controller configuration, the configuration software of the upper computer should be checked for the configuration of the controller: whether the security mode such as encryption is adopted for remote management of the equipment.

If the results show that the device or system undergoing test is remotely managed in an encrypted manner, the result is conforming; otherwise it is non-conforming or partially conforming.

[L3-CES1-04/ L4-CES1-04 Interpretation and Description]

The evaluation index "should adopt cryptography, cryptography technology, biological technology and other two or more combination authentication technologies to authenticate the user's identity, and at least one authentication technology should be implemented with cryptography technology". The evaluation targets are network equipment (including virtualized network equipment), security equipment (including virtualized security equipment), operation system in workstation and server equipment (including host computer and virtual machine operation system), control equipment, control equipment management system, business application system, database system, middleware, system management software and system design document, etc.

The implementation points of the evaluation include: check whether the identity authentication function of dual-factor identity authentication is adopted; check whether two or more identification technologies such as one-time cryptography, digital certificate, biotechnology and device fingerprint are adopted to authenticate the user's identity; check whether one of the authentication technologies is realized by the cryptography technology; if the evaluation target is a controller device, if the control equipment is deployed in a complete set with the upper unit state software, and the configuration software is used to issue the controller configuration, the configuration software of the upper computer should be checked for the configuration of the controller: whether the identity authentication function of two-factor identity authentication is adopted.

If the results show that the equipment or system undergoing test uses more than two combined authentication technologies to authenticate the users, and at least one of the authentication technologies is realized by using the cryptographic techniques, the result is conforming; otherwise it is non-conforming or partially conforming.

2. Access Control

[Standard Requirements]

Level III of the control point includes evaluation units L3-CES1-05, L3-CES1-06, L3-CES1-07, L3-CES1-08, L3-CES1-09, L3-CES1-10, L3-CES1-11, and Level IV includes evaluation units L4-CES1-05, L4-CES1-06, L4-CES1-07, L4-CES1-08, L4-CES1-09, L4-CES1-10, L4-CES1-11.

[L3-CES1-05/ L4-CES1-05 Interpretation and Description]

The evaluation targets of "account and authority distribution for logged-in users" are network equipment (including virtualized network equipment), security equipment (including virtualized security equipment), operation system in workstation and server equipment (including host computer and virtual machine operation system), control equipment, control equipment management system, business application system, database system, middleware, system management software and system design document, etc.

The key points of evaluation implementation include: check user account and permission settings, testing whether to design access to system functions and data according to different user permissions, and verify whether access permissions of anonymous and default accounts have been disabled or restricted. If the business application system is logged in with B/S architecture, test whether the system functions and data can be accessed without logging in the system; if the evaluation target is the controller equipment, if the control equipment is deployed in a complete set with the upper unit state software and the configuration software is used to issue the controller configuration, the configuration software of the upper computer should be checked for the configuration of the controller: whether the access to the system functions and data is designed according to different user permissions.

If the test results show that the system undergoing test has allocated the account and authority to the logged-in user, and the default account does not exist, the result is conforming; otherwise it is non-conforming or partially conforming.

【L3-CES1-06/ L4-CES1-06 Interpretation and Description】

The targets of the evaluation index "it is necessary to rename or delete the default user and modify the default cryptography of the default account" are the network equipment (including the virtualized network equipment), the security equipment (including the virtualized security equipment), the operation system in the workstation and server equipment (including the host computer and virtual machine operation system), the control equipment, the management system of the control device, the business application system, the database system, the middleware, the system management software and the system design document, etc.

The key implementation points of the evaluation include: check whether the default account does not exist or the default account has been renamed; check whether the default cryptography of the default account has been modified; if the evaluation target is the controller device, verify whether the programming cryptography has changed the default password; if the evaluation target is the controller device, and if the control device is deployed in a complete set with its upper unit software and use the configuration software to issue the controller configuration, it is necessary to check the configuration software of the upper computer for the controller: whether the default account does not exist or the default account has been renamed, and whether the default password has been changed.

If the results show that the device or system undergoing test does not have the default account and the default cryptography, the unit determines that the result is consistent; otherwise the device or the system doe not comply with the default cryptography.

【L3-CES1-07/ L4-CES1-07 Interpretation and Description】

The targets of the evaluation index "redundant and expired accounts should be deleted or disabled in time to avoid the existence of shared accounts" are network equipment (including virtualized network equipment), security equipment (including virtualized security equipment), operation system in workstation and server equipment (including host computer and virtual machine operation system), control equipment, management system of control equipment, business application system, database system, middleware, system management software and system design document, etc.

The key points for evaluation implementation include: check whether there is a shared account number and whether the administrator user and account are one-to-one correspondence; check and test whether the redundant and expired accounts are deleted or disabled; if the evaluation target is a controller device, if the control device is deployed in a complete set with the upper unit software, and the configuration software is used to issue the controller configuration, the configuration software of the upper computer should be checked for the configuration of the controller: whether there is a shared, redundant or expired account number.

If the results show that the system undergoing test does not have a shared, redundant or expired account, the result is conforming; otherwise it is non-conforming or partially conforming.

[L3-CES1-08/ L4-CES1-08 Interpretation and Description]

The targets of the evaluation index "the minimum authority required by the authorized management user to realize the authority separation of the management user" are network equipment (including virtualized network equipment), security equipment (including virtualized security equipment), operation system (including host computer and virtual machine operation system), control equipment, management system of control equipment, business application system, database system, middleware, system management software and system design document in the industrial control system.

The key points of evaluation implementation include: especially check whether different administrator users are established, such as system administrator, security administrator and audit administrator; check whether role division and responsibility separation are performed for administrators; verify whether the authority of management user is separated, whether it is the minimum authority required for their work tasks; check access control policy to verify whether the authority of management user has been separated; focus on verify whether the authority of application software user in engineer station and operator station in industrial control system is separated; if the evaluation target is controller equipment, if the control equipment is deployed in a complete set with the software of upper unit, and the configuration software is used to issue controller configuration, the configuration software of the upper computer should be checked for the configuration of controller: whether the role is divided, whether it is the minimum authority required for the working task of the administrator.

If the results show that the system undergoing test has divided the administrator authority according to the principle of minimum authorization, the unit judges that the result is conform; otherwise the system does not conform or partially conforms.

[L3-CES1-09/ L4-CES1-09 Interpretation and Description]

The targets of the evaluation index "the authorized subject should configure the access control policy, and the access control policy specifies the access rules of the subject to the object" are the network equipment (including the virtualized network equipment), the security equipment (including the virtualized security equipment), the operation system in the workstation and server equipment (including the host computer and virtual machine operation system), the control equipment, the management system of the control equipment, the business application system, the database system, the middleware, the system management software and the system design document, etc.

The key points of evaluation implementation: whether the interview system administrator performs different access control policy configuration for different user subjects of each manufacturer to avoid unauthorized access; check whether different access control policies are configured for different operation users of third-party operation and maintenance manufacturers to verify whether the access control policy is effective; check whether the access control policy is configured by the management user; test whether the low-privilege user accesses the system function module of the high-authority user; if the evaluation target is a controller device, if the control device is deployed in a complete set with the software of the host unit, and the configuration software is used to issue the controller configuration, the configuration software of the host computer should be checked for the configuration of the controller; whether the access rules of the subject to the object are configured according to the security policy.

If the test results indicate that the management user of the system undergoing test configures the access control rule according to the security policy, the result is conforming; otherwise it is non-conforming or partially conforming.

【L3-CES1-10/ L4-CES1-10 Interpretation and Description】

The evaluation index "the subject of access control should be user-level or process-level, and the object should be file-level and database table-level" should be the network equipment (including virtualized network equipment), security equipment (including virtualized security equipment), operation system in workstation and server equipment (including host computer and virtual machine operation system), control equipment, control equipment management system, business application system, database system, middleware, system management software and system design document, etc.

The key points of evaluation implementation include: check whether the authorization subject (such as the management user) is responsible for configuring the access control policy; check whether the authorization subject configures the access rules of the subject to the object according to the security policy; if the evaluation target is a server device, check whether the subject user-level access control to the object file level is realized; if the evaluation target is a database system, check whether the subject user-level access control to the database table level is realized; if the evaluation target is a controller device, if the control device is deployed in a complete set with its unit state software, and use the configuration software to issue the controller configuration, it is necessary to check the configuration software of the host computer for the controller; whether the upper-level access control policy is configured, and check the granularity of its access control.

If the results show that the access control authority granularity of the system undergoing test reaches the user level or the process level of the main body and the object is the file and database table level, the unit judges that the result is consistent; otherwise it does not conform or partially conforms.

【L3-CES1-11 Interpretation and Description】

The targets of the evaluation index of "setting security marks for important subjects and objects, and controlling the access of subjects to information resources with security marks" are network equipment (including virtualized network equipment), security equipment (including virtualized security equipment), operation system in workstation and server equipment (including host computer and virtual machine operation system), control equipment, control equipment management system, business application system, database system, middleware, system management software and system design document, etc.

The key points of evaluation implementation include: check whether the security label is set for the subject and object according to the security policy; test and verify whether the mandatory access control policy of the subject to the object access is controlled according to the security label of the subject and object; if the evaluation target is the controller device, if the control device is deployed in a complete set with the upper unit software, and the configuration software is used to issue the controller configuration, the configuration software of the upper computer should be checked for the configuration of the controller: whether the security mark policy is set.

If the results show that the system undergoing test has set a security flag on the sensitive resource and the access control policy is configured, the unit determines that the result is accord; otherwise it is not or partially conforming.

[L4-CES1-11 Interpretation and Description]

The targets of the evaluation index "set security marks for subjects and objects, and determine subject-to-object access according to security marks and mandatory access control rules" are network devices (including virtualized network devices), security devices (including virtualized security devices), operation systems (including host and virtual machine operation systems), control devices, control device management systems, business application systems, database systems, middleware, system management software and system design documents in the industrial control system.

The key points of evaluation implementation include: check whether the security label is set for the subject and object according to the security policy; test and verify whether the mandatory access control strategy for the access of the subject to the object is controlled according to the security label of the subject and object; if the evaluation target is the controller device, if the control device is deployed in a complete set with the upper unit software, and the configuration software is used to issue the controller configuration, the configuration software of the upper computer should be checked for the configuration of the controller: whether the security mark policy is set or not.

If the results show that the system undergoing test has set a security flag on the sensitive resource and the forced access control policy is configured, the result is conforming; otherwise it is non-conforming or partially conforming.

3. Security Audit

[Standard Requirements]

Level Ⅲ of the control point includes evaluation units L3-CES1-12, L3-CES1-13, L3-CES1-14, L3-CES1-15, and Level Ⅳ includes evaluation units L4-CES1-12, L4-CES1-13, L4-CES1-14 and L4-CES1-15.

[L3-CES1-12/ L4-CES1-12 Interpretation and Description]

The targets of the evaluation index "the security audit function should be enabled, the audit should cover each user, and the audit of important user behaviors and important security events" should be network equipment (including virtualized network equipment), security equipment (including virtualized security equipment), operation system in workstation and server equipment (including host computer and virtual machine operation system), control equipment, control equipment management system, business application system, database system, middleware, system management software and system design document, etc.

The key points of evaluation implementation include: check whether the security audit function is provided and enabled; check whether the security audit scope covers each user; check whether important user behaviors and important security events are audited; if the evaluation target is controller equipment, check whether the operation log, operation log and error log are provided and enabled; if the evaluation target is controller equipment, if the control device is deployed in a complete set with its host unit software and the configuration software is used to issue the configuration of the controller, it is necessary to check the configuration software of the upper computer for the configuration of the controller: whether the security audit function is provided and enabled, whether the security audit scope covers each user, and whether the important user behavior and important security events are audited.

If the test results show that the tested equipment or system has enabled the audit function covering all users and can audit the important user behaviors and important security events, the result is conforming; otherwise it is non-conforming or partially conforming.

[L3-CES1-13 Interpretation and Description]

The targets of the evaluation index "the audit record should include the date and time of the event, user, time type, success of the event and other audit-related information" are network equipment (including virtualized network equipment), security equipment (including virtualized security equipment), operation system in workstation and server equipment (including host computer and virtual machine operation system), control equipment, control equipment management system, business application system, database system, middleware, system management software and system design documents, etc.

The implementation points of the evaluation include: check whether the audit record information includes the date and time of the event, user, event type, whether the event is successful and other audit-related information; check whether the audit record is kept for at least 6 months according to the requirements of the network security law; if the evaluation target is the controller equipment, check whether the audit record information includes the date and time of the event, user, event type, success of the event and other audit-related information; if the evaluation target is the controller device, if the control device is deployed in a complete set with the software of the upper unit, its log information may need to be viewed in the upper computer.

If the test results show that the audit record of the EUT or system includes the date and time of the event, the user, the type of time, the success of the event and other audit-related information, and the record is kept for at least 6 months, the result is conforming; otherwise it is non-conforming or partially conforming.

【L4-CES1-13 Interpretation and Description】

The targets of the evaluation index "the audit record should include the date and time of the event, event type, subject identification, object identification and result" are network equipment (including virtualized network equipment), security equipment (including virtualized security equipment), operation system in workstation and server equipment (including host computer and virtual machine operation system), control equipment, control equipment management system, business application system, database system, middleware, system management software and system design document, etc.

The key points of evaluation implementation include: check whether the audit record information includes the date and time of the event, event type, subject identification, object identification and result; check whether the audit record is kept for at least 6 months according to the requirements of the network security law; if the evaluation target is the controller equipment, check whether the audit record information includes the date and time of the event, event type, subject identification, object identification and result, etc.; if the evaluation target is the controller device, if the control device is deployed in a complete set with the software of the upper unit, its log information may need to be checked in the upper computer.

If the test results show that the content of the audit log of the tested equipment or system is satisfactory and the records are retained for at least 6 months, the result is conforming; otherwise it is non-conforming or partially conforming.

【L3-CES1-14/ L4-CES1-14 Interpretation and Description】

The targets of the evaluation index "audit records should be protected, backed up regularly to avoid unexpected deletion, modification or coverage" mainly include network equipment (including virtualized network equipment), security equipment (including virtualized security equipment), operation system in workstation and server equipment (including host computer and virtual machine operation system), control equipment, control equipment management system, business application system, database system, middleware, system management software and system design documents, etc.

The implementation points of the evaluation include: check whether protective measures are taken to protect the audit records, such as configuring the log server; check whether technical measures have been taken for regular backup of audit records and its backup strategy; The log keeping cycle should take into account the storage requirements of 6 months for log audit records in the Network Security Law of the People's Republic of China and 12 months for the log audit records in the Basic Requirements for Network Security Protection of Critical Information Infrastructure; if the evaluation target is the controller equipment, if the control equipment is deployed in complete set with the upper unit state software and the configuration software is used to issue the controller configuration, it is necessary to check the configuration software of the upper computer for the configuration of the controller: whether protective measures are taken to protect the audit records, whether the technical measures are taken to backup the audit records regularly, and check the storage time of the records; check whether there are relevant access control policies for the storage disk location of the audit log of the key control device and the log storage disk location of the application business system.

If the test results show that the equipment or system undergoing test is protected by audit records and the records are retained for at least 6 months, the result is conforming; otherwise it is non-conforming or partially conforming.

[L3-CES1-15/ L4-CES1-15 Interpretation and Description]

The targets of the evaluation index "the audit process should be protected against unauthorized interruption" are network equipment (including virtualized network equipment), security equipment (including virtualized security equipment), operation system in workstation and server equipment (including host computer and virtual machine operation system), control equipment, control equipment management system, business application system, database system, middleware, system management software and system design documents.

The key points of evaluation implementation include: use the account of non-audit administrator to try to interrupt the audit process and verify whether the audit process is protected; if the evaluation target is the controller device, and the logs of the key control equipment on the industrial site are received by the upper computer, it should be noted that the response protection mechanism should be provided for the receiving log audit process of the upper computer.

If the test results show that the audit process of the tested equipment or system has an interruption protection mechanism, the unit determines that the result is compliant; otherwise it is not or partially conforming.

4. Intrusion Prevention

[Standard Requirements]

Level Ⅲ of the control point includes evaluation units L3-CES1-17, L3-CES1-18, L3-CES1-19, L3-CES1-20, L3-CES1-21, L3-CES1-22, and Level Ⅳ includes evaluation units L4-CES1-16, L4-CES1-17, L4-CES1-18, L4-CES1-19, L4-CES1-20, L4-CES1-21.

[L3-CES1-17/ L4-CES1-16 Interpretation and Description]

The "minimum installation principle should be followed, and only the required components and application programs should be installed" is the evaluation targets of the operation system (including host computer and virtual machine operation system), application program and control equipment in the workstation and server equipment in the industrial control system.

The key points of evaluation implementation include: check whether the minimum installation principle is followed; whether non-essential components and applications, such as non-business programs such as QQ, are not installed; if the evaluation target is controller equipment, check the product design documents, and check whether the modules and components are minimally tailored according to the task requirements.

If the test results show that only the required components and applications are installed in the equipment or system under test, the result is conforming; otherwise it is non-conforming or partially conforming.

[L3-CES1-18/ L4-CES1-17 Interpretation and Description]

The targets of the evaluation index "Unneeded system services, default sharing and high-risk ports should be closed" are network devices (including virtualized network devices), security devices (including virtualized security devices) and operation systems (including host and virtual machine operation systems) in the industrial control system. For the field control layer, this evaluation applies to the control equipment connected to the upper computer through Ethernet/industrial Ethernet, but not to the control equipment connected to the upper computer through field bus or hard wiring.

The key points of evaluation implementation include: check whether unnecessary system services and default sharing are closed; check whether there are no unnecessary high-risk ports; and perform vulnerability inspection by category when the business system is down when the tool is used for vulnerability inspection. The vulnerability scanning tools should adopt the lightweight scanning mode; when the industrial site needs to apply for the construction point/skylight point for vulnerability inspection, the inspection should be completed within 30 minutes before the end of the construction point/skylight point, so as to reserve time to recover the business application system; if the evaluation target is the controller equipment in the Ethernet/industrial Ethernet environment, it is required to check whether there is no unnecessary high-risk port, and the industrial control miss-scan tool can be used for vulnerability inspection.

If the test results show that the device or system undergoing test does not open redundant system services, default sharing and high-risk ports, the result is conforming; otherwise it is non-conforming or partially conforming.

【L3-CES1-19/ L4-CES1-18 Interpretation and Description】

The targets of the evaluation index that "the terminal managed through the network should be restricted by setting the terminal access mode or network address range" is the network equipment (including the virtualized network equipment), the security equipment (including the virtualized network equipment), the operation system in the workstation and server equipment (including host computer and virtual machine operation system) in the industrial control system. For the field control layer, this evaluation applies to the control equipment connected to the upper computer through Ethernet/ industrial Ethernet, but not to the control equipment connected to the upper computer through field bus or hard wiring.

The key points of evaluation implementation include: check the configuration file or parameter to determine whether the access range of the terminal is limited; interview the administrator to see whether there is an independent and controllable safe operation and maintenance area, and check whether the technical measures of the safe operation and maintenance area can meet the security requirements for the management terminal; if the evaluation target is the controller equipment in Ethernet/ industrial Ethernet environment, it is necessary to check whether the network access range is limited.

If the results show that the device or system undergoing test has limited the access range of the terminal, the unit determines that the result is accord; otherwise it does not conform or partially complies.

【L3-CES1-20/ L4-CES1-19 Interpretation and Description】

The targets of the evaluation index "should provide the data validity test function to ensure that the contents input through the man-machine interface or through the communication interface conform to the system setting requirements" are the control equipment management system, business application system, middleware, system management software and system design documents, etc. in the industrial control system.

The key points of evaluation implementation include: check whether the content of the system design document includes the content or module of the data validity verification function.

If the results show that the system undergoing test provides the data validity check function, the result is conforming; otherwise it is non-conforming or partially conforming.

【L3-CES1-21/ L4-CES1-20 Interpretation and Description】

The targets of the evaluation index that "it is necessary to find possible known vulnerabilities and repair them in time after full test and evaluation" are network devices (including virtualized network devices), security devices (including virtualized security devices), operation systems in workstations and server devices (including host and virtual machine operation systems), control devices, control device management systems, business application systems, database systems, middleware and system management software, etc. For the field control layer, this evaluation applies to the control equipment connected to the upper computer through Ethernet/ industrial Ethernet, but not to the control equipment connected to the upper computer through field bus or hard wiring.

The key points of evaluation implementation include: to check whether there is no high-risk vulnerability by means of vulnerability scanning and penetration test; to interview the security administrator to understand the vulnerability inspection measures and to verify whether the vulnerability is repaired in a timely manner after full test and evaluation; if the evaluation target is the controller equipment in Ethernet/ industrial Ethernet environment, the industrial control scanning tool can be used to check whether there is no high-risk vulnerability; core controller equipment: some key control devices in the industrial scenario are of foreign brands, so the vulnerability should be emphasized; security equipment: security equipment is the main guarantee means for the information security of the industrial control system, and it is not allowed to introduce or have high-risk vulnerabilities by itself; key business application system: most of the key business systems in the industrial control system are independently researched and developed at home or introduced from abroad, and there may be high-risk vulnerabilities with serious consequences.

If the test results show that the tested equipment or system has a mechanism to periodically check and fix the vulnerability, the result is conforming; otherwise it is non-conforming or partially conforming.

[L3-CES1-22/ L4-CES1-21 Interpretation and Description]

The evaluation targets of "should be able to detect the intrusion of important nodes and provide alarm in case of serious intrusion event" are mainly the network equipment (including the virtualized network equipment), the security equipment (including the virtualization security equipment), the operation system in the workstation and server equipment (including host computer and virtual machine operation system) in the industrial control system.

The key points of evaluation implementation include: check whether there are intrusion detection measures; check whether the alarm is provided in case of serious intrusion events. In the industrial scenario, each node generally does not be able to independently discover intrusion behaviors, so it generally adopts the third-party technical measures for implementation, and checks whether the deployed third-party technical measures are effective.

If the results show that the equipment or system undergoing test has taken intrusion detection measures and is configured with an alarm, the result is conforming; otherwise it is non-conforming or partially conforming.

5. Malicious Code Prevention

[Standard Requirements]

Level Ⅲ of the control point includes the evaluation unit L3-CES1-23, and Level Ⅳ includes the evaluation unit L4-CES1-22.

[L3-CES1-23 Interpretation and Description]

The evaluation targets of "take technical measures to prevent malicious code attack or active immune trusted verification mechanism timely to identify intrusion and virus behaviors and effectively block them" are network equipment (including virtualized network equipment), security equipment (including virtualized security equipment), operation system in workstation and server equipment (including host computer and virtual machine operation system) in industrial control system.

The key points of evaluation implementation include: check whether the anti-malicious code software or the software with corresponding functions is installed, upgrade and update the anti-malicious code database regularly (including the network level and mobile media local ferry); check whether the technical measures to prevent the malicious code attack or the active immune trusted verification technology are adopted to timely identify the intrusion and virus behavior; check whether the intrusion and virus behavior are effectively blocked when identifying the intrusion and virus behavior.

If the test results show that the tested equipment or system adopts the technical measures to prevent the malicious code attack or the active immune trusted verification to detect and block the virus, the result is conforming; otherwise it is non-conforming or partially conforming.

[L4-CES1-22 Interpretation and Description]

The targets of the evaluation index "active immune trusted verification mechanism should be adopted to timely identify intrusion and virus behaviors and effectively block them" are network equipment (including virtualized network equipment), security equipment (including virtualized security equipment), operation system in workstation and server equipment (including host computer and virtual machine operation system) in the industrial control system.

The key points of evaluation implementation include: check whether anti-malicious code software or software with corresponding function is installed, upgrade and update anti-malicious code database regularly (including network level regular upgrade and mobile media local ferry); check whether active immune trusted verification technology is adopted to timely identify intrusion and virus behavior; check whether intrusion and virus behavior are effectively blocked when identifying intrusion and virus behavior.

If the results show that the equipment or system undergoing test adopt the active immune trusted verification technology to detect and block the virus, the result is conforming; otherwise it is non-conforming or partially conforming.

6. Trusted Verification

[Standard Requirements]

Level Ⅲ of the control point includes the evaluation unit L3-CES1-24, and Level Ⅳ includes the evaluation unit L4-CES1-23.

[L3-CES1-24 Interpretation and Description]

The evaluation targets of "can conduct trusted verification on the system boot program, system program, important configuration parameters and application program of the computing device based on the root of trust, and conduct dynamic trusted verification in the key execution links of the application program, give an alarm after the credibility is detected to be damaged, and form the verification result into audit record and send it to the security management center" are network equipment (including virtualized network equipment), security devices (including virtualized security devices), operation systems (including host and virtual machine operating systems) in the industrial control system, control equipment, business application system, database system, middleware, system management software and system design documents and others.

The key points of evaluation implementation include: check whether the above evaluation targets conduct trusted verification on the system boot program, system program, important configuration parameters and application program based on the trusted root (such as TCM security chip); check whether the evaluation target conducts dynamic trusted verification in the key execution link of the application program; verify whether to give an alarm when the credibility is damaged; and check whether the trusted verification result is sent to the security management center in the form of audit record.

If the results show that the system undergoing test adopts the trust verification technology and is configured with an alarm, and the trusted verification result is sent to the security management center, the result is conforming; otherwise it is non-conforming or partially conforming.

[L4-CES1-23 Interpretation and Description]

The targets of "can conduct trusted verification on the system boot program, system program, important configuration parameters and application program of the computing device based on the root of trust, conduct dynamic trusted verification in all execution links of the application program, give an alarm when the credibility is detected to be damaged, and send the verification result into audit record to the security management center, and conduct dynamic association perception" are network equipment (including virtualized network equipment), security equipment (including virtualized security equipment), operation system (including host and virtual machine operation system) in workstation and server equipment in industrial control system, control equipment, business application system, database system, middleware and system management software and system design document.

The key points of evaluation implementation include: check whether the above-mentioned evaluation targets conduct trusted verification on the system boot program, system program, important configuration parameters and application program based on the trusted root (such as TCM security chip); check whether the evaluation target conducts dynamic trusted verification in all the execution links of the application program; verify whether the alarm is given when the credibility of the evaluation target is detected to be damaged; check whether the trusted verification result is sent to the security management center in the form of audit record; check whether the trusted verification result is dynamically associated and perceived.

If the results show that the equipment or system undergoing test adopts the trust verification technology and is configured with an alarm, and the trusted verification result is sent to the security management center, and the dynamic association perception is performed, the result is conforming; otherwise it is non-conforming or partially conforming.

7. Data Integrity

[Standard Requirements]

Level III of the control point includes evaluation units L3-CES1-25, L3-CES1-26, and Level IV includes evaluation units L4-CES1-24, L4-CES1-25 and L4-CES1-26.

[L3-CES1-25/ L4-CES1-24 Interpretation and Description]

The evaluation targets of "verification technology or cryptography technology should be adopted to ensure the integrity of important data during transmission, including but not limited toauthentication data, important business data, important audit data, important configuration data, important video data and important personal information" are network equipment (including virtualized network equipment), security equipment (including virtualized security equipment), operation system in workstation and server equipment (including host computer and virtual machine operation system), control equipment, control equipment management system, business application system, database system, data security protection system, middleware and system management software and system design documents.

The key points of evaluation implementation include: check the system design documents, verify whether the authentication data, important business data, important audit data, important configuration data, important video data and important personal information are used in the transmission process to ensure the integrity; verify whether the system can detect that the integrity of the data is damaged in the transmission process and can timely recover after the authentication data, important business data, important audit data, important configuration data, important video data and important personal information are tampered in the transmission process; if the evaluation target is the controller equipment, check the network transmission between the upper station and the controller, and whether the verification technology or cryptography technology is used for data integrity protection.

If the test results show that the tested equipment or system adopts the verification technology or the cryptographic techniques to ensure the integrity of the important data in the transmission process, the result is conforming; otherwise it is non-conforming or partially conforming.

[L3-CES1-26/ L4-CES1-25 Interpretation and Description]

The evaluation targets of "verification technology or cryptography technology should be adopted to ensure the integrity of important data in the storage process, including but not limited to authentication data, important business data, important audit data, important configuration data, important video data and important personal information" are network equipment (including virtualized network equipment), security equipment (including virtualized security equipment), operation system in workstation and server equipment (including host computer and virtual machine operation system), control equipment, control equipment management system, business application system, database system, data security protection system, middleware and system management software and system design documents.

The key points of evaluation implementation include: check system design documents, verify whether verification technology or cryptography technology is used in the storage of authentication data, important business data, important audit data, important configuration data, important video data and important personal information to ensure the integrity; check whether technical measures (such as data security protection system, etc.) are adopted to ensure the integrity of authentication data, important business data, important audit data, important configuration data, important video data and important personal information in the storage process; testing and verify that authentication data, important business data, important audit data, important configuration data, important video data and important personal information are tampered in the storage process, and whether the integrity of the data in the storage process can be detected and recovered in time.

If the results show that the equipment or system undergoing test adopts the verification technology or the cryptographic techniques to ensure the integrity of the important data in the storage process, the result is conforming; otherwise it is non-conforming or partially conforming.

[L4-CES1-26 Interpretation and Description]

The targets of "In the applications that may involve legal liability identification, cryptography techniques should be adopted to provide original data evidence and data reception evidence to realize the non-repudiation of the data original behaviour and the non-repudiation of the data receiving behaviour" are the business application system, system management software and system design document which may be involved in the legal liability determination in the industrial control system.

The key points of evaluation implementation include: interview the administrator to understand the operation purpose, background and process of the system, and verify whether the system application may involve the legal responsibility identification; check the design documents to check whether the cryptography technology is adopted to ensure the non-repudiation of data sending and data receiving operations; and verify whether the system realizes non-repudiation by using digital certificate and other technologies.

If the results show that the system adopts the cryptographic techniques to realize non-repudiation, the unit determines that the result is accord; otherwise the system does not conform or partially conforms.

8. Data Confidentiality

[Standard Requirements]

Level Ⅲ of the control point includes evaluation units L3-CES1-27 and L3-CES1-28, and Level Ⅳ includes evaluation units L4-CES1-27 and L4-CES1-28.

[L3-CES1-27/ L4-CES1-27 Interpretation and Description]

The targets of the evaluation index "cryptography technology should be adopted to ensure the confidentiality of important data during transmission, including but not limited to authentication data, important business data and important personal information" include network equipment (including virtualized network equipment), security equipment (including virtualized security equipment), operation system (including host computer and virtual machine operation system) in workstation and server equipment, control equipment, control equipment management system, business application system, database system, data security protection system, middleware and system management software and system design documents.

The implementation points of the evaluation include: check whether the system design documents, authentication data, important business data and important personal information are transmitted to ensure the confidentiality with the cryptography technology; retrieve the data packets during the transmission process by sniffing, and verify whether the authentication data, important business data and important personal information are encrypted during the transmission process; if the evaluation target is the controller equipment, check the network transmission between the upper engineer station and the controller, and whether the cryptography technology is used to protect the data confidentiality.

If the results show that the system undergoing test adopts the cryptographic techniques to ensure the confidentiality of its important data in the transmission process, the result is conforming; otherwise it is non-conforming or partially conforming.

[L3-CES1-28/ L4-CES1-28 Interpretation and Description]

The evaluation targets of "cryptography technology should be adopted to ensure the confidentiality of important data in the storage process, including but not limited to authentication data, important business data and important personal information" are network equipment (including virtualized network equipment), security equipment (including virtualized security equipment), operation system in workstation and server equipment (including host computer and virtual machine operation system), control equipment, control equipment management system, business application system, database system, data security protection system, middleware and system management software and system design documents.

The key points of evaluation implementation include: check whether the system design documents, authentication data, important business data and important personal information are stored with cryptography technology to ensure confidentiality; check whether technical measures (such as data security protection system) are adopted to ensure the confidentiality of authentication data, important business data and important personal information in the storage process; test and verify whether the specified data is encrypted.

If the results show that the system undergoing test adopts the cryptographic techniques to ensure the confidentiality of the important data in the storage process, the result is conforming; otherwise it is non-conforming or partially conforming.

9. Data Backup and Recovery

[Standard Requirements]

Level III of the control point includes evaluation units L3-CES1-29, L3-CES1-30, L3-CES1-31, and Level IV includes evaluation units L4-CES1-29, L4-CES1-30, L4-CES1-31, L4-CES1-32.

[L3-CES1-29/ L4-CES1-29 Interpretation and Description]

The targets of the evaluation index "Local data backup and recovery function of important data should be provided" are network equipment (including virtualized network equipment), security equipment (including virtualized security equipment), operation system in workstation and server equipment (including host computer and virtual machine operation system), control equipment, control equipment management system, business application system, database system, data security protection system, middleware and system management software and system design documents.

The key points of evaluation implementation include: check whether the local backup is carried out according to the backup strategy formulated in the management system; check whether the backup strategy is reasonable and the backup configuration is correct; check whether the backup result is consistent with the backup strategy; check whether the recent recovery test records can be recovered normally.

If the results show that the system undergoing test has the function of local data backup and recovery, the result is conforming; otherwise it is non-conforming or partially conforming.

【L3-CES1-30/ L4-CES1-30 Interpretation and Description】

The targets of the evaluation index "the remote real-time backup function should be provided, and the important data should be backed up to the backup site in real time through the communication network" are the network equipment (including the virtualized network equipment), the security equipment (including the virtualized security equipment), the operation system in the workstation and server equipment (including host computer and virtual machine operation system), the control equipment, the management system of the control equipment, the business application system, the database system, the data security protection system, the middleware and the system management software and the system design documents.

The key points of evaluation implementation include: check whether the configuration data and business data have a remote real-time backup mechanism; check whether the backup site meets the requirements for remote backup; check whether the configuration data and business data can be backed up to the remote backup site through the network; check the data backup records to confirm whether the real-time backup mechanism is effectively implemented.

If the results show that the system undergoing test provides the remote real-time backup function, the result is conforming; otherwise it is non-conforming or partially conforming.

【L3-CES1-31/ L4-CES1-31 Interpretation and Description】

The targets of the evaluation index "hot redundancy of important data processing system should be provided to ensure the high availability of the system" are network equipment (including virtualized network equipment), security equipment (including virtualized security equipment), operation system in workstation and server equipment (including host computer and virtual machine operation system), control equipment, management system of control equipment, business application system, database system, data security protection system, etc. in the industrial control system.

The key points of evaluation implementation include: check whether the important data processing system of the production management layer adopts the hot redundancy deployment mode; check whether the monitoring server and history server of the process monitoring layer adopt the hot redundancy deployment mode; check whether the important control equipment of the field control layer adopts the hot redundancy deployment mode.

If the results show that the important data processing system undergoing test adopts the hot redundancy and other deployment mode to ensure the high availability of the system, the unit determines that the result is consistent; otherwise the data processing system does not conform to or partially complies with the requirement.

【L4-CES1-32 Interpretation and Description】

The targets of the evaluation index "remote disaster backup center should be established to provide real-time switching of business applications" are the disaster backup center and relevant components in the industrial control system, including network equipment (including virtualized network equipment), security equipment (including virtualized security equipment), operation system in workstation and server equipment (including host computer and virtual machine operation system), control equipment, control equipment management system, business application system, database system, middleware and system management software, etc.

The key points of evaluation implementation include: check whether a remote disaster backup center is established and equipped with communication lines, network equipment, security equipment and data processing equipment required for disaster recovery; check whether real-time switching function of business applications is provided; focus on systems with high requirements for business continuity, such as intelligent traffic control system of transportation industry, meteorological acquisition system of meteorological industry, and system of power industry, etc.

If the results show that the remote disaster backup center has been establish and the real-time switching of the business application is provided, the result is conforming; otherwise it is non-conforming or partially conforming.

10. Residual Information Protection

[Standard Requirements]

Level III of the control point includes evaluation units L3-CES1-32 and L3-CES1-33, and Level IV includes evaluation units L4-CES1-33 and L4-CES1-34.

[L3-CES1-32/ L4-CES1-33 Interpretation and Description]

The targets of the evaluation index "should ensure that the storage space in which the authentication information is located is completely cleared before being released or re-allocated" are network equipment (including virtualized network equipment), security equipment (including virtualized security equipment), operation system (including host computer and virtual machine operation system) in workstation and server equipment, control equipment, management system of control equipment, field device, business application system, database system, middleware, system management software and system design document, etc.

The key points of evaluation implementation include: check whether the system or equipment of the production management layer, process monitoring layer, site control layer and field device layer, its configuration information or system design document, and the storage space in which the user's authentication information is located is completely cleared before being released or re-allocated.

If the results show that the storage space of the device or system undergoing test has been completely cleared after the release or clearing of the authentication information of the device or system undergoing test, the result is conforming; otherwise it is non-conforming or partially conforming.

【L3-CES1-33/ L4-CES1-34 Interpretation and Description】

The targets of the evaluation index "should ensure that the storage space with sensitive data is completely cleared before being released or re-allocated" are network equipment (including virtualized network equipment), security equipment (including virtualized security equipment), operation system in workstation and server equipment (including host computer and virtual machine operation system), control equipment, management system of control equipment, field device, business application system, database system, middleware, system management software and system design document, etc.

The key points of evaluation implementation include: check the system of the production management layer, process monitoring layer, site control layer and field device layer, check their configuration information or system design documents, and verify whether the storage space in which the sensitive information of the user is located is completely cleared before being released or re-allocated.

If the test results indicate that the storage space of the tested equipment or system has been completely cleared after the release or removal of the sensitive information, the result is conforming; otherwise it is non-conforming or partially conforming.

11. Personal Information Protection

【Standard Requirements】

Level Ⅲ of the control point includes evaluation units L3-CES1-34 and L3-CES1-35, and Level Ⅳ includes evaluation units L4-CES1-35 and L4-CES1-36.

【L3-CES1-34/ L4-CES1-35 Interpretation and Description】

The targets of the evaluation index "only collect and save the user's personal information necessary for the business" are the control equipment management system, business application system, database system, system management software, etc. in the industrial control system.

The implementation points of the evaluation include: verify whether the personal information collected by the production management layer or the process monitoring layer is necessary for the business application; and verify whether the management system and process related to the protection of the user's personal information have been formulated.

If the results show that the system undergoing test only collects and stores the personal information of the user necessary for the business, and the management system and flow for the protection of the personal information of the user are formulated, the result is conforming; otherwise it is non-conforming or partially conforming.

【L3-CES1-35/ L4-CES1-36 Interpretation and Description】

The evaluation targets of "Unauthorized access to and illegal use of user's personal information should be prohibited" are control equipment management system, business application system, database system, system management software, etc. in the industrial control system.

The key points of the evaluation include: check the management system documents, check whether the management system and process for the protection of user's personal information are formulated, especially the contents that prohibit unauthorized access to the user's personal information; check whether the technical measures are adopted to limit the access and use of the user's personal information.

If the results show that the system adopt technical measures to prohibit unauthorized access and illegal use of the user's personal information and formulate the management system and flow for the protection of the user's personal information, the result is conforming; otherwise it is non-conforming or partially conforming.

12. Security of Control Equipment

[Standard Requirements]

Level III includes evaluation units L3-CES5-01, L3-CES5-02, L3-CES5-03, L3-CES5-04, L3-CES5-05, and Level IV includes evaluation units L4-CES5-01, L4-CES5-02, L4-CES5-03, L4-CES5-04 and L4-CES5-05.

[L3-CES5-01/ L4-CES5-01 Interpretation and Description]

The evaluation targets of "the control equipment should realize the security requirements such as identity authentication, access control and security audit provided by the general security requirements of the corresponding level. If the control equipment cannot realize the above requirements due to the conditions, the upper control or management equipment should realize the same function or be controlled by management means" are the DCS control unit, PLC controller, RTU, process controller, paperless recorder, measurement and control device of the field control layer. Suppose the field control layer equipment cannot directly log in and verify the security functions such as identity authentication, access control, and security audit. In that case, the evaluation target should be extended to the upper control or management equipment of the control equipment, such as the engineer station, operator station. If the relevant security requirements of the control equipment are not realized by technical means, the management means should be selected as the evaluation target. This indicator applies to the asset components of the equipment in the field control layer. During the evaluation implementation, attention should be paid to realizing the security function of the control equipment itself.

The key points of evaluation implementation include: because the industrial control system focuses on the real-time requirement, part of the security control measures in the general requirements do not apply to the control equipment. The evaluation here is implemented in combination with the technical requirements for the design of the security computing environment for the third-level industrial control system in GB/ T 25070. The evaluation implementation of the identity authentication control point of the control device should focus on whether the program running on it and the corresponding data set have a unique identification, to prevent unauthorized modification. The evaluation implementation of access control should focus on whether the control device can check whether the upper program or the user that issues the operation command has the authority to execute the operation after receiving the operation command, and the evaluation implementation of security audit should focus on whether the control device adopts the real-time audit tracking technology to ensure the timely capture of network security event information. When it is necessary to check the upper control or management equipment, attention should be paid to how the upper control or management equipment realizes the functions of identity authentication, access control and security audit for the control equipment, rather than these security control measures of the upper control or management equipment itself. If the unit undergoing test realizes the identity identification, access control, security audit and other security control points of the control equipment through management measures, it must carefully check the manning, implementation, execution records and video recording of the management measures, so as to ensure that the security management measures can realize the security control effect with the same effect.

If the test results show that the security functions provided by the equipment undergoing test such as identity authentication, access control and security audit are correctly configured, or the upper control or management equipment realizes the same function or is controlled by management means and has standard management records, the result is conforming; otherwise it is non-conforming or partially conforming.

【L3-CES5-02/ L4-CES5-02 Interpretation and Description】

The targets of the evaluation index "patch update and firmware update should be carried out on the control equipment without affecting the safe and stable system operating after full test and evaluation" mainly include DCS control unit, PLC controller, RTU, process controller, paperless recorder, measurement and control device, etc.

The key points of evaluation implementation include: interview the administrator responsible for the daily update of the control equipment to control whether the equipment has the conditions for patch upgrade and firmware upgrade; the vulnerability discovery of the control device can be conducted by means of vulnerability scanning; pay attention to the vulnerability release platform or tracking official patch release, etc. ; pay attention to the method to test and evaluate the impact of the update on the security and stability of the system before the update of the device version, patch and firmware, such as test verification on the standby control equipment or in the built test environment, and view the test verification records.

Taking the PLC controller of urban gas SCADA system as an example, the implementation steps mainly include: interview the equipment administrator or security administrator to control whether the equipment is subject to patch update and firmware update regularly or irregularly; check whether there is security test report, vulnerability scan or evaluation record for PLC controller; check whether the impact of upgrade and update on the security and stability of PLC controller and on-site equipment is fully tested before the equipment version, patch and firmware are updated, and check the traces of test record or test environment.

If the results show that the unit undergoing test performs sufficient test before upgrading and updating the PLC controller to evaluate the possible influence brought by the upgrade and update, the result is conforming; otherwise it is non-conforming or partially conforming.

[L3-CES5-03/ L4-CES5-03 Interpretation and Description]

The targets of the evaluation index of "should close or remove diskette drive, optical disk drive, USB interface, serial port or redundant network interface of the control equipment, and perform strict monitoring and management through relevant technical measures if necessary" are DCS control unit, PLC controller, RTU, process controller, paperless recorder, measurement and control device, etc. This index applies to the equipment of the field control layer, and the equipment that can detect and control the industrial production process and device should be selected for evaluation according to the business function of the industrial control system during the evaluation implementation.

The key points of evaluation implementation include: interview the security administrator or equipment administrator, and check which hardware interfaces or drivers are provided for the control equipment, such as floppy disk drive, optical disk drive, USB interface, serial port or redundant network interface, which interfaces or drivers must be reserved for use; the reasons for retaining the interface or driver should be sufficient, and the purpose should be recorded; the monitoring and management measures include the control equipment is locked in the protection box and equipped with damage alarm device, or the operation and maintenance monitoring system is used to monitor and alarm the use of the interface or driver.

If the test results show that the unitundergoing test has closed or removed the floppy disk drive, optical disk drive, USB interface, serial port or redundant network interface of the relevant control equipment, or the interface that must be reserved is monitored through technical measures and the monitoring records are kept, the result is conforming; otherwise it is non-conforming or partially conforming.

[L3-CES5-04/ L4-CES5-04 Interpretation and Description]

The targets of the evaluation index "use specialized equipment and specialized software to update the control equipment" are DCS control unit, PLC controller, RTU, process controller, paperless recorder, measurement and control device, etc. This index applies to the equipment of the field control layer, and the equipment that can detect and control the industrial production process and device should be selected for evaluation according to the business function of the industrial control system during the evaluation implementation.

The key points of evaluation implementation include: interview the security administrator or equipment administrator, update the brand model and software version of the special equipment and special software used for the control equipment, which are usually realized by the service program in the upper management or control equipment; check the update records of the special equipment and special software to confirm their normal operation.

If the results show that the unit undergoing test updates the control equipment with special equipment and special software, the result is conforming; otherwise it is non-conforming or partially conforming.

[L3-CES5-05/ L4-CES5-05 Interpretation and Description]

The targets of the evaluation index "the control equipment should be ensured to pass the security detection before being on-line, to avoid malicious code programs in the firmware of the control equipment" are DCS control unit, PLC controller, RTU, process controller, paperless recorder, measurement and control device, etc. This index applies to the equipment of the field control layer, and the equipment that can detect and control the industrial production process and device should be selected for evaluation according to the business function of the industrial control system during the evaluation implementation.

The key points of evaluation implementation include: the possibility that the firmware of the control equipment is preset with malicious code program; the special security test and verification should be conducted in the offline environment before the control equipment is online, wherein, the offline environment refers to the environment that is physically isolated from the production environment, and the organization implementing the security inspection should be the detection institution with international authorization or accreditation, such as the detection institution with CNAS qualification; pay attention to the detection part of the malicious code of the control equipment firmware in the security inspection report.

If the results show that the control device undergoing test has passed the security detection before being online and there is no malicious code, the result is conforming; otherwise it is non-conforming or partially conforming.

7.2.5 Security Management Center

Please refer to the general security evaluation requirement of this book for its application and interpretation.

7.2.6 Security Management System

Please refer to the general security evaluation requirement of this book for its application and interpretation.

7.2.7 Security Management Organization

Please refer to the general security evaluation requirement of this book for its application and interpretation.

7.2.8　Security Management Personnel

Please refer to the general security evaluation requirement of this book for its application and interpretation.

7.2.9　Security Development Management

This chapter interprets the extended security evaluation requirements of security development management of the industrial control system.

1. Product Procurement and Use

[Standard Requirements]

Level Ⅲ of the control point includes the evaluation unit L3-CMS5-01, and Level Ⅳ includes the evaluation unit L4-CMS5-01.

[L3-CMS5-01/ L4-CMS5-01 Interpretation and Description]

The targets of the evaluation index "important equipment of industrial control system can be purchased and used only after passing the security test of professional organization" are the responsible person for security system construction of industrial control system, inspection report document and qualification certificate of testing institution.

The key points of evaluation implementation include: check whether the important equipment of industrial control system and network security special products used in the system have passed the security test of professional organization, and whether there is security inspection report issued by professional organization; check whether the testing organization meets the requirements of national regulations or the regulations of relevant departments.

If the test results show that the important equipment of the industrial control system of the tested unit has the inspection report of the professional organization, and the professional organization has the testing qualification, the result is conforming; otherwise it is non-conforming or partially conforming.

2. Outsourcing Software Development

[Standard Requirements]

Level Ⅲ of the control point includes the evaluation unit L3-CMS5-02, and Level Ⅳ includes the evaluation unit L4-CMS5-02.

[L3-CMS5-02/ L4-CMS5-02 Interpretation and Description]

Outsourcing contract is the target of specifying the restrictive terms for the development unit and supplier in the outsourcing development, including the contents related to the confidentiality, prohibition of critical technology diffusion and equipment industry-specific aspects of the equipment and system within the life cycle.

The key points of evaluation implementation include: check the outsourcing development contract, whether the contract stipulates the restrictive terms for the development unit and supplier, including the contents related to the confidentiality of the equipment and system within the life cycle, prohibition of the proliferation of key technologies and the exclusive use of the equipment industry.

If the test results show that the outsourcing development contract of the unit undergoing test contains restrictive clauses for the development unit and supplier, including the contents related to the confidentiality of the equipment and system, prohibition of the diffusion of key technologies and the special purpose of the equipment industry within the life cycle, the result is conforming; otherwise it is non-conforming or partially conforming.

7.2.10 Security Operation and Maintenance Management

Please refer to the general security evaluation requirement of this book for its application and interpretation.

Chapter 8 Application and Interpretation of the Extended Security Evaluation Requirement of Big Data

8.1 Basic Concepts

8.1.1 Big Data

Big data generally refers to the data with huge quantity, various sources, fast generation, and changeable characteristics, which is difficult to deal with effectively by traditional data architecture. [GB/T 35295—2017, Definition 2.1.1]

Big data is a huge and complex data set that can not be extracted, stored, searched, shared, analyzed and processed by existing software tools. The industry typically uses "4 V" to characterize big data:

(1) Huge data volume. The capacity of a typical computer hard disk is Terabytes (TB), while the amount of data in some large enterprises approaches Exabyte (EB). Massive data is liable to the risk of data leakage, tampering and loss, so it is necessary to strengthen the security and protection capability of storage, isolation, query and clearing of mass data to ensure the confidentiality and integrity of data transmission and storage, and the backup and recovery capability of data.

(2) Data sources variety. Data may come from multiple data warehouses, data domains, or multiple data types, including structured and unstructured data. Unstructured data includes web logs, audio, video, pictures, geographic information, and so on. Once the data has problems, the diversified data sources are easy to cause the difficulty of responsibility identification, so it is necessary to strengthen the security ability of data traceability, guarantee the protection of the rights and interests of individuals and enterprises, and trace and identify the legal liabilities.

(3) Fast processing velocity. In the face of massive data, the efficiency of data processing is the life of an enterprise. During data processing, it is necessary to ensure the security and stability of the operation process, and it is easy to cause security risks, such as misbehaving and operation interruption. It is necessary to strengthen the security capabilities of network communication between components, network boundary protection, identity authentication, data processing security audit, data cleaning and conversion, to ensure the

efficiency, security and stability of mass data processing.

(4) Variability. The amount and type of big data are changing and developing all the time, which hinders the data processing and effective management. In the process of data processing, it is easy to cause security risks such as the leakage of value data or personal information, so it is necessary to strengthen the security capabilities of data classification and classification, data access control, desensitization and de-identification, so as to prevent personal sensitive information from being disclosed.

8.1.2 Targets of Big Data Classification Protection

1. Big Data Platform

Supporting integrated environment that provides resources and services for big data applications, including part or all of the functions of infrastructure layer, data platform layer, computing and analysis layer and big data management platform. The infrastructure layer provides physical or virtual computing, network and storage capabilities; the data platform layer provides the physical storage and logical storage capacity of structured and unstructured data; the computing and analysis layer provides the analytical and computational capability to process large, high-speed, diverse and variable data; and the big data management platform provides the auxiliary service capability of the big data platform. The big data platform can provide services for multiple big data applications and big data resources.

The big data platform mainly includes: file system (such as HDFS), data storage (such as HBASE), memory technology (such as Ignite, GemFire, GridGain), data collection, message system, data processing (such as SparkHPCCp, Spark Streaming, Storm, HaLoop, Yahoo S4, Flink, Trident), query engine (such as Presto, Drill, Phoenix, Pig, Hive, SparkSQL, Stinger, Impala, Shark), analysis and reporting tools (such as Kylin, Druid), Cascading and management services (such as YARN, Mesos, Ambari, ZooKeeper, Thrift, Chukwa), machine learning and developing platforms (such as Lumify, etc.).

The Hadoop big data platform is taken as an example:

① Infrastructure layer:

The infrastructure layer consists of two parts: the Zookeeper cluster and the Hadoop cluster. It provides infrastructure services for the basic platform layer, such as naming services, distributed file systems, MapReduce/ YARN, etc.

② Data platform layer:

The basic platform layer consists of three parts: task scheduling console, HBase, and Hive. It provides the basic service invocation interface for the computing and analysis layer.

③ Computing analysis layer:

The computing analysis layer is used to provide characterized calling interface and user identity authentication for the end users, and is the only visible big data platform operation portal for users. Only through the interface provided by the computing and analysis layer can the end user interact with the big data platform.

2. Big Data Application

Implement data processing activities related to data life cycle based on big data platform, generally including data acquisition, data transmission, data storage, data processing (such as calculation, analysis, visualization, etc.), data exchange, data destruction and other data activities.

Common cases of big data application include text and image search system, video search system of TV media, automatic parking system, supermarket article placement analysis system, map navigation system, online education system, translation system, product push advertising system, etc.

3. Big Data Resources

Big data resources mainly refer to data resources, that is, the data with huge quantity, various sources, fast generation, and changeable characteristics, which are difficult to deal with effectively by traditional data architecture.

Since different operators are solely responsible for security, from the point of view of the responsible subject of the grading target, the big data resource can be regarded as the grading target independently or combined with the big data platform or the big data application as the grading target. When the big data resources are independently regarded as the grading targets, the carriers (such as storage, server, database system, etc.) carrying the data should also be included in the classification objects of the big data resources. If big data resources provide external data resource services, the security protection software and hardware should also be included in the classification objects of big data resources.

8.2 Extended Security Requirements and Best Practices

Annex H of GB/T 22239—2019 provides the big data application scenario description. In order to guide the evaluation of network security level protection of big data system more effectively, this chapter gives the comparison between Annex H of GB/T 22239—2019 and the best practice of security protection of big data system (as shown in Table 8.1). The following chapters 8.3—8.4 add the interpretation of best practice of big data security protection on the basis of the application and interpretation of GB/T 22239—2019 for readers' reference.

Chapter 8 Application and Interpretation of the Extended Security Evaluation Requirement of Big Data

Table 8.1 Comparison Table Between Annex H of GB/T 22239—2019 and Best Practice of Big Data

CATEGORY	Annex H of GB/T 22239—2019				Big Data System Protection Best Practices	
	Label No.	Control Point	Requirement Item	Corresponding Level	Requirement Item	Corresponding Level
Security Physical Environment	H.3.1 H.4.1 H.5.1	Infrastructure Location	Equipment room for big data storage, processing and analysis shall be ensured to be located in China.	2,3,4	Equipment room for big data storage, processing and analysis shall be ensured to be located in China.	2,3,4
Security Communication Network	H.2.1 H.3.2 H.4.2 H.5.2	Network Architecture	a) It shall be ensured that the big data platform does not bear big data applications higher than its security protection level;	1,2,3,4	a) It shall be ensured that the big data platform does not bear big data applications higher than its security protection level;	1,2,3,4
	H.4.2 H.5.2		b) The management flow of the big data platform shall be separated from the system business flow.	3,4	b) The management flow of the big data platform shall be separated from the system business flow.	3,4
Security Computing environment	H.2.2 H.3.3 H.4.3 H.5.3	Identification	a) The big data platform shall implement identity authentication for the use of data acquisition terminal, data import service component, data export terminal and data export service component;	1,2,3,4	a) The use of data acquisition terminal, data import service component, data export terminal and data export service component shall be authenticated;	1,2,3,4
			b) Big data platform shall be able to identify and identify big data applications of different customers;	2,3,4	b) The big data platform shall be able to identify the big data applications of different customers;	1,2,3,4
					c) Big data resource shall authenticate the object calling its function;	1,2,3,4
					d) Important external calling interfaces provided by big data platform shall be authenticated.	2
					e) Various external calling interfaces provided by the big data platform shall carry out corresponding strength identity authentication according to the operation authority of the calling subject.	3,4

(Continued)

CATEGORY	Annex H of GB/T 22239—2019				Big Data System Protection Best Practices	
	Label No.	Control Point	Requirement Item	Corresponding Level	Requirement Item	Corresponding Level
Security Computing environment	H.3.3 H.4.3 H.5.3	Access Control	c) The big data platform, platform or third party providing services to the outside can access, use and manage the data resources of the big data application only when the big data application is authorized;	2,3,4	a) The big data platform, platform or third party providing services to the outside can access, use and manage the data resources of the big data application only when the big data application is authorized;	2,3,4
			d) The big data platform shall provide data classification and hierarchical security management function for big data application to take different security protection measures according to data of different categories and levels;	3,4	b) The data shall be classified and managed;	2
					c) Big data platform shall provide data classification and grading identification function;	3,4
	H.4.3 H.5.3		e) Big data platform shall provide the function of setting data security tags, authorization and access control measures based on security tags to meet the requirements of fine-grained authorization access control management capability;	3,4	d) Big data platform shall have the function of setting data security tags and access control based on security tags;	3,4
			f) The big data platform shall support the classification and hierarchical disposal of data in all links such as data acquisition, storage, processing and analysis, and ensure the consistency of security protection strategies;	3,4	e) Data collection, transmission, storage, processing, exchange and destruction shall be disposed differently according to data classification and grade identification. Relevant protection measures for the highest level data shall not be lower than Level Ⅲ safety requirements, and safety protection strategies shall be consistent in each link; f) Data collection, transmission, storage, processing, exchange and destruction shall be disposed differently according to data classification and grade identification. Relevant protection measures for the highest level data shall not be lower than Level Ⅳ safety requirements, and safety protection strategies shall be consistent in each link;	3

Chapter 8　Application and Interpretation of the Extended Security Evaluation Requirement of Big Data

(Continued)

CATEGORY	Annex H of GB/T 22239—2019				Big Data System Protection Best Practices	
	Label No.	Control Point	Requirement Item	Corresponding Level	Requirement Item	Corresponding Level
Security Computing environment	H.5.3	Access Control	g) Big data platform shall be able to distinguish and dispose data of different categories and levels throughout the life cycle;	4		4
	H.4.3 H.5.3		h) Access control shall be implemented for calls involving important data interfaces and important service interfaces, including but not limited to data processing, use, analysis, export, sharing, exchange and other related operations;	3,4	a) The big data platform shall implement access control on the invocation of various interfaces provided by it, including but not limited to data acquisition, processing, use, analysis, export, sharing, exchange and other related operations;	3,4
					b) Technical means shall be taken to restrict the use of data acquisition terminals, data import service components, data export terminals and data export service components;	2
					c) The operating authority of various interfaces shall be minimized;	2,3,4
					d) Data sets used, analyzed, exported, shared and exchanged shall be minimized;	2,3,4
					e) The big data platform shall provide the ability to isolate the data resources of different customers' applications.	3,4
					f) Technical means shall be adopted to limit the output of important data at the terminal.	4
	H.4.3 H.5.3	Security Audit	i) The big data platform shall ensure that the audit data of big data applications of different customers are stored separately, and provide the ability to collect, summarize and centralize the audit data of different customers;	3,4	a) The big data platform shall ensure that the audit data of big data applications of different customers are stored separately, and can provide different customers with the ability to collect and summarize relevant audit data through interface call.	3,4

(Continued)

CATEGORY	Annex H of GB/T 22239—2019				Big Data System Protection Best Practices	
	Label No.	Control Point	Requirement Item	Corresponding Level	Requirement Item	Corresponding Level
Security Computing environment		Security Audit			b) Big data platform shall audit the invocation of important interfaces provided by it;	2
					c) The big data platform shall audit the invocation of various interfaces provided by it;	3,4
					d) It shall be ensured that the big data platform service provider's operation on the data of service customers can be audited by service customers.	2,3,4
		Intrusion Prevention			a) Data collected from import or other data collection methods shall be detected to avoid malicious data input;	3,4
		Data Integrity			b) Data integrity test shall be conducted for data exchange process by technical means;	1,2,3,4
					c) The integrity protection of data in stored process shall meet the security protection requirements of data source system.	1,2,3,4
	H.3.3 H.4.3 H.5.3	Data Confidentiality	j) Big data platform shall provide static desensitization and de-identification tool or service component technology;	2,3,4	a) Big data platform shall provide static desensitization and de-identification tool or service component technology;	2,3,4
					b) Data shall be statically desensitized and de-identified according to relevant security policies;	3,4
					c) Data shall be subjected to static desensitization and de-identification according to relevant security policy and data classification and classification identification;	1,2
					d) The confidentiality protection of data during storage shall meet the security protection requirements of data source system.	2,3,4

Chapter 8 Application and Interpretation of the Extended Security Evaluation Requirement of Big Data

(Continued)

CATEGORY	Annex H of GB/T 22239—2019				Big Data System Protection Best Practices	
	Label No.	Control Point	Requirement Item	Corresponding Level	Requirement Item	Corresponding Level
Security Computing environment		Data Backup Recovery			a) The backup data shall be provided with safety protection measures consistent with the original data;	2,3,4
					b) The big data platform shall ensure that there are several available copies of user data, the contents of each copy shall be consistent, and the copies shall be verified regularly;	3,4
					c) Backup of key traceability data shall be provided.	3,4
		Residual Information Protection			a) In the process of overall data migration, data residues shall be eliminated;	2,3,4
					b) The big data platform shall be able to carry out data destruction according to the data destruction requirements and methods proposed by the big data application;	2,3,4
					c) Big data application shall specify data destruction requirements and methods based on data classification and hierarchical protection strategy.	3,4
		Personal Information Protection			a) The collection, processing, use, transfer, sharing and disclosure of personal information shall be within the scope of authorization and consent for personal information processing;	2,3,4
					b) Measures shall be taken to prevent identification of personally identifiable information during data processing, use, analysis, export, sharing, exchange, etc.	2,3,4

(Continued)

CATEGORY	Annex H of GB/T 22239—2019				Big Data System Protection Best Practices	
	Label No.	Control Point	Requirement Item	Corresponding Level	Requirement Item	Corresponding Level
Security Computing environment	H.4.3 H.5.3	Data Traceability	k) The process of data collection, processing, analysis and excavation shall be tracked and recorded to ensure that the traceability data can reproduce the corresponding process and that the traceable data meet the compliance audit requirements;	3,4	a) The process of data collection, processing, analysis and excavation shall be tracked and recorded to ensure that the traceable data can reproduce the corresponding process;	3,4
					b) Traceability data shall meet data service requirements and compliance audit requirements;	3,4
			l) Important data shall be protected during data cleaning and conversion, so as to ensure the consistency after cleaning and conversion of important data, avoid data distortion, and be able to restore and recover effectively in case of problems;	3,4	c) Important data shall be protected during data cleaning and conversion, so as to ensure the consistency after cleaning and conversion of important data, avoid data distortion, and be able to restore and recover effectively in case of problems;	3,4
					d) Technical means shall be adopted to ensure the authenticity and credibility of data sources;	3,4
					e) Technical means shall be adopted to ensure the authenticity and confidentiality of traceability data;	4

Chapter 8 Application and Interpretation of the Extended Security Evaluation Requirement of Big Data

(Continued)

CATEGORY	Annex H of GB/T 22239—2019				Big Data System Protection Best Practices	
	Label No.	Control Point	Requirement Item	Corresponding Level	Requirement Item	Corresponding Level
Safety Management Center	H.3.3	Systems Management	a) The big data platform shall provide the big data application with the ability to control the usage of its computing and storage resources;	2.	a) The big data platform shall provide the big data application with the ability to manage the usage of its computing and storage resources;	2
	H.3.3 H.4.3 H.5.3		b) The big data platform shall effectively manage the auxiliary tools or service components provided by it;	2,3,4	b) The big data platform shall effectively manage the auxiliary tools or service components provided by it;	2,3,4
			c) Big data platform shall shield failure of computing, memory and storage resources to ensure normal operation of business;	2,3,4	c) The big data platform shall shield the failure of computing, memory and storage resources to ensure the normal operation of business;	2,3,4
					d) In case of system maintenance and online capacity expansion, the big data platform shall ensure the normal business processing capacity of big data application.	2,3,4
	H.4.3 H.5.3	Centralized Management and Control	d) The big data platform shall provide the big data application with the ability to centrally control the usage of its computing and storage resources;	3,4	a) The big data platform shall provide the big data application with the ability to centrally manage its computing and storage resource usage;	3,4
					b) The use of various interfaces provided by the big data platform shall be subject to centralized audit and monitoring.	3,4

355

(Continued)

CATEGORY	Annex H of GB/T 22239—2019				Big Data System Protection Best Practices	
	Label No.	Control Point	Requirement Item	Corresponding Level	Requirement Item	Corresponding Level
Safety Management Organization		Authorization and Approval			a) Data collection shall be authorized by the data source manager to ensure minimum data collection;	1,2,3,4
					b) The authorization and approval control process for data integration, analysis, exchange, sharing and disclosure shall be established, and relevant control shall be implemented according to the process and the process shall be recorded;	3,4
					c) A cross-border data assessment, approval and regulatory control process should be established and implemented and documented in accordance with the process.	3,4
Safety Construction Management	H.2.3 H.3.4 H.4.4 H.5.4	Selection of Big Data Service Provider	a) A safe and compliant big data platform shall be selected, and the big data platform service provided shall provide corresponding level of security protection capability for the big data application carried by it;	1,2,3,4	a) A safe and compliant big data platform shall be selected, and the big data platform service provided shall provide corresponding level of security protection capability for the big data application carried by it;	1,2,3,4
	H.3.4 H.4.4 H.5.4		b) The authority and responsibility of the big data platform provider, various service contents and specific technical indicators shall be agreed in writing, especially the security service contents;	2,3,4	b) The authority and responsibility of the big data platform provider, various service contents and specific technical indicators shall be agreed in writing, especially the security service content.	1,2,3,4
	H.4.4 H.5.4	Supply Chain Management	c) The responsibility of the receiver for data exchange and sharing shall be clearly restricted, and the receiving party shall be ensured to have sufficient or equivalent security protection capability.	3,4	a) The selection of suppliers shall be ensured in accordance with relevant national regulations;	1,2,3,4
					b) The data protection responsibilities of the data exchange and sharing receiver shall be agreed in writing, and the data security protection requirements shall be specified; meanwhile, the supply chain security event information or security threat information shall be timely transmitted to the data exchange and sharing receiver.	3,4
		Data Source Management			All types of data should be obtained through legitimate and legitimate sources.	1,2,3,4

Chapter 8 Application and Interpretation of the Extended Security Evaluation Requirement of Big Data

(Continued)

CATEGORY	Annex H of GB/T 22239—2019				Big Data System Protection Best Practices	
	Label No.	Control Point	Requirement Item	Corresponding Level	Requirement Item	Corresponding Level
Safety Operation and Maintenance Management	H.3.5 H.4.5 H.5.5	Asset Management	a) The security management strategy of digital assets shall be established to specify the operation specifications, protection measures and responsibilities of management personnel in the whole life cycle of data, including but not limited to the processes of data acquisition, storage, processing, application, flow and destruction;	2,3,4	a) The safety management strategy of data assets shall be established, and the operation specifications, protection measures and responsibilities of management personnel in the whole life cycle of data shall be specified, including but not limited to data acquisition, transmission, storage, processing, exchange, destruction and other processes;	2,3,4
			b) Data classification and hierarchical protection strategy shall be formulated and implemented, and different safety protection measures shall be formulated for data of different categories and levels;	3,4	b) Data classification and hierarchical protection strategy shall be formulated and implemented, and safety protection requirements of corresponding strength shall be formulated for data of different categories and levels.	3,4
			c) The scope of important digital assets shall be divided on the basis of data classification and classification, and the use scenarios and business processing flow for automatic desensitization or de-identification of important data shall be specified;	3,4	a) Data assets shall be registered and managed, and a list of data assets shall be established;	2
					b) Data assets and external data interface shall be registered and managed, and corresponding assets list shall be established;	3,4
			d) The category and level of data shall be reviewed regularly. If the category or level of data needs to be changed, the change shall be implemented in accordance with the change approval process.	3,4	c) The category and level of data shall be reviewed regularly. If it is necessary to change the category or level to which the data belongs, the change shall be implemented according to the change approval process.	3,4
		Media Management			a) Data shall be removed or destroyed in China.	2,3,4
		Network and System Security Management			a) The external data interface security management mechanism shall be established, and all interface calls shall be authorized and approved.	2,3,4

8.3 Application and Interpretation of the Extended Security Evaluation Requirements of Big Data at Level III and Level IV

8.3.1 Security Physical Environment

This chapter interprets the extended security evaluation requirement of security physical environment of big data system.

Infrastructure Location

[Standard Requirements]

The Level III control point includes the evaluation unit L3-PES6-01, and Level IV includes the evaluation unit L4-PES6-01.

[L3-CNS6-01/ L4-PES6-01 Interpretation and Description]

The targets of "should ensure that the equipment room carrying big data storage, processing and analysis is located in China" are the physical machine room location description document issued by the public cloud computing platform or the physical machine room storing and processing data resources, etc. This evaluation index applies to the grading targets that include big data platform, big data application or big data resources.

The implementation points of the evaluation include: interview the administrator of the big data platform to understand the physical location of the big data infrastructure, storage node, processing node, analysis node, big data management platform and other hardware and software rooms where big data services and data are carried. Check whether the big data platform storage node, processing node, analysis node, big data management platform, and other hardware and software supporting big data service and data are all located in China. If the big data platform is built on the third-party infrastructure, check whether the third-party infrastructure machine room is located in China; check whether the physical infrastructure such as big data platform server, storage device and other physical infrastructures are located in China.

If the test results show that the computer room of the deployed big data platform environment, all kinds of servers such as storage, processing and analysis, and the software and hardware of big data service and data such as big data management platform are located in China, the result is conforming; otherwise it is non-conforming or partially conforming.

8.3.2 Security Communication Network

This chapter interprets the extended security evaluation requirements of security communication network of big data system.

According to the characteristics of big data system, this chapter also lists the personalized interpretation contents of some general security evaluation requirements of big data environment.

1. Communication Transmission

[Standard Requirements]

The Level III control point includes the evaluation units L3-CNS1-01 and L3-CNS1-02, and Level IV includes the evaluation units L4-CNS1-01, L4-CNS1-02, L4-CNS1-03, L4-CNS1-04.

[L3-CNS1-01/ L4-CNS1-01 Interpretation and Description]

The targets of "adopt verification technology or cryptography technology to ensure the integrity of data during communication" are the data communicated between big data platform, big data application and big data resources as well as the equipment/ components for data transmission among big data platform, big data application and big data resources. The communication data includes but is not limited to identification data, important business data, important audit data, important configuration data, important video data and important personal information, etc., and focuses on the big data handled by the big data platform (or application); the equipment/ components for data transmission include but are not limited to VPN, message middleware, API interface, WEB page, etc. This evaluation index applies to the grading targets that include big data platform, big data application or big data resources.

The implementation points of the evaluation include: check whether the verification technology or cryptography technology is used to ensure the integrity of the data during the communication process. Extract the character code of the data to be transmitted by checking technology or cryptography technology, and output the characteristic code with fixed length, which is irreversible. The common algorithms include SM3, MD5, SHA1, SHA256, SHA512 and CRC-32, and SM3 algorithm is recommended. Send the data characteristic code and source data to the receiver, and the receiver calculates the characteristic code according to the source data and compares it with the received characteristic code to verify the communication integrity; if the data is the same, the data is not modified, otherwise, the data has been modified; or check whether the equipment/component performing data transmission adopts HTTPS, TLS or SSL protocol to ensure the integrity of data transmission in the communication channel.

If the results show that the big data platform adopts verification technology or cryptography technology for authentication data, important business data, important audit data, important configuration data, important video data and important personal information to ensure the integrity of data in the communication process, the result is conforming; otherwise it is non-conforming or partially conforming.

[L3-CNS1-02/ L4-CNS1-02 Interpretation and Description]

　　The targets of "should adopt cryptographic technology to ensure the confidentiality of data during communication" are the data communicated between big data platforms, big data application and big data resources as well as the equipment/components for data transmission among big data platforms, big data application and big data resources. The communication data includes but is not limited to identification data, important business data, audit data, configuration data, video data and personal information, and focuses on the big data handled by the big data platform (or application); the equipment/ components for data transmission include but are not limited to VPN, message middleware, API interface, WEB page and the like. This evaluation index applies to the grading targets that include big data platforms, applications, or big data resources.

　　The implementation points of the evaluation include: to prevent the information from being eavesdropping, verify whether the technical means are adopted to encrypt the sensitive information field or the whole message during the communication process, symmetric encryption andadopt asymmetric encryption to realize the confidentiality of data. Or check whether HTTPS, TLS or SSL protocols are adopted by the equipment/ components for data transmission to ensure the confidentiality of data transmission in the communication channel. During the verification, attention should be paid to whether the adopted version has disclosed security risks.

　　If the results show that the big data platform adopts cryptographic technology for the authentication data, important service data, important audit data, important configuration data, important video data and important personal information to ensure the confidentiality of data in the communication process, and the packet grabbing tool displays that the transmitted information is an encrypted message, the result is conforming; otherwise it is non-conforming or partially conforming.

[L4-CNS1-03 Interpretation and Description]

　　The targets of "verify and authenticate the both parties based on cryptographic technology before communication" are equipment/components for data transmission between big data platform, big data application and big data resources to provide cryptographic technology functions, including but not limited to kerberos authentication equipment/ components, digital certificate authentication devices/components, etc. This evaluation index applies to the grading targets that include big data platform, big data application or big data resources.

　　The implementation points of the evaluation include: in order to prevent the information from being eavesdropped, check whether the cryptography technology is used to conduct session initialization verification or authentication before the communication parties establish the connection, such as kerberos authentication, digital certificate authentication, etc.; check whether the authentication mechanism is reasonable; verify the management of the key used in the authentication to the life cycle; and whether the adopted cryptographic technology has disclosed security risks.

If the results show that the both parties of communication use the cryptographic technology to perform session initialization verification or authentication before establishing the connection, and the authentication mechanism or relevant configuration is reasonable, the result is conforming; otherwise it is non-conforming or partially conforming.

[L4-CNS1-04 Interpretation and Description]

The targets of "should perform cryptography calculation and management for important communication processes based on hardware cryptography module" are the hardware cryptography module in the equipment/ components for data transmission between big data platform, big data application and big data resources, including but not limited to hardware cryptography module in kerberos authentication device/component, hardware cryptography module in digital certificate authentication device/component, etc. This evaluation index applies to the grading targets that include big data platform, big data application or big data resources.

In order to prevent the information from being eavesdropping, check whether the cryptography technology can be used for session initialization verification or authentication before the communication parties establish the connection, such askerberos authentication, digital certificate authentication, etc.; verify whether the cryptography is generated and calculated based on the hardware cryptography module; check whether the relevant products have obtained the effective test report or cryptography product model certificate specified by the national cryptography management department.

If the results show that the communication parties use the cryptographic technology to perform session initialization verification or authentication before establishing the connection, and generate cryptographys based on the hardware cryptographic module and carry out cryptography operation, and the key is reasonably managed for the life cycle, and at the same time, an effective test report or cryptography product model certificate specified by the national cryptography management department is obtained, the result is conforming; otherwise it is non-conforming or partially conforming.

2. Network Architecture

[Standard Requirements]

The Level III control point includes evaluation units L3-CNS6-01 and L3-CNS6-02, and Level IV includes evaluation units L4-CNS6-01 and L4-CNS6-02.

[L3-CNS6-01/ L4-CNS6-01 Interpretation and Description]

The targets of "ensure that the big data platform should not bear big data application higher than its security protection level" are the big data platform, application system, rating materials and filing certificate of big data resources. This evaluation index applies to the grading targets that include big data platform, big data application or big data resources.

The key points of evaluation implementation include: define the security level of each grading target in the Big Data System by viewing the filing certificate and other filing materials of grading targets. When the big data platform, big data application and big data resource are regarded as the grading targets independently or arbitrarily combined as the grading targets, the level of big data application and big data resource cannot be higher than the security level of big data platform; the level of big data resource cannot be higher than the security level of big data application and big data platform.

If the test results show that the level of the big data application is not higher than the level of the big data platform where the big data application is located, the result is conforming; otherwise it is non-conforming or partially conforming.

【L3-CNS6-02/ L4-CNS6-02 Interpretation and Description】

The targets of "ensure the separation of management traffic and system business traffic on the big data platform" are the devices and components that realize out-of-band management, VLAN, VPC, VxLAN and other mechanisms in the big data platform, such as switches, security management and control platform, and dual network card servers in the big data platform. This evaluation index applies to the grading targets that include big data platform.

The implementation points of the evaluation include: check whether the physical separation between the device management traffic and the business traffic is realized through out-of-band management. On this set of dedicated management network channels independent of the data network, conduct centralized monitoring, management and maintenance of network devices (switches, routers and the like) and security devices (firewall, intrusion prevention devices, etc.) through the Console port and operate and manage the server devices through KVM. Check whether the devices and components capable of implementing mechanisms such as VLAN, VPC and VxLAN are used, such as switches, and whether the policy is configured to realize logical network structure partition to ensure logical channel isolation of service flow and device management traffic. Verify whether dual network card configuration is used to isolate device management traffic and service flow of virtual machine on the physical server; try to capture traffic flow in management domain of big data platform or penetrate business management domain using intranet penetration test.

If the results show that the network architecture in which the measure big data platform is located can separate the management traffic and the service traffic by adopting out-of-band management or policy configuration and other modes through reliable technical means, and after testing, the service data cannot be accessed in the management domain or penetrate into the service domain, the result is conforming; otherwise it is non-conforming or partially conforming.

8.3.3 Security Area Boundary

According to the characteristics of big data system, this chapter also lists the personalized interpretation contents of some general security evaluation requirements of big data environment.

1. Intrusion Prevention

[Standard Requirements]

The Level Ⅲ control point includes the evaluation unit L3-ABS1-01, and Level Ⅳ includes the evaluation unit L4-ABS1-01.

[L3-ABS1-01/ L4-ABS1-01 Interpretation and Description]

The targets of "detect, prevent or limit the external network attack behavior at key network nodes" are network equipment, security equipment, etc. This evaluation index applies to the grading targets that include big data platform, big data application or big data resources.

The implementation points of the evaluation include: during the evaluation, attention should be paid to whether the system undergoing test actively monitors at key network nodes to check whether intrusion and attack have occurred; for large data systems, attention should also be paid to whether to detect data disclosure, data theft, etc. and whether to monitor illegal outflow of data.

If the test results show that the system undergoing test or the equipment can detect the external network attack behavior, and no exploitable vulnerability is found through penetration test, the configuration information and the rules formulated in the security policy cover the IP addresses of key nodes of the system, and the monitored attack log information is consistent with the security policy, the result is conforming; otherwise it is non-conforming or partially conforming.

2. Security Audit

[Standard Requirements]

The Level Ⅲ control point includes the evaluation unit L3-ABS1-02, and Level Ⅳ includes the evaluation unit L4-ABS1-02.

[L3-ABS1-02/ L4-ABS1-02 Interpretation and Description]

The targets of "audit equipment and audit system at network boundary and important network nodes, and cover each user and audit important user behaviors and important security events" are audit devices and audit systems. This evaluation index applies to the grading targets that include big data platform, big data application or big data resources.

The implementation points of evaluation include: for big data system, in addition to the security audit behaviors and events that should be paid attention to by general information system, attention should also be paid to whether the audit scope covers the access and operation behaviors during the whole life cycle of important data.

If the results show that the system undergoing test deploys the audit equipment at the network boundary and the important network node, and it can cover each user, and the audit event covers important user behavior and important security event audit (including data access and operation), the result is conforming; otherwise it is non-conforming or partially conforming.

8.3.4 Security Computing Environment

This chapter interprets the extended security evaluation requirements of security computing environment of big data system.

1. Identification

[Standard Requirements]

The Level III control point includes the evaluation units L3-CES6-01, L3-CES6-02, L3-CES6-03, L3-CES6-04, and Level IV includes the evaluation units L4-CES6-01, L4-CES6-02, L4-CES6-03, L4-CES6-04.

[L3-CES6-01/ L4-CES6-01 Interpretation and Description]

The targets of "authenticate the identify for the use of data acquisition terminal, data import service component, data export terminal and data export service component" are data acquisition terminal, data import service component, data management system, system management software, external database management system, external file server, business application system, etc. This evaluation index applies to the grading targets that include big data platform, big data application or big data resources.

The implementation points of the evaluation include: which data acquisition terminals, data import service components, data export terminals and data export service components exist for big data platform, big data application and big data resources, whether identity authentication measures are implemented for various components and services; whether the big data platform, big data application and big data resources implement identity authentication measures for various components and services; whether the identity authentication measures can be bypassed.

If the results show that the measure big data platform or application takes identity authentication measures for the use of the data acquisition terminal, data import service component, data export terminal and data export service component, the result is conforming; otherwise it is non-conforming or partially conforming.

[L3-CES6-02/ L4-CES6-02 Interpretation and Description]

The targets of "big data platform should be able to identify the big data applications of different customers" are big data platform, big data platform management system and big data application system. This evaluation index applies to the grading targets that include big data platform.

The implementation points of the evaluation include: interview the administrator of the big data platform to verify whether the big data platform can authenticate the identity of all customer big data applications; whether the identity authentication measures of the big data platform can be bypassed.

If the test results show that the big data platform can authenticate the big data applications of different customers, and the identity authentication measures of the big data platform cannot be bypassed, the result is conforming; otherwise it is non-conforming or partially conforming.

【L3-CES6-03/ L4-CES6-03 Interpretation and Description】

The targets of "Big data resource should identify the objects calling its functions. (best practice)" are the external data management system, application system and data export service components that call its functions. This evaluation index applies to grading target containing big data resources.

The implementation points of the evaluation include: sort out which external entities call the big data resources; sort out the identity authentication measures implemented by the big data resources to all external entities calling their functions; and the identity authentication measures should not be bypassed.

If the test results show that the big data resource implements identity authentication measures on the external data management system, application system and data export service component calling its functions, and the identity authentication measures cannot be bypassed through the test verification, the result is conforming; otherwise it is non-conforming or partially conforming.

【L3-CES6-04/ L4-CES6-04 Interpretation and Description】

The targets of "authenticate the identity of all kinds of external calling interfaces provided by the big data platform according to the operation authority of the calling subject. (best practice)" are the interfaces provided by the big data platform for external services and the external entities calling the interfaces (including database management system, business system, system management software, file system, etc.). This evaluation index applies to the grading targets that include big data platform.

The key points for implementation of the evaluation include: sort out the interfaces provided by the big data platform for external calling; check whether the identity authentication measures are implemented for the relevant interfaces when calling the external entities; the interface should implement identity authentication measures with different strengths according to different calling authorities of the external entities, such as the authentication method based on account number and key; attention should be paid to whether the key length, key algorithm type, key algorithm strength, key life cycle management mode and process are different; the external entity with the highest calling authority should meet the three-level identity authentication strength; and test whether the identity authentication measures can be bypassed.

Taking the big data platform providing big data service as an example, the implementation steps of the evaluation mainly include: interview the administrator of the big data platform, which interfaces the big data platform provides for external invocation, whether the interface implements identity authentication measures on the calling subject when being invoked, and whether the identity authentication strength differs according to the authority of the calling entity.

If the results show that the interface provided by the big data platform undergoing test implements identity authentication measure when the external entity calls, and the identity authentication measure cannot be bypassed through test verification, the result is conforming; otherwise it is non-conforming or partially conforming.

2. Access Control

[Standard Requirements]

The Level Ⅲ control point includes evaluation units L3-CES6-05, L3-CES6-06, L3-CES6-07, L3-CES6-08, L3-CES6-09, L3-CES6-10, L3-CES6-11, L3-CES6-12, L3-CES6-13, L3-CES6-14; Level Ⅳ includes evaluation units L4-CES6-05, L4-CES6-06, L4-CES6-07, L4-CES6-08, L4-CES6-09, L4-CES6-10, L4-CES6-11, L4-CES6-12, L4-CES6-13, L4-CES6-14, L4-CES6-15, L4-CES6-16.

[L3-CES6-05/ L4-CES6-05 Interpretation and Description]

The targets of "big data platform, platform or third party providing services externally can access, use and manage data resources of big data application only when big data application is authorized" are big data platform, big data application system, big data platform application client, data management system and system design document, etc. This evaluation index applies to the grading targets that include big data platform or big data application.

The implementation points of the evaluation include: whether the big data platform provides the access control function for the big data application data resources; the big data application should be able to use the big data platform access control function to implement access control over its data resources; the big data application should configure the access control policy; the big data platform or the third party cannot access the data resources of the big data application without authorization.

If the results show that the big data platform or the third party cannot access the data resource of the big data application without authorization, if the access control rule and the verification test are configured, the result is conforming; otherwise it is non-conforming or partially conforming.

[L3-CES6-06/ L4-CES6-06 Interpretation and Description]

The targets of "big data platform should provide data classified and categorized security management function for big data application to adopt different security protection measures according to different types and levels of data" are big data platform, big data application system, data management system and system design document, etc. This evaluation index applies to the grading targets that include big data platform.

The implementation points of the evaluation include: the big data platform should formulate the classification strategy according to the national, local and industrial standards and regulations; the big data platform should provide the function of providing data identification based on the classified and categorized strategy; the big data platform should provide the function of selecting different security protection measures according to the data identification.

If the results show that the big data platform has the function of classification and category identification, and the data are classified and categorized according to the classification and category strategy, and a variety of data security protection measures are available for selection, the result is conforming; otherwise it is non-conforming or partially conforming.

[L3-CES6-07/ L4-CES6-07 Interpretation and Description]

The targets of "big data platform should provide the function of setting data security tags, authorization and access control measures based on security tags, and meet the requirements of fine-grained authorization access control management capability" are big data platform, data management system and system design documents. This evaluation index applies to the grading targets that include big data platform or big data resources.

The implementation points of the evaluation include: the big data platform should formulate the data security tag security strategy and provide the function of data security tag; the object granularity of access control based on security token should reach the file level or database list level; the subject granularity of access control should reach the big data application level or the big data platform user level.

If the results show that the measure big data platform sets the security mark on the data according to the security policy, and provide the access control authorization capability based on the security mark, the result is conforming; otherwise it is non-conforming or partially conforming.

[L3-CES6-08/ L4-CES6-08 Interpretation and Description]

The targets of "the big data platform should support the classified and categorized disposal of data in each link such as data acquisition, transmission, storage, processing and analysis, and ensure the consistency of security protection strategy" mainly include big data platform, big data application system, big data platform application client, data management system and system design document, etc. This evaluation index applies to the grading targets that include big data platform, big data application or big data resources.

The key points of evaluation implementation include: sort out data protection and disposal measures for each link such as data acquisition, transmission, storage, processing and analysis; classified and categorized disposal of data in each link; sorting out security protection strategies for each link of data acquisition, transmission, storage, processing and analysis to check whether they are consistent.

If the results show that the measure big data platform has taken disposal measures for each link such as data acquisition, transmission, storage, processing, exchange and destruction according to the data classification and grading identification, and has carried out categorized protection for the data according to the classified and categorized strategy, the result is conforming; otherwise it is non-conforming or partially conforming.

【L3-CES6-09/ L4-CES6-09 Interpretation and Description】

The targets of "conduct access control for the calls involving important data interface and important service interface, including but not limited to data processing, use, analysis, export, sharing, exchange and other related operations" are big data platform, big data application system, data management system and system management software, etc. This evaluation index applies to the grading targets that include big data platform, big data application or big data resources.

The implementation points of the evaluation include: interview the administrator of the big data platform to understand various interfaces existing in the big data platform, big data application and big data resources; check whether the big data platform, big data application and big data resource have established access control security policies; verify whether the access control measures of the test interface can be bypassed.

Taking the big data platform providing big data service as an example, the implementation steps of evaluation mainly include: interview the administrator of big data platform to understand whether the big data platform or application system provides access control measures for important data interfaces and important service interfaces.

If the results show that the big data platform or big data application system takes access control measures for calling important data interfaces and important service interfaces, and the access control measures include but not limited to related operations such as data processing, use, analysis, export, sharing, exchange. The result is conforming; otherwise it is non-conforming or partially conforming.

【L4-CES6-10 Interpretation and Description】

The targets of "big data platform should be capable of distinguishing and disposing data of different categories and levels in the whole life cycle" are big data platform, big data application system, big data platform application client, data management system and system design document, etc. This evaluation index applies to the grading targets that include big data platform, big data application or big data resources.

The key points of evaluation implementation include: interview the administrator, consult the system design documents, understand the data protection and disposal measures for each link such as data collection, transmission, storage, processing, exchange and destruction; check whether corresponding security protection measures are taken for different types and levels of data in each link.

If the test results show that the tested big data platform has taken corresponding disposal measures for each link such as data acquisition, transmission, storage, processing, exchange and destruction according to the data classified and categorized identification, and the verification test shows that different disposal measures can be taken according to the categorized identification and the link where the same data is collected, transmitted, stored, processed, exchanged and destroyed, the result is conforming; otherwise it is non-conforming or partially conforming.

[L3-CES6-10/ L4-CES6-11 Interpretation and Description]

The targets of "big data platform should implement access control for the callings of various interfaces provided by it, including but not limited to relevant operations such as data acquisition, processing, use, analysis, export, sharing and exchange; (best practice)" mainly include big data platform, big data application system, data management system and system management software, etc. This evaluation index applies to the grading targets that include big data platform, big data application or big data resources.

The implementation points of the evaluation include: interview the administrator of the big data platform to understand various interfaces existing in the big data platform, big data application and big data resources; check whether the big data platform and big data resources have established access control security policies; test and verify whether the access control measures can be bypassed.

If the results show that the tested big data platform or application system provide effective access control measures for calling important data interfaces and important service interfaces, and the verification test and access control measures cannot be bypassed, the result is conforming; otherwise it is non-conforming or partially conforming.

[L3-CES6-11/ L4-CES6-12 Interpretation and Description]

The targets of "process the data differently according to their identifier of classifications and categories in each link such as data acquisition, transmission, storage, processing, exchange and destruction, and the relevant protection measures for the highest-level data should not be lower than Level III security requirements, and the security protection strategies should be consistent in each link; (best practice L3-CES6-11)" and "process the data differently according to their identifier of classifications and categories in each link such as data acquisition, transmission, storage, processing, exchange and destruction, and the relevant protection measures for the highest-level data should not be lower than Level IV security requirements, and the security protection strategies should be consistent in each link; (best practice L4-CES6-12)" mainly include big data platform, big data application system, big data platform application client, data management system and system design documents. This evaluation index applies to the grading targets that include big data platform, big data application or big data resources.

The key points for implementation of the evaluation include: interview the administrator, consult the system design documents to understand the data protection and disposal measures for each link such as data acquisition, transmission, storage, processing, exchange and destruction; check whether each link has the capability of meeting the data protection and treatment measures of Level III (if Level IV protection object, Level IV should be satisfied); check whether the security protection policies of data acquisition, transmission, storage, processing, exchange and destruction are consistent.

If the test results show that the big data platform undergoing test has taken disposal measures for each link such as data acquisition, transmission, storage, processing, exchange and destruction according to the data classification and grading identification, and through test verification, different disposal measures can be taken for each link such as collection, transmission, storage, processing, exchange and destruction of the same data according to the categorized identification, and the protection measures for the highest level data are not lower than Level III (if the protection object of Level IV is not lower than Level IV), the result is conforming; otherwise it is non-conforming or partially conforming.

【L3-CES6-12/ L4-CES6-13 Interpretation and Description】

The targets of "minimize all kinds of interface operation authority. (best practice)" are big data platform, big data application system, data management system and system management software, etc. This evaluation index applies to the grading targets that include big data platform, big data application or big data resources.

The key points for implementation of the evaluation include: interview the administrator of big data platform, consult the security policy documents, and understand the authority regulations of various interfaces.

If the results show that the measure big data platform formulates the security strategy and specifies the operation authority required by various interfaces, the result is conforming; otherwise it is non-conforming or partially conforming.

【L3-CES6-13/ L4-CES6-14 Interpretation and Description】

The targets of "minimize the data sets of data using, analysis, export, sharing and exchange; (best practice)" are big data platform, big data application system, data management system and system management software. This evaluation index applies to the grading targets that include big data platform, big data application or big data resources.

The implementation points of the evaluation include: according to a certain type of business, check whether the data set in each link of data use, analysis, export, sharing and exchange complies with relevant security policies and whether all kinds of data in the data set are approved by the approval process; spot test the data in each link of data use, analysis, export, sharing and exchange to check whether the operations of various data have been approved by the approval process and whether they comply with relevant security policies.

If the results show that the data set of the measure big data platform does not contain unnecessary data when the data is used, analyze, exported, shared and exchanged, the result is conforming; otherwise it is non-conforming or partially conforming.

[L3-CES6-14/ L4-CES6-15 Interpretation and Description]

The targets of "big data platform should provide the ability to isolate the application data resources of different customers (best practice)." mainly includes big data platform, big data application system, data management system and system management software. This evaluation index applies to the grading targets that include big data platform.

The implementation points of the evaluation include: interview the administrator of the big data platform, understand the user list of the big data platform and the data isolation mode of the big data platform, check the isolation methods used by each user; the isolation measures cannot be bypassed.

If the test results show that the big data platform should provide the ability to isolate the application data resources of different customers, and the isolation measures cannot be bypassed through the test and verification, the result is conforming; otherwise it is non-conforming or partially conforming.

[L4-CES6-16 Interpretation and Description]

The targets of "adopt technical means to restrict the output of important data on the terminal (best practice)." are big data platform, big data application system, data management system, system management software, terminal and terminal operating system. This evaluation index applies to the grading targets that include big data platform, big data application or big data resources.

The implementation points of the evaluation include: interview the administrator, consult the system design documents, learn about the usage of big data platform, big data application, special terminal and non-special terminal of big data resource; check whether the restrictive measures taken by big data platform, big data application and big data resource are effective at the terminal.

If the results show that the big data platform adopt relevant technical measures to restrict the output of important data such as authentication data, important business data, important audit data, important configuration data, important video data and important personal information in the terminal, the result is conforming; otherwise it is non-conforming or partially conforming.

3. Security Audit

[Standard Requirements]

The Level Ⅲ control point includes evaluation units L3-CES6-15, L3-CES6-16, L3-CES6-17, and Level Ⅳ includes evaluation units L4-CES6-17, L4-CES6-18, L4-CES6-19.

[L3-CES6-15/ L4-CES6-17 Interpretation and Description]

The main evaluation object of the evaluation index "big data platform should ensure that the audit data of big data applications of different customers are stored separately and can provide collection and summary of audit data related to interface call for different customers" are the audit module provided by the big data application and the storage module of audit records. Under special circumstances, the audit function may also be divided into multiple systems or components, i. e. , the evaluation objects include log collection system, log storage system, log analysis system, etc. This evaluation index applies to the grading targets that include big data platform.

The implementation points of the evaluation include: interview the developers of the big data platform and the auditing administrator of big data to confirm whether the record of interactive data between the platform and big data application is recorded by the big data platform, and the platform provides an interface for big data application to invoke audit data; verify whether the stored audit data is segregated and stored separately according to the client roles of different applications; check whether the isolation measures and access control measures can be bypassed.

If the results show that the measure big data platform provides an isolated storage function for the audit data of different client big data applications, and provides an interface for the big data application to call the audit data, the result is conforming; otherwise it is non-conforming or partially conforming.

[L3-CES6-16/ L4-CES6-18 Interpretation and Description]

The targets of "big data platform should audit the callings of various interfaces provided by it; (best practice)" are various interfaces and audit modules provided by the platform. This evaluation index applies to the grading targets that include big data platform.

The implementation points of the evaluation include: interview the administrator of the big data platform (application) to understand the access control policy of the data service interface type and the audit content of the audit module; conduct operation test for each type of interface service, and check the log record in the audit module.

If the test results show that the tested big data platform can have access control measures for the callings of various interfaces provided by the platform, and audit the call of interfaces, and the audit records are complete and valid, the result is conforming; otherwise it is non-conforming or partially conforming.

[L3-CES6-17/ L4-CES6-19 Interpretation and Description]

The evaluation index "should ensure that the big data platform service provider's operation on the service customer data can be audited by the service customer (best practice)." mainly refers to the audit module provided by the big data platform for the service customer or the message notification mode of other service customers, such as platform configuration change announcement. This evaluation index applies to the grading targets that include big data platform or big data application.

The implementation points of the evaluation include: interview the management personnel of the big data platform to understand the auditing situation of the big data platform service provider on the operation of the service customer data; check the operation history of the big data platform service provider on the service customer data that can be seen by the customer side.

If the test results show that the audit module on the client side can consult the important configuration changes of the platform on the client side or publish the records of important changes of the platform through the announcement process, the result is conforming; otherwise it is non-conforming or partially conforming.

4. Intrusion Prevention

[Standard Requirements]

The Level Ⅲ control point includes the evaluation unit L3-CES6-18, and Level Ⅳ includes the evaluation unit L4-CES6-20.

[L3-CES6-18/ L4-CES6-20 Interpretation and Description]

The targets of "inspect the data imported or the data collected by other ways to avoid malicious data input. (best practice)" are data acquisition, data import system and data cleansing module. This evaluation index applies to the grading targets that include big data platform or big data application.

The implementation points of the evaluation include: interview the administrator of big data platform, understand the handling strategy of non-compliant data, and test and verify the effectiveness of data filtering module.

If the results show that the measure big data platform has a non-compliant data processing strategy, and through field test verification, it has relevant function module to filter the input data, then the result is conforming; otherwise it is non-conforming or partially conforming.

5. Data Integrity

[Standard Requirements]

The Level Ⅲ control point includes evaluation units L3-CES6-19 and L3-CES6-20, and Level Ⅳ includes evaluation units L4-CES6-21 and L4-CES6-22.

[L3-CES6-19/ L4-CES6-21 Interpretation and Description]

The targets of "take technical means to test the data integrity during data exchange process; (best practice)" are the products or components related to the integrity verification technology adopted by the system at the data exchange stage. This evaluation index applies to the grading targets that include big data platform, big data application or big data resources.

The implementation points of the evaluation include: interview the administrator of the big data platform to understand the data exchange process and the integrity verification technology adopted; the data concerned include system business data, system configuration parameter data, mirror snapshot data, stored backup data, log records, etc.; check the feedback behavior and handling strategy adopted by the system if the integrity verification fails in the data exchange.

If the test results show that the big data platform adopts relevant technical measures such as cryptography and can detect the integrity of data in the data exchange process through spot test verification, the result is conforming; otherwise it is non-conforming or partially conforming.

[L3-CES6-20/ L4-CES6-22 Interpretation and Description]

The targets of "the integrity protection of data in the storage process should meet the security protection requirements of the data source system" are the integrity verification technical products or components adopted by the system in the storage stage. This evaluation index applies to the grading targets that include big data platform, big data application or big data resources.

The key points of evaluation implementation include: interview the developer of big data system whether there is a data source system; if it does not exist, it is inapplicable; if yes, inquire whether the data source system has data storage integrity protection requirements; determine the security protection level of the data source system.

If the test results show that the integrity protection measures (such as the adopted cryptographic technology and cryptographic algorithm configuration parameters) during the storage process meet the protection requirements for the source data system, the result is conforming; otherwise it is non-conforming or partially conforming.

6. Data Confidentiality

[Standard Requirements]

The Level III control point includes evaluation units L3-CES6-21, L3-CES6-22, L3-CES6-23, and Level IV includes evaluation units L4-CES6-23, L4-CES6-24, L4-CES6-25.

[L3-CES6-21/ L4-CES6-23 Interpretation and Description]

The targets of "big data platform should provide static desensitization and de-identification tool or service component technology" is data cleaning module or other data de-identification and desensitization components. This evaluation index applies to the grading targets that include big data platform.

The key points of evaluation implementation include: check whether tools or service components for static desensitization and de-identification are provided; desensitization and de-identification principles of recording components, not limited to Tokenization, masking, encryption, suppression, data discovery, TDE encryption, etc.

If the results show that the big data platform has the function of data de-identification and desensitization process, the result is conforming; otherwise it is non-conforming or partially conforming.

[L3-CES6-22/ L4-CES6-24 Interpretation and Description]

The targets of "The data should be statically desensitized and de-identified according to relevant security policies and data classification and identification. (best practice)" are data cleaning module or other data de-identification and desensitization components. This evaluation index applies to the grading targets that include big data platform, big data application or big data resources.

The key points of evaluation implementation include: interview big data system administrator, understand the system or component of data de-identification and desensitization processing, understand relevant security policy and specific content of data classification and grading identification, and verify whether thedeidentified and desensitized data can be restored to source data information by means of social workers, etc.

If the results show that it has processing rule of data classification dei-dentification and desensitization, and the source data information can not be restored and obtained by means such as social worker after the data de-identification and desensitization, the result is conforming; otherwise it is non-conforming or partially conforming.

[L3-CES6-23/ L4-CES6-25 Interpretation and Description]

The targets of "confidentiality protection of data during storage process should meet the security protection requirements of data source system (best practice)" are the confidential storage technology products or components adopted in each stage of the system or the data management system independently providing data storage. This evaluation index applies to the grading targets that include big data platform, big data application or big data resources.

The implementation points of the evaluation include: confirm the security protection level of the data source system; understand the storage control system used in the data storage process, the confidential storage technology used by the storage control system and the industry standard referenced; compare the data storage protection confidentiality requirements adopted by this system and the data source system.

If the results show that the measure big data platform has taken technical measures (such as cryptography) for the confidentiality of the data in the storage process, and the data stored in encryption cannot obtain plaintext data and plaintext file data through field test verification, the result is conforming; otherwise it is non-conforming or partially conforming.

7. Data Backup Recovery

[Standard Requirements]

The Level III control point includes evaluation units L3-CES6-24, L3-CES6-25, L3-CES6-26, and Level IV includes evaluation units L4-CES6-26, L4-CES6-27, L4-CES6-28.

[L3-CES6-24/ L4-CES6-26 Interpretation and Description]

The targets of "backup data should adopt security protection measures consistent with the original data (best practice)" are the data backup module and corresponding data backup strategy, and the data management system that provides independent data storage. This evaluation index applies to the grading targets that include big data platform, big data application or big data resources.

The implementation points include: interview the administrator of the big data platform to understand the backup data to be protected and corresponding backup strategy; record the implementation mode of backup, the implementation mode of data integrity protection, whether reliable integrity cryptography algorithm is adopted and stored as HASH list; if the standard cryptography technology is adopted, check the corresponding standard document. If the non-standard cryptographic technology is used and a single algorithm is used, the algorithm type and algorithm parameters should be recorded; if the network environment between the source data and the backup data storage point is not controllable, it is necessary to confirm whether the backup transmission mode is at risk of unauthorized acquisition. For example, the transmission channel integrity protection and special line transmission protection should be adopted for safe methods.

If the test results show that there is an effective backup data integrity detection strategy to make the data have confidentiality and integrity check protection during the whole process from the source to the backup point, and reliable encryption technology or reliable encryption algorithm and reliable integrity cryptographic algorithm are adopted, and the permissions of the HASH field table are protected, then the result is conforming; otherwise, if the standard cryptographic technology is not used and a single unreliable integrity algorithm such as SHA1, MD5 or a single unreliable encryption algorithm DES is used, it is regarded as non-conformance or partial compliance.

[L3-CES6-25/ L4-CES6-27 Interpretation and Description]

The targets of "The big data platform should ensure that there are several available copies of user data, and the contents of each copy should be consistent, and the copies should be verified regularly (best practice)" are the data replica storage component in the big data system and its strategy. This evaluation index applies to the grading targets that include big data platform.

The implementation points of the evaluation include: interview the administrator of big data platform to understand whether there are several available copies of user data and the storage mode of the copies, check the cryptographic techniques and algorithms to ensure the consistency and integrity of multiple data backup copies, and verify the effectiveness of the consistency detection strategy of duplicate data.

If the results show that the measure big data platform has a backup copy data integrity detection strategy, adopts a reliable integrity cipher algorithm, has data backup copy availability test record, and there are reliable technical measures to regenerate a new copy when the replica availability test fail, the result is conforming; otherwise it is non-conforming or partially conforming.

【L3-CES6-26／L4-CES6-28 Interpretation and Description】

The targets of "should provide backup of key traceability data (best practice)" are data backup module or independent audit system. This evaluation index applies to the grading targets that include big data platform or big data resources.

The key points of evaluation implementation include: interview the administrator of big data platform to understand the type of key traceability data suitable for the system, confirm whether the key data used for traceability includes data circulation events such as user management, interface call, own collection, processing and analysis, and data operation events such as cleaning and conversion; record the backup storage mode and storage duration of traceable key data, check whether the backed-up data covers all traceable key data; check whether the backup traceability data is available; record the traceability process and check the archiving of historical traceability events.

If the test results show that the backup data covers all traceability key data and the traceable backup data is valid, the result is conforming; otherwise it is non-conforming or partially conforming.

8. Residual Information Protection

【Standard Requirements】

The Level III control point includes evaluation units L3-CES6-27, L3-CES6-28, L3-CES6-29, and Level IV includes evaluation units L4-CES6-29, L4-CES6-30, L4-CES6-31.

【L3-CES6-27／L4-CES6-29 Interpretation and Description】

The targets of "in the process of overall data migration, residual data should be avoided. (best practice)" are related objects involved in the process of data migration such as infrastructure layer, data platform layer, calculation and analysis layer and big data management platform, including operating system (including host and virtual machine operating system), network equipment (including virtual network equipment), security equipment (including virtual security equipment), business application system, database management system, middleware, system management software and system related documents, etc. in devices such as terminal and server. This evaluation index applies to the grading targets that include big data platform, big data application or big data resources.

The key points for implementation of the evaluation include: interview the administrator of the big data platform, understand the relevant management specifications of the undergoing test on data migration, check whether relevant specifications have been made on the residual data during the data migration process; understand whether the big data platform has conducted data migration. If there is data migration, check the process document and migration scheme of the data migration at that time, understand the original storage location of platform data, the process of migration and the processing method of the original storage space, and view the corresponding codes and cases according to the processing mode of the organization undergoing test; if there is no data migration, pay attention to the system development scheme and other processing methods related to data residues.

If the test results show that the data migration management system is formulated, the documents related to historical migration are complete and consistent with the management system, and the original storage space is destroyed without data residue, the result is conforming; otherwise it is deemed as non-conformance or partial compliance; if there is no overall data migration, it is mainly focused on whether there is a treatment method for data residue in the system development scheme. If the method is proper, the result is conforming; otherwise it is non-conforming or partially conforming.

【L3-CES6-28/ L4-CES6-30 Interpretation and Description】

The targets of "big data platform should be able to carry out data destruction according to the data destruction requirements and methods proposed by the big data application. (best practice)" are mainly evaluated by big data platform, big data application system, data management system, design and development documents, agreements and contracts, etc. This evaluation index applies to the grading targets that include big data platform.

The key points for implementation of the evaluation include: interview the administrator of the big data platform, consult the design and development documents of the big data platform, understand the types of data destruction methods (such as index table deletion, disk erasing, overwrite erasing, disk destruction, etc.); consult the service agreement or contract between the big data platform and big data application to check whether there are corresponding data destruction requirements and methods. If yes, check whether the data destruction mode of the big data platform meets the requirements of big data application, and verify whether the actual data destruction mode meets the requirements of big data application.

If the results show that the measure big data platform can meet the destruction requirements and method proposed by the big data application and carry out data destruction, the result is conforming; otherwise it is non-conforming or partially conforming.

【L3-CES6-29/ L4-CES6-31 Interpretation and Description】

The targets of the evaluation index "big data application should be based on data classified and categorized protection strategy, and requirements and methods for data destruction (best practice)" are security policy documents and data destruction records. This evaluation index applies to rated objects containing big data applications.

The key points of evaluation implementation include: interview the administrator of big data platform, consult data classified and categorized management system, check the data destruction method formulated according to classified and categorized protection strategy; check the corresponding data destruction records of the tested system, check whether the list of recovered destruction equipment and destruction operation records are consistent, and

Chapter 8　Application and Interpretation of the Extended Security Evaluation Requirement of Big Data

check whether the data destruction measures at different levels are consistent with the management system.

If the results show that the measure big data platform formulates the corresponding data destruction requirement and mode based on the data classification categorized protection strategy, and the data is destroyed according to the security policy, the result is conforming; otherwise it is non-conforming or partially conforming.

9. Personal Information Protection

【Standard Requirements】

The Level Ⅲ control point includes evaluation units L3-CES6-30 and L3-CES6-31, and Level Ⅳ includes evaluation units L4-CES6-32 and L4-CES6-33.

【L3-CES6-30/ L4-CES6-32 Interpretation and Description】

The targets of "collection, processing, use, transfer, sharing and disclosure of personal information should be within the scope of authorization and consent of personal information processing (best practice)" are data management system, system management software, big data platform, big data application system and system management software. This evaluation index applies to the grading targets that include big data platform, big data application or big data resources.

The implementation points of the evaluation include: interview the administrator of the big data platform, understand all processing procedures such as collection and use of personal information, consult the registration agreement and personal privacy protection agreement, and check whether the authorized scope of personal information processing covers all processes of personal information processing.

If the results show that all the processing of personal information relate to the big data platform undergoing test are within the authorized and approved scope of personal information processing, the result is conforming; otherwise it is non-conforming or partially conforming.

【L3-CES6-31/ L4-CES6-33 Interpretation and Description】

The targets of "take measures to prevent the identification of personally identifiable information during data processing, use, analysis, export, sharing, exchange (best practice)" mainly refer to the objects involved in the process of data processing, use, analysis, export, sharing and exchange of infrastructure layer, data platform layer, computing analysis layer and big data management platform, etc., including operating system (including host computer and virtual machine operating system), business application system, database management system, middleware, system management software and system management software and system related documents in devices such as terminal and server. This evaluation index applies to the grading targets that include big data platform, big data application or big data resources.

The key points for implementation of the evaluation include: interview the administrator of big data platform to understand the specific measures to prevent personal information from being identified in each link of data processing, such as desensitization and encryption of data; test or check the effectiveness of preventive measures.

If the results show that the measure big data platform takes relevant measures to prevent the identification of personal identification information in each link of data processing, the unit evaluation result is judged as conforming; otherwise, the result is conforming; otherwise it is non-conforming or partially conforming.

10. Data Traceability

[Standard Requirements]

The Level III control point includes evaluation units L3-CES6-32, L3-CES6-33, L3-CES6-34, and Level IV includes evaluation units L4-CES6-34, L4-CES6-35, L4-CES6-36, L4-CES6-37.

[L3-CES6-32/ L4-CES6-34 Interpretation and Description]

The targets of "track the process of data collection, processing, analysis and mining to ensure that the traceability data can reproduce the corresponding process, and the traceable data meet the compliance audit requirements" are those objects involved in the process of data processing, using, analysis, export, sharing and exchange of infrastructure layer, data platform layer, calculation and analysis layer and big data management platform, including the operating systems (host and virtual machine operating systems), network equipment (virtual network equipment), security equipment (including virtual security equipment), business application system, database management system, system management software and system management software and system-related documents in the infrastructure layer, data platform layer, calculation and analysis layer and big data management platform. This evaluation index applies to the grading targets, including big data platforms, big data applications, or big data resources.

The implementation points: interview the administrator of big data platform to understand the process of data acquisition, processing, analysis and mining of the system undergoing test, check the tracking measures and record documents of the process of data acquisition, processing, analysis and mining. Select traceability data, trace back relevant processes, and check whether the process can be reproduced. Check whether relevant logs are kept and whether they meet the requirements of enterprise audit system, etc.

If the results show that the measure big data platform tracks and record the data acquisition, processing, analysis, mining and other processes, can reproduce the corresponding process according to the traceable data, and meets the data service requirements and compliance audit requirements, the result is conforming; otherwise it is non-conforming or partially conforming.

[L3-CES6-33/ L4-CES6-35 Interpretation and Description]

The targets of "protect important data during data cleaning and conversion to ensure the consistency after cleaning and conversion of important data, avoid data distortion, and effectively restore and restore in case of problems" are objects involved in data processing, use, analysis, export, sharing and exchange of infrastructure layer, data platform layer, computing analysis layer and big data management platform, including operating system (host computer and virtual machine operating system), network equipment (virtual network equipment), security equipment (virtual security equipment), business application system, database management system, middleware, system management software and system management software and system related documents in infrastructure layer, data platform layer, computing and analysis layer and big data management platform. This evaluation index applies to the grading targets that include big data platform or big data resources.

The key points of evaluation implementation include: interview the administrator of big data platform to understand the protection measures for important data in the process of data cleaning and conversion, test and check the effectiveness of specific protection measures, and consult relevant records of restoration and recovery.

If the test results show that the big data platform to be tested has taken technical measures to ensure the consistency after cleaning and conversion of important data, and can be restored and restored when problems occur, the result is conforming; otherwise it is non-conforming or partially conforming.

[L3-CES6-34/ L4-CES6-36 Interpretation and Description]

The targets of "take technical measures to ensure the authenticity and credibility of data source (best practice)" mainly refers to the objects involved in the process of data processing, use, analysis, export, sharing and exchange of infrastructure layer, data platform layer, computing analysis layer and big data management platform, etc., including operating system (host computer and virtual machine operating system), network equipment (virtual network equipment), security device (virtual security device), business application system, database management system, middleware and system management software as well as system-related documents and data source procurement contracts in terminals and servers. This evaluation index applies to the grading targets that include big data platform, big data application or big data resources.

The implementation points of the evaluation include: interview the administrator of the big data platform to understand the technical measures or means for the big data platform to be tested to ensure the integrity and confidentiality of the data source. If the verification code technology or cryptography technology is used to realize the implementation, the completeness of the key used by the check code technology and cryptography technology should be ensured; if the data is provided by a third party, the identity of the third party should be confirmed, and the data format transmitted by the third party should be strictly limited, whether the contract has been signed with the third party, and whether the contract restricts the content, scope and legitimacy of the data source.

If the results show that the measure big data platform system has data source integrity protection measures, the result is conforming; otherwise it is non-conforming or partially conforming.

[L4-CES6-37 Interpretation and Description]

The targets of "take technical means to ensure the authenticity and confidentiality of traceability data (best practice)" are the processes of data processing, use, analysis, export, sharing and exchange of infrastructure layer, data platform layer, computing analysis layer and big data management platform, including operating system (host computer and virtual machine operating system), network equipment (virtual network equipment), security equipment (virtual security equipment), business application system, database management system, middleware, system management software and system related documents in devices such as terminal and server. This evaluation index applies to the grading targets that include big data platform, big data application or big data resources.

The implementation points of the evaluation include: interview the administrator of the big data platform to understand the technical measures or means of the big data platform to be tested to ensure the authenticity and confidentiality of the traceable data; if the verification code technology or cryptography technology is used to realize the implementation, the completeness of the key used by the verification code technology and cryptography technology should be ensured; check whether the traceability data of the big data platform to be tested is managed by special personnel.

If the test results show that the big data platform to be tested is equipped with a specially-assigned person to manage the traceability data, and the corresponding check code technology and cryptography technology are adopted to ensure the authenticity and confidentiality of the traceability data, the result is conforming; otherwise it is non-conforming or partially conforming.

8.3.5 Security Management Center

This chapter is the interpretation of "Security Management Center" required for extended security evaluation of big data system.

1. System Management

[Standard Requirements]

The Level III control point includes the evaluation units L3-SMC6-01, L3-SMC6-02, L3-SMC6-03, and Level IV includes the evaluation units L4-SMC6-01, L4-SMC6-02, L4-SMC6-03.

[L3-CNS6-01/ L4-SMC6-01 Interpretation and Description]

The targets of "big data platform should implement effective management on the auxiliary tools or service components provided by it" are the management platform/component providing auxiliary tools or service components. This evaluation index applies to the grading targets that include big data platform.

The key points of evaluation implementation include: check whether the provided auxiliary tools or service components have security measures such as identity authentication, access control, administrator authority management, log audit, etc.; if the provided auxiliary tools or service components have security management measures such as installation, deployment, upgrade and uninstall, check whether the function is effective; test whether horizontal authority override or vertical authority override can be realized among accounts through penetration testing.

If the test results show that the tested big data platform can effectively manage the auxiliary tools or service components provided by it, such as identity authentication, access control, administrator authority management, log audit, etc., as well as security management for its installation, deployment, upgrade and unloading, and there is no horizontal or vertical authority override between accounts through penetration test, the result is conforming; otherwise it is non-conforming or partially conforming.

【L3-CNS6-02/ L4-SMC6-02 Interpretation and Description】

The targets of "big data platform should shield the failure of computing, memory and storage resources to ensure normal operation of business" are relevant design documents of calculation, memory and storage resource deployment mode; alarm record of computing, memory and storage resource operation; and operation record of big data platform and big data application. This evaluation index applies to the grading targets that include big data platform.

The implementation points of the evaluation include: interview the administrator of the big data platform, consult the design documents or development documents, and understand the measures and technical means for shielding the failure of computing, memory and storage resources, such as memory failure transfer mode, calculation mode, storage mode, hard disk failure recovery mechanism, etc.; understand the monitoring measures for the use of computing, memory and storage resources, such as how to alarm when the storage resources are exhausted, broken disks, storage disks are damaged, and how to give alarms when calculation is abnormal; check the alarm records or historical records; close a single computing node or storage node on site to check whether the normal operation of the big data platform and big data application business is affected.

If the results show that the big data platform undergoing test has measures to shield the failure of calculation, memory and storage resources, and has measures and alarm records for monitoring the usage of calculation, memory and storage resources, and the normal operation of the big data platform and big data application business is not affected by the failure of calculation, memory and storage resources, the result is conforming; otherwise it is non-conforming or partially conforming.

【L3-CNS6-03/ L4-SMC6-03 Interpretation and Description】

The targets of "big data platform should guarantee the normal business processing capability of big data application during system maintenance and online capacity expansion (best practice)" are the design scheme of online capacity expansion, and the process system and relevant records of major changes and batch changes, and the emergency response plan. This evaluation index applies to the grading targets that include big data platform.

The key points of evaluation implementation include: interview the administrator of the big data platform, consult the online capacity expansion design scheme, understand the specific measures and technical means for shielding the impact of capacity expansion, check the implementation process records and result records of capacity expansion; check whether the emergency response plan for online expansion has been formulated and consult specific measures; check whether the system documents related to the change process include the definition of major change, change mechanism, change approval process and operation process, and check the change records.

If the test results show that the big data platform undergoing test is equipped with measures and technical means to shield the big data platform's online expansion, relevant system documents and change records for major changes to the big data platform, and emergency response plan for online capacity expansion and system change of the big data platform, the result is conforming; otherwise it is non-conforming or partially conforming.

2. Centralized Control

[Standard Requirements]

The Level Ⅲ control point includes the evaluation units L3-SMC6-04 and L3-SMC6-05, and Level Ⅳ includes the evaluation units L4-SMC6-04 and L4-SMC6-05.

[L3-CNS6-04/ L4-SMC6-04 Interpretation and Description]

The targets of "big data platform should provide big data application with the ability to centrally control its usage of computing and storage resources" is the system/component that big data platform provides centralized management and control of computing and storage resources to big data application, or the system/component that cloud computing platform provides centralized management and control of computing and storage resources to big data application. This evaluation index applies to the grading targets that include big data platform.

The implementation points of the evaluation include: interview the administrator of the big data platform, understand whether the big data platform has deployed the centralized management system, and monitor and manage the computing resources and storage resources used by the big data application. Check whether the centralized management system can perform centralized and real-time monitoring on the operation status of computing resources and storage resources, such as the space occupied by resources, resource utilization rate, service status and the like. The centralized management system can perform centralized and real-time management on computing resources and storage resources, such as expansion, reduction, activation and deactivation of resources; business management personnel should establish big data application test accounts for test verification, and consult historical alarm records and management operation records.

If the results show that the measure big data platform has the function of monitoring and managing the computing resource and the storage resource of the big data application, the unit evaluation result is judged as conforming; otherwise it is regarded as non-conforming or partially conforming.

[L3-CNS6-05/ L4-SMC6-05 Interpretation and Description]

The targets of "conduct centralized audit and monitoring on the use of various interfaces provided by the big data platform (best practice)" are the system/component for centralized audit and monitoring of the usage of various interfaces provided by the platform, and various API application interfaces provided by the big data platform, and the using process documents of various interfaces provided by the big data platform, and the monitoring records and audit records of interface usage. This evaluation index applies to the grading targets that include big data platform.

The implementation points of the evaluation include: check whether the big data platform has deployed centralized management systems/components, audit and monitor the usage of various interfaces provided by the big data platform, such as interface application, activation, use, cancellation, authority management, etc.; check whether the use process documents and monitoring records of various interfaces are complete and effective; check whether the use conditions of various interfaces provided by the big data platform are centralized and audited in real time; check whether audit records are available, including the account number, time and operation of interface access, etc.

If the test results show that the tested big data platform has the functions of centralized monitoring and auditing on the service conditions of various interfaces provided, and has the use process documents, monitoring records and audit records of various interfaces, the result is conforming; otherwise it is non-conforming or partially conforming.

8.3.6 Security Management System

According to the characteristics of big data system, this chapter lists the personalized interpretation contents of some general requirements for security evaluation under big data environment.

1. Security Policy

[Standard Requirements]

The Level Ⅲ control point includes the evaluation unit L3-PSS1-01, and Level Ⅳ includes the evaluation unit L4-PSS1-01.

[L3-PSS1-01/ L4-PSS1-01 Interpretation and Description]

The targets of "formulate overall policy and security strategy of network security work, and clarify the overall objective, scope, principle and security framework of the organization's security practices" are the overall policy and security strategy documents of network security work, and the documents specifying the overall strategy of network security work may be in the form of a single document or a set of documents. This evaluation index applies to the grading targets that include big data platform, big data application or big data resources.

The implementation points of the evaluation include: the overall security policy and policy documents should include the big data security objectives, principles, security framework and the overall security strategy to be followed. The security strategy should include all key security management activities in the whole life cycle of the big data system such as data acquisition, transmission, storage, processing (such as calculation, analysis, visualization, etc.), exchange and destruction, such as physical environment of data, data traceability, data authorization, output control of sensitive data, event handling and emergency response of data, etc.

If the test results show that the organization undergoing test has formulated the network security policy and policy document covering the security policy and security policy of big data security work or has the security policy and security policy with independent big data security work, and the contents such as the goal, principle, security framework and overall strategy to be followed of big data security work are specified in the document, the result is conforming; otherwise it is non-conforming or partially conforming.

2. Management System

[Standard Requirements]

The Level III control point includes the evaluation unit L3-PSS1-02, and Level IV includes the evaluation unit L4-PSS1-02.

[L3-PSS1-02/ L4-PSS1-02 Interpretation and Description]

The targets of "establish security management system for various management contents in security management activities" are the series of security management systems established for security management activities. It may take the form of a number of institutional components, or a number of fascicles. This evaluation index applies to the grading targets that include big data platform, big data application or big data resources.

The implementation points of the evaluation include: verify whether the security management system includes the management system related to the data life cycle of the big data platform; verifying whether the management system covers the security management activities (such as: data acquisition, transmission, storage, processing, exchange, destruction, etc.) involved in the data life cycle of the big data platform, or establish an independent big data management system according to the characteristics of the big data system.

If the test results show that the organization undergoing test has established the management system related to the data life cycle of the big data platform, and the security management activities (such as data acquisition, transmission, storage, processing, exchange, destruction, etc.) involved in the data life cycle of the big data platform are specified in the system, then the result is conforming; otherwise it is non-conforming or partially conforming.

8.3.7 Security Management Organization

This chapter interprets the extended security evaluation of security management organization of big data system security evaluation.

According to the characteristics of big data system, this chapter also lists the personalized interpretation contents of some general security evaluation requirements of big data environment.

1. Post Setting

[Standard Requirements]

The Level III control point includes the evaluation unit L3-ORS1-01, and Level IV includes the evaluation unit L4-ORS1-01.

[L3-ORS1-01/ L4-ORS1-01 Interpretation and Description]

The targets of "set up system administrator, audit administrator, security administrator and other posts should be set, and responsibilities of departments and various posts should be defined" are security supervisor, post responsibility document, etc. This evaluation index applies to the grading targets that include big data platform, big data application or big data resources.

The key points for implementation of evaluation include: for big data system, roles or posts responsible for data security during data acquisition, transmission, storage, processing, exchange and destruction should be specified, and security responsibilities of data security roles or posts during data collection, transmission, storage, processing, exchange and destruction should be specified.

If the test results show that the organization undergoing test has established the roles or posts of system administrator, network administrator, security administrator and administrator in charge of data security, and the work responsibilities of each post are specified, including the responsibilities of roles or posts in charge of data security during data acquisition, transmission, storage, processing, exchange and destruction, etc., the result is conforming; otherwise it is non-conforming or partially conforming.

2. Authorization and Approval

[Standard Requirements]

The Level III control point includes the evaluation units L3-ORS1-02, L3-ORS6-01, L3-ORS6-02, L3-ORS6-03, and Level IV includes the evaluation units L4-ORS1-02, L4-ORS6-01, L4-ORS6-02, L4-ORS6-03.

[L3-ORS1-02/ L4-ORS1-02 Interpretation and Description]

The targets of "establish approval procedures for such items as system changes, important operations, physical access and system access, and execute the approval process according to the approval procedures, and the set up level-by-level approval system for important activities" are security management system, operating procedure documents, items approval records and level-by-level approval record forms. This evaluation index applies to the grading targets that include big data platform, big data application or big data resources.

The key points of evaluation implementation include: for big data system, establish approval procedures for data authorization, data desensitization and de-identification processing, sensitive data output control, data classification and identification storage, and classified destruction of data in the process of data acquisition, transmission, storage, processing, exchange and destruction to determine the approval department and approver of the items, and specify which items need to be approved level by level; the approval records should be consistent with the approval procedures, and the approval procedures should be consistent with the requirements of corresponding management systems. Check relevant documents or records to verify whether management system or control measures have been established for the above contents.

If the test results show that the organization undergoing test has formulated the approval procedures for system changes, important operations, physical access and system access for the big data platform, and the approval records are complete and effective, the result is conforming; otherwise it is non-conforming or partially conforming.

[L3-ORS6-01/ L4-ORS6-01 Interpretation and Description]

The targets of "data collection should be authorized by the data source manager to ensure the principle of minimizing data collection (best practice)" are the authorization agreement or contract signed by the data source manager for data collection, the big data platform development scheme or system requirement analysis design scheme and other documents. This evaluation index applies to the grading targets that include big data platform, big data application or big data resources.

The key points of evaluation implementation include: for big data system, data collection of big data system should sign authorization agreement or contract with data source manager, define authorized scope of data collection and data use to ensure legitimacy of data source; check whether the data used is collected beyond the scope according to business carried out by big data system, and ensure that only the minimum data range meeting clear business purpose and business scenario is used. Review the relevant documentation to verify that the above requirements are met.

If the results show that the organization undergoing test signs an authorization agreement or contract with the data source manage, collects data according to the authorized

Chapter 8 Application and Interpretation of the Extended Security Evaluation Requirement of Big Data

content, and the collected data follows the minimum principle (data necessary for business development), the result is conforming; otherwise it is non-conforming or partially conforming.

【L3-ORS6-02/ L4-ORS6-02 Interpretation and Description】

The targets of "establish authorization and approval control process for data integration, analysis, exchange, sharing and publication, and relevant control should be implemented according to the process and the process should be recorded (best practice)" mainly include the security management system, operating procedures, data authorization approval control process, approval control process record, etc. This evaluation index applies to the grading targets that include big data platform, big data application or big data resources.

The key points of evaluation implementation include: for big data system, establish the approval control process for the authorized matters of data integration, analysis, exchange, sharing and disclosure during data collection, transmission, storage, processing, exchange and destruction, and determine the approval department and approver for authorized approval items; the approval records should be consistent with the approval procedures, and the approval procedures should be consistent with the requirements of corresponding management systems. Review the relevant documentation to verify that the above requirements are met.

If the test results show that the organization undergoing test has established an approval control process for the authorization matters of data integration, analysis, exchange, sharing and disclosure in the process of data collection, transmission, storage, processing, exchange, destruction, etc., the authorization control process has complete records, the records are consistent with the approval procedures, and the approval procedures are consistent with the requirements of corresponding management systems, the result is conforming; otherwise it is non-conforming or partially conforming.

【L3-ORS6-03/ L4-ORS6-03 Interpretation and Description】

The targets of "establish the evaluation, approval and regulatory control process for cross-border data, and implement relevant control and record the process according to the process (best practice)" are documents related to cross-border data security management, documents related to development scheme or system deployment, and documents related to record forms. This evaluation index applies to the grading targets that include big data platform, big data application or big data resources.

The key points of evaluation implementation include: interview the system administrator to understand the requirements of cross-border data flow; if there is a need for cross-border data flow, check whether the cross-border data evaluation, approval and supervision control process has been established; consult the cross-border data evaluation results, approval and regulatory control process records, etc., check whether approvers at all levels and audit departments sign/seal, check whether the records of control process are consistent with the approval and supervision control process, and check whether the approval process is consistent with the requirements of corresponding management systems.

If the test results show that the organization undergoing test has established the cross-border data evaluation, approval and supervision control process, and the document of the implementation process of cross-border data evaluation, approval and supervision control process is complete and effective, the result is conforming; otherwise it is non-conforming or partially conforming.

3. Audit and Inspection

[Standard Requirements]

The Level III control point includes the evaluation unit L3-ORS1-03, and Level IV includes the evaluation unit L4-ORS1-03.

[L3-ORS1-03/ L4-ORS1-03 Interpretation and Description]

The targets of "conduct security inspection on a regular basis, including the daily operation of the system, system loopholes and data backup" are the security management system and routine security inspection records. This evaluation index applies to the grading targets that include big data platform, big data application or big data resources.

The key points of evaluation implementation include: check whether the organization undergoing test regularly conducts routine security inspection. The security inspection mainly includes data authorization, data desensitization or de-identification processing during data acquisition, transmission, storage, processing, exchange and destruction, output control of sensitive data, audit of important operations, destruction of data by classified and categorized, data traceability, etc.; verification of routine security inspection records.

If the test results show that the organization undergoing test carries out regular security inspection regularly, including data authorization, data desensitization or de-identification processing, sensitive data output control, audit of important operations, destruction by classified and categorized of data and data traceability during data acquisition, transmission, storage, processing, exchange and destruction, etc., and the inspection records are complete and effective, the result is conforming; otherwise it is non-conforming or partially conforming.

8.3.8 Security Management Personnel

According to the characteristics of big data system, this chapter lists the personalized interpretation contents of some general security evaluation requirements of big data environment.

1. Personnel Recruitment

[Standard Requirements]

The Level III control point includes the evaluation unit L3-HRS1-01, and Level IV includes the evaluation unit L4-HRS1-01.

【L3-HRS1-01/ L4-HRS1-01 Interpretation and Description】

The targets of "sign confidentiality agreement with employed personnel and post responsibility agreement with key personnel" are confidentiality agreement, post responsibility document, post responsibility agreement, etc. This evaluation index applies to the grading targets that include big data platform, big data application or big data resources.

The implementation points of the evaluation include: for big data system, spot check the signed confidentiality agreement to make clear the signed confidentiality agreement; pay special attention to the personnel of key posts related to data security management, sign the post responsibility agreement in data acquisition, transmission, storage, processing, exchange, destruction and other links, and ensure that the job responsibility agreement specifies the security responsibility of key posts related to data, the validity period of the agreement.

If the test results show that the organization undergoing test has signed confidentiality agreements with all employed personnel, and at the same time, the post responsibility agreement has been signed with the personnel of key posts related to data security management in data acquisition, transmission, storage, processing, exchange and destruction, etc., the result is conforming; otherwise it is non-conforming or partially conforming.

2. Personnel Leaving the Post

【Standard Requirements】

The Level III control point includes the evaluation unit L3-HRS1-02, and Level IV includes the evaluation unit L4-HRS1-02.

【L3-HRS1-02/ L4-HRS1-02 Interpretation and Description】

The targets of "terminate all access rights of the off-duty personnel in time, and retrieve back all kinds of identity documents, keys, badges and software and hardware equipment provided by the organization" are the person in charge of personnel, department head, security management system, personnel leaving records, asset registration form, etc. This evaluation index applies to the grading targets that include big data platform, big data application or big data resources.

The key points of evaluation implementation include: for big data system, besides retrieving various identification documents, keys, badges and software and hardware equipment (including data storage media) provided by the organization, all access rights of off-duty personnel should be terminated in time, data access and use rights involved in corresponding data acquisition, transmission, storage, processing, exchange and destruction should be withdrawn, and important data should be withdrawn. Check personnel management system and relevant records and forms, etc.

If the test results show that the asset return records of the off-duty personnel of the organization undergoing test are complete, and all the big data access rights of the off-duty personnel are terminated or revoked in time, the result is conforming; otherwise it is non-conforming or partially conforming.

3. Security Awareness Education and Training

[Standard Requirements]

The third level includes the evaluation unit L3-HRS1-03, and Level IV includes the evaluation unit L4-HRS1-03.

[L3-HRS1-03/ L4-HRS1-03 Interpretation and Description]

The targets of "formulate different training plans according to different posts, and conduct training about basic security knowledge and post operation procedures" are security management documents, training plans, security education and training records. This evaluation index applies to the grading targets that include big data platform, big data application or big data resources.

The key points of evaluation implementation include: check whether relevant training plans are formulated for big data security management personnel, and whether there are training records of relevant personnel.

If the test results show that the organization undergoing test has specified the roles or positions related to big data security or data security, and has training plans and relevant training records for big data security or data security, the result is conforming; otherwise it is non-conforming or partially conforming.

8.3.9 Security Development Management

This chapter interprets the extended security evaluation requirements of security development management of big data system.

According to the characteristics of big data system, this chapter also lists the personalized interpretation contents of some general security evaluation requirements of big data environment.

1. Security Scheme Design

[Standard Requirements]

The Level III control point includes the evaluation unit L3-PES6-01, and Level IV includes the evaluation unit L4-PES6-01.

[L3-CMS1-01/ L4-CMS1-01 Interpretation and Description]

The targets of "select basic security measures according to the security protection level, and supplement and adjust security measures according to the results of risk analysis" are security planning and design documents. This evaluation index applies to the grading targets that include big data platform, big data application or big data resources.

The key points of evaluation implementation include: select basic security measures

Chapter 8 Application and Interpretation of the Extended Security Evaluation Requirement of Big Data

based on the security protection level of protection targets, focus on the management and control of security risks at each stage of big data life cycle, and supplement and adjust security measures according to the results of risk analysis. Check whether security planning and design documents design corresponding security protection measures according to the particularity of big data, such as data desensitization, data security mark.

If the test results show that the selection of basic security measures of the organization undergoing test is consistent with the security protection level, the contents of security planning and design documents cover all aspects, and corresponding security protection measures are designed according to the particularity of big data, the result is conforming; otherwise it is non-conforming or partially conforming.

2. Testing Acceptance

[Standard Requirements]

The Level Ⅲ control point includes the evaluation unit L3-CMS1-02, and Level Ⅳ includes the evaluation unit L4-CMS1-02.

[L3-CMS1-02/ L4-CMS1-02 Interpretation and Description]

The targets of "conduct security test before going online, and issue security test report, which should include relevant contents of cryptography application security test" are the security test scheme, security test report, security test process record and other relevant documents. This evaluation index applies to the grading targets that include big data platform, big data application or big data resources.

The key points of implementation of the evaluation include: as for the security test before going online, in addition to cryptography application security test, it is also necessary to test the defects or vulnerabilities that may exist in the process of data acquisition, transmission, storage, processing, exchange and destruction, and test the security of data security components such as data desensitization component, data de-identification component and relevant interfaces. Check the security test scheme, report and test process record, check whether it includes the security test of data acquisition, transmission, storage, processing, exchange and destruction, check whether the security test has been carried out for the security components such as data desensitization component and data de-identification component, and check whether the cryptography application security test has been conducted.

If the test results show that the security test scheme, report and test process records before launch are complete and effective, and the security test contents cover all aspects, the result is conforming; otherwise it is non-conforming or partially conforming.

3. Selection of Big Data Service Provider

[Standard Requirements]

The Level Ⅲ control point includes the evaluation units L3-CMS6-01 and L3-CMS6-02, and Level IV includes the evaluation units L4-CMS6-01 and L4-CMS6-02.

【L3-CMS6-01/ L4-CMS6-01 Interpretation and Description】

The targets of "select safe and compliant big data platform, and the big data platform service provided by it should provide corresponding level of security protection capability for the big data application it carrying" mainly include the person in charge of big data application development, big data resource manager, big data platform service contract, big data platform level evaluation report, big data platform qualification and security service capability report. This evaluation index applies to the grading targets that include big data applications or big data resources.

The key points of evaluation implementation include: interview the person in charge of big data application development to understand the security protection level of big data application and the security protection level of the selected big data platform; if big data application is built on the third-party big data platform, check the qualification documents of the third party big data platform, check whether the service provider meets the national regulations, check whether the big data platform meets the requirements of laws, regulations and relevant standards and whether the rating evaluation is carried out regularly. If it is not deployed on that third-party big data platform, check whether the private big data platform meets the requirements of laws, regulations and relevant standards and whether the level evaluation is carried out regularly. Check the evaluation report or result of the big data platform and check whether the selected big data platform service can provide the application with corresponding security protection capability. Check the relevant qualification and security capability report of the selected big data platform, and check whether it has the security protection capability corresponding to or higher than the big data application.

If the test results show that the selected big data platform is safe and compliant and provides corresponding level of security protection capability, the result is conforming; otherwise it is non-conforming or partially conforming.

【L3-CMS6-02/ L4-CMS6-02 Interpretation and Description】

The targets of "specify the authority and responsibility of the big data platform provider in written, various service contents and specific technical indicators, especially the security service content" are the service contract, agreement or service agreement, security statement, etc. This evaluation index applies to the grading targets that include big data applications or big data resources.

The key points of evaluation implementation include: authority and responsibility of big data platform provider (such as management scope, division of responsibilities, access authorization, privacy protection, code of conduct, liability for breach of contract, etc.), various service contents and specific technical indicators should be clearly specified in service agreement or contract. The service should include security services, such as interface security management, resource assurance, fault shielding. Check whether the service agreement or contract clearly defines the authority and responsibilities of the big data platform provider, including the management scope, division of responsibilities, access authorization, privacy protection, code of conduct, liability for breach of contract. Verify whether the service agreement, service contract and security statement clearly specify various service contents and specific technical indicators provided by the big data platform provider.

If the test results show that the organization undergoing test has signed a service agreement or contract with the big data platform service provider, and the service agreement or contract clearly specifies the rights and responsibilities of the big data platform provider, various service contents and specific technical indicators, etc., the result is conforming; otherwise it is non-conforming or partially conforming.

4. Supply Chain Management

[Standard Requirements]

The Level Ⅲ control point includes the evaluation units L3-CMS6-03 and L3-CMS6-04, and Level Ⅳ includes the evaluation units L4-CMS6-03 and L4-CMS6-04.

[L3-CMS6-03/ L4-CMS6-03 Interpretation and Description]

The targets of "restrain the protection responsibility for the receiver of data exchange and sharing, and ensure that the receiver has sufficient or equivalent security protection capability" are data exchange, sharing strategy, service agreement or contract, level evaluation report of the receiver, etc. This evaluation index applies to the grading targets that include big data platform, big data application or big data resources.

The key points of evaluation implementation include: the data sending party should sign a contract or agreement with the receiving party, in which the responsibilities and obligations of both parties are specified. The receiving party should provide relevant qualification or certification of capability to ensure thatit has the corresponding data security protection capability. Check the service agreement or contract to verify whether the data receiving party has specified the data protection responsibility; check whether the data receiver has corresponding data security protection capability, such as grade evaluation report of corresponding level.

If the test results show that the organization undergoing test has signed a service agreement or contract with the data receiver, specifying the data protection responsibility of the receiver, and the receiving party has corresponding data security protection capability, the result is conforming; otherwise it is non-conforming or partially conforming.

[L3-CMS6-04/ L4-CMS6-04 Interpretation and Description]

The targets of "ensure that the supplier selection complies with relevant national regulations (best practice)" are the responsible person of development, bidding documents, relevant contracts, qualification certificates, sales licenses. This evaluation index applies to the grading targets that include big data platform, big data application or big data resources.

The key points of evaluation implementation include: the selection of suppliers (such as big data platform supplier, security service provider, big data platform infrastructure, etc.) should comply with relevant national laws, regulations and standard specifications, such as meeting the requirements of Network Security Law of the People's Republic of China, GB/T36637—2018 Information Security Technology ICT Supply Chain Security Risk Management Guidelines, having corresponding qualification certificates and sales licenses. Review relevant documents such as bidding documents, supplier qualification certificates, sales licenses, relevant contracts, etc. to check whether the above requirements are met.

If the test results show that the supplier selection complies with relevant national regulations, the result is conforming; otherwise it is non-conforming or partially conforming.

5. Data Source Management

[Standard Requirements]

The Level III control point includes the evaluation unit L3-CMS6-05, and Level IV includes the evaluation unit L4-CMS6-05.

[L3-CMS6-05/ L4-CMS6-05 Interpretation and Description]

The targets of "obtain all kinds of data through legal and proper channels (best practices)" are data administrators, authorization documents or records, and related contracts and agreements. This evaluation index applies to the grading targets that include big data platform, big data application or big data resources.

The key points of implementation of the evaluation include: check whether the data source is a legitimate and legitimate channel, whether a contract or agreement has been signed with the provider or has been authorized to use the data, and checking whether the data acquisition method violates the principle of legality and legality.

If the test results show that the channels through which the big data platform/application system obtains all kinds of data are legitimate and authorized, the result is conforming; otherwise it is non-conforming or partially conforming.

8.3.10 Security Operation and Maintenance Management

This chapter is the interpretation of "safe operation and maintenance management" required for extended security evaluation of big data system.

According to the characteristics of big data system, this chapter also lists the personalized interpretation contents of some general security evaluation requirements of big data environment.

1. Security Incident Handling

[Standard Requirements]

The Level III control point includes the evaluation unit L3-MMS1-01, and Level IV

Chapter 8　Application and Interpretation of the Extended Security Evaluation Requirement of Big Data

includes the evaluation unit L4-MMS1-01.

【L3-MMS1-01/ L4-MMS1-01 Interpretation and Description】

The targets of "formulate security incident report and disposal management system, and specify the process for reporting, disposal and response process of different security incidents, and clarify the management responsibilities for on-site handling, incident report and post-recovery of security incidents" are security incident report and disposal management system, disposal record, and post-incident recovery records. This evaluation index applies to the grading targets that involve big data platform, big data application or big data resources.

The implementation points of the evaluation include: check the security incident report and disposal management system, check whether the reporting, disposal and response processes of different security events are specified, and check whether the management responsibilities for on-site handling, incident reporting and post-recovery of security incidents are specified; check the security incident report and disposal management system to see whether the reporting, disposal and response processes have been established for security incidents such as different data destruction, big data information disclosure, data abuse, data manipulation. If any security incidents occur, view relevant security event handling records, security event report records, security incident post-recovery records.

If the test results show that the organization undergoing test has formulated the security incident (including the security incident related to big data) report and disposal management system, and the relevant records and forms are complete and effective, the result is conforming; otherwise it is non-conforming or partially conforming.

2. Emergency Plan Management

【Standard Requirements】

The Level Ⅲ control point includes the evaluation unit L3-MMS1-02, and Level Ⅳ includes the evaluation unit L4-MMS1-02.

【L3-MMS1-02/ L4-MMS1-02 Interpretation and Description】

The targets of "formulate emergency plan for important security events, including emergency handling process, system recovery process and other contents" are emergency response plan (special emergency plan for big data important security incidents). This evaluation index applies to the grading targets that include big data platform, big data application or big data resources.

The key points for implementation of the evaluation include: check whether the emergency plan documents include important events related to big data, or whether special emergency plans have been formulated for important security events of big data, such as data disclosure, major data abuse or major data manipulation; verify whether the emergency plan for important events includes emergency handling process, data recovery process, etc.

If the test results show that the organization undergoing test has formulated the special emergency plan for important events (including but not limited to the special emergency plan for important events of big data) and the emergency response process and data recovery process are specified in the emergency plan, the result is conforming; otherwise it is non-conforming or partially conforming.

3. Asset Management

[Standard Requirements]

The Level III control point includes the evaluation units L3-MMS6-01, L3-MMS6-02, L3-MMS6-03, L3-MMS6-04, and Level IV includes the evaluation units L4-MMS6-01, L4-MMS6-02, L4-MMS6-03, L4-MMS6-04.

[L3-MMS6-01/ L4-MMS6-01 Interpretation and Description]

The targets of "establish digital asset security management strategy, and specify the operation specifications, protection measures and responsibilities of management personnel in the whole life cycle of data, including but not limited to data acquisition, storage, processing, application, flow, destruction and other processes" are the management policy of data asset security. This evaluation index applies to the grading targets that include big data platform, big data application or big data resources.

The key points of evaluation implementation include: check whether there are documents related to security management strategy of digital assets; verify whether security management objectives, principles and scope of digital assets are specified in relevant documents; verify whether relevant documents specify operation specifications (including but not limited to data acquisition, storage, processing, application, flow, destruction, etc.) and protective measures in the whole life cycle of data; checking whether there are relevant operation record forms including but not limited to data acquisition, storage, processing, application, flow, destruction. Verify whether responsibilities of management personnel are specified in relevant documents of security management strategy of digital assets.

If the test results show that the big data platform/application system undergoing test has formulated the data asset security management strategy, and the policy content includes the provisions such as operation specifications, protective measures, responsibilities of management personnel, etc. formulated according to different data life cycles, the result is conforming; otherwise it is non-conforming or partially conforming.

[L3-MMS6-02/ L4-MMS6-02 Interpretation and Description]

The targets of "formulate and implement data classified and categorized protection strategy, and formulate different security protection measures for data of different categories and levels" are the strategy documents and the result documents of data classification and categorization. This evaluation index applies to the grading targets that include big data platform, big data application or big data resources.

The key points for implementation of the evaluation include: interview with the security administrator to determine whether the data classified and categorized protection strategy has been formulated; check whether the data classified and categorized protection strategy document specifies the data classified and categorized method; check whether different security protection measures (such as security mark, encryption, desensitization) are formulated for data of different categories and levels in the data classified and categorized protection policy document; verify whether different security protection measures are taken according to the classified and categorized protection policy; check whether relevant data classification and grading result documents are available, and check whether the data classification and grading results are consistent with the data classified and categorized protection strategy.

If the test results show that the big data platform/ application system undergoing test has formulated the data classified and categorized protection strategy, the content of the strategy includes formulating different security protection measures for different levels of data and implementing the data classification categorized protection strategy, the result is conforming; otherwise it is non-conforming or partially conforming.

[L3-MMS6-03/ L4-MMS6-03 Interpretation and Description]

On the basis of data classified and categorized, the evaluation index should divide the scope of important digital assets and define the use scenarios and business processing flow for automatic desensitization or de-identification of important data. The main evaluation objects are data asset list and management documents related to data desensitization or de-identification requirements. This evaluation index applies to the grading targets that include big data platform, big data application or big data resources.

The key points of evaluation implementation include: interview data administrator, check management of data assets and external data interface, whether data assets and data interface assets list are established, such as data asset manager, list of software and hardware assets, interface name, interface parameters, interface security requirements, importance degree of assets, etc. ; check whether relevant management system documents have corresponding requirements and descriptions for data or scenarios requiring automatic desensitization or de-identification.

If the test results show that the big data platform/application system registers and manages the data assets and external data interfaces, and establishes an asset list containing the importance degree of assets, and there are relevant requirements documents for desensitization and de-identification, the result is conforming; otherwise it is non-conforming or partially conforming.

[L3-MMS6-04/ L4-MMS6-04 Interpretation and Description]

The targets of "review the data category and level regularly; if it is necessary to change the category or level of data, the change should be implemented according to the change approval process" are the security management system, periodic review records of data classification and grading and change records of data category level. This evaluation index applies to the grading targets that include big data platform, big data application or big data resources.

The key points of evaluation implementation include: check the security management system, check whether there is regular review content of data category and level, and whether there is approval process for change of data category or level; check the regular review record of data classified and categorized, check whether the regular review record of data classification and grade is consistent with the requirements of security management system; check relevant data category and level change records, and check whether the change is implemented according to the approval process of data category level change.

If the test results show that the big data platform/ application system undergoing test has formulated relevant systems for data management, and the data review, change approval process is specified, and there are records such as review records of regular review data category and level and review and approval process, the result is conforming; otherwise it is non-conforming or partially conforming.

4. Media Management

[Standard Requirements]

The Level Ⅲ control point includes the evaluation unit L3-MMS6-05, and Level Ⅳ includes the evaluation unit L4-MMS6-05.

[L3-MMS6-05/ L4-MMS6-05 Interpretation and Description]

The targets of "remove or destroy the data within the territory of China (best practice)" are the strategies or management systems relating to data removal or destruction, records of data removal and destruction. This evaluation index applies to the grading targets that include big data platform, big data application or big data resources.

The key points of evaluation implementation include: interview system administrator or security administrator of big data platform to understand whether relevant systems have been established and make clear that data clearing or destruction should be in China; verify relevant data clearing or destruction strategies or management systems, and verify data clearing and destruction records.

If the test results show that the organization undergoing test has established relevant data elimination or destruction strategies or management systems, and stipulates the data clearance and destruction within the territory of China, the result is conforming; otherwise it is non-conforming or partially conforming.

5. Network and System Security Management

[Standard Requirements]

The Level Ⅲ control point includes the evaluation unit L3-MMS6-06, and Level Ⅳ includes the evaluation unit L4-MMS6-06.

【L3-MMS6-06/ L4-MMS6-06 Interpretation and Description】

The targets of "establish external data interface security management mechanism, and all interface calls should be authorized and approved (best practice)" are the management system documents, authorization approval records. This evaluation index applies to the grading targets that include big data platform, big data application or big data resources.

The key points of evaluation implementation include: interview the system administrator or security administrator of big data platform, check whether security management mechanism of external data interface is specified in security management system; verify whether external data interface is authorized and approved by data interface manager; check whether there is authorization and approval record of external data interface.

If the results show that the big data platform undergoing test has established the security management mechanism of the external data interface, and the external data interface has obtained the authorization and approval of the big data platform, and has the authorization and approval record of the external data interface, the result is conforming; otherwise it is non-conforming or partially conforming.